T0374458

The Board of Longitude

In the first book-length history of the Board of Longitude, a distinguished team of historians of science brings to life one of Georgian Britain's most important scientific institutions. Having developed in the eighteenth century following legislation that offered rewards for methods to determine longitude at sea, the Board came to support the work of navigators, instrument makers, clockmakers and surveyors, and assembled the *Nautical Almanac*. The authors use the archives and records of the Board, which they have recently digitised, to shed new light on the Board's involvement in colonial projects and in Pacific and Arctic exploration, as well as on innovative practitioners whose work would otherwise be lost to history. This is an invaluable guide to science, state and society in Georgian Britain, a period of dramatic industrial, imperial and technological expansion.

ALEXI BAKER operates and curates the History of Science and Technology collection at Yale Peabody Museum.

RICHARD DUNN is Keeper of Technologies and Engineering at the Science Museum, London.

REBEKAH HIGGITT is Principal Curator of Science at National Museums Scotland.

SIMON SCHAFFER is Emeritus Professor of the History of Science at the University of Cambridge and was Principal Investigator on the Arts and Humanities Research Council project on the Board of Longitude.

SOPHIE WARING is Director of the Old Operating Theatre Museum and Herb Garret, London.

The Board of Longitude

Science, Innovation and Empire

Alexi Baker
Yale Peabody Museum

Richard Dunn
Science Museum, London

Rebekah Higgitt
National Museums Scotland

Simon Schaffer
University of Cambridge

Sophie Waring
Old Operating Theatre Museum and Herb Garret

CAMBRIDGE
UNIVERSITY PRESS

Shaftesbury Road, Cambridge CB2 8EA, United Kingdom

One Liberty Plaza, 20th Floor, New York, NY 10006, USA

477 Williamstown Road, Port Melbourne, VIC 3207, Australia

314–321, 3rd Floor, Plot 3, Splendor Forum, Jasola District Centre, New Delhi – 110025, India

103 Penang Road, #05–06/07, Visioncrest Commercial, Singapore 238467

Cambridge University Press is part of Cambridge University Press & Assessment, a department of the University of Cambridge.

We share the University's mission to contribute to society through the pursuit of education, learning and research at the highest international levels of excellence.

www.cambridge.org
Information on this title: www.cambridge.org/9781009602525

DOI: 10.1017/9781009602556

When citing this work, please include a reference to the DOI 10.1017/9781009602556

First published 2025

Cover image: William Chevasse, illustration of a marine chair for observing Jupiter's satellites, 1813, CUL RGO 14/36, p. 51r. Reproduced by kind permission of the Syndics of Cambridge University Library.

A catalogue record for this publication is available from the British Library

A Cataloging-in-Publication data record for this book is available from the Library of Congress

ISBN 978-1-009-60252-5 Hardback
ISBN 978-1-009-60250-1 Paperback

In memory of Jim Bennett (1947–2023)

Contents

Figures

Tables

Boxes

Acknowledgements

This book is the result of two major projects, the funders of which deserve our particular thanks. 'The Board of Longitude 1714–1828: Science, Innovation and Empire in the Georgian World', funded by the Arts and Humanities Research Council, was jointly based at the University of Cambridge and the National Maritime Museum, Greenwich, in 2010–2015. The Principal Investigator was Simon Schaffer, Co-Investigators Richard Dunn and Rebekah Higgitt, postdoctoral researchers Alexi Baker and Nicky Reeves, doctoral students Katy Barrett, Eóin Phillips and Sophie Waring, and Engagement Officer Katherine McAlpine. Its aim was to produce the first comprehensive history of the British Board of Longitude, examining its changing role as an influential player in Georgian culture. The Huntington Library was a project partner, for which we thank Robert C. Ritchie, Steve Hindle and colleagues. Alongside the project members who have written the chapters in this book, we warmly thank Katy, Katherine, Eóin and Nicky, who have all contributed invaluably and whose own work has been credited. In addition, we received enormous help during the project and in the later writing of this book from our advisory board: Jim Bennett, Jonathan Betts (whose helpful comments on several draft chapters should also be acknowledged), Richard Drayton, John Gascoigne, John McAleer, Anita McConnell, David Philip Miller, Alison Morrison-Low, Nigel Rigby, Nicholas Rodger, Martina Schiavon and Larry Stewart. Thanks as well to Tamara Hug in Cambridge and Sally Archer in Greenwich for tireless logistical support throughout.

Members of the project also took part in a project funded by the Joint Information Systems Committee, 'Navigating 18th-Century Science and Technology: the Board of Longitude', with colleagues at Cambridge University Library. The project digitised the Board of Longitude archive held in Cambridge and related material from the National Maritime Museum and other archives, making them publicly available with additional research aids through the Cambridge Digital Library. It is impossible to overstate how valuable this was in the later stages of writing the book, when the benefits of remote access to high-quality images of primary resources became so apparent. It would not have been possible without many staff at Cambridge University

Library, notably Adam Perkins, Grant Young, Huw Jones, Emma Saunders, Robert Steiner and Karen Davies, as well as Megan Barford and James Poskett, who wrote introductory essays, Lucinda Blaser and Gareth Bellis, who oversaw the digitisation of material at Royal Museums Greenwich, and Mary Ferguson, engagement officer for the project, not to mention the band of volunteers she recruited.

Many of those already named spoke to the research team and took part in workshops, panels and conferences during the project, and have helped in countless ways since. Thank you again. Many others also kindly spoke at project events or helped in other ways as we explored the many narratives in which the Board of Longitude was entwined. For this we wish to thank: Emily Akkermans, Will Ashworth, Guy Boistel, Michael Bravo, David Bryden, Jenny Bulstrode, Joe Cain, Joyce Chaplin, Andreas Christoph, Gloria Clifton, Gareth Cole, Andrew Cook, Adriana Craciun, Mary Croarken, James Davey, Karel Davids, Suzanne Débarbat, Nicolàs de Hilster, Louise Devoy, Nick Dew, Graham Dolan, Danielle Fauque, Héloïse Finch-Boyer, Debra Francis, John Gascoigne, Tim Hitchcock, Caitlin Homes, Gillian Hutchinson, Richard Johns, Stephen Johnston, Michael Kershaw, Andrew King, Roger Knight, Wolfgang Köberer, Albert Krayer, Greg Lynall, Christine Macleod, Margaret Makepeace, Rory McEvoy, Amy Miller, Willem Mörzer Bruyns, Maya Nuku, Jacob Orrje, Joseph Payne, Juan Pimentel, Jo Poppleton, Stephen Pumfrey, Georgina Rannard, Frank Reed, Martin Rees, Neil Safier, Hannah Salisbury, Martin Salmon, Margaret Schotte, Jared Shurin, Ivan Tafteberg Jakobsen, Ilya Vinkovetsky, Simon Werrett, Jane Wess, Diederick Wildeman and Charlie Withers. We can only apologise if we have omitted anyone from this list.

Timeline

Note: Only some rewards paid out by the Board of Longitude are included below. For a full list, see Derek Howse, 'Britain's Board of Longitude: The Finances, 1714–1828', *The Mariner's Mirror*, 84 (1998), 400–417.

1675	Foundation of the Royal Observatory, Greenwich, for 'the finding out of the longitude of places for perfecting navigation and astronomy' John Flamsteed appointed first Astronomer Royal
1675–1676	Henry Bond rewarded for his magnetic work and publishes *Longitude Found*
1691	Thomas Axe dies, establishing a longitude reward in his will
1698–1700	Edmond Halley commands two Atlantic voyages to map magnetic variation
1701	Halley creates the first published isogonic maps with lines indicating magnetic variation
1712	John Hutchinson petitions the House of Commons for protection for his design for a longitude timekeeper
1713	William Whiston and Humphry Ditton begin lobbying for money for a longitude scheme
1714	Whiston and Ditton petition Parliament, followed by a petition from 'several Captains of her Majesty's Ships, Merchants of *London*, and Commanders of Merchant-men' Discovery of Longitude at Sea Act (12 Anne c 15) passed, constituting up to twenty-three Commissioners who can assess schemes and approve rewards
1716–1728	Henry Sully invents and trials sea clocks, presenting designs to the Académie Royale des Sciences in 1716 and 1723 and also attempting to arouse interest in Britain
1720	A bill to allow the Commissioners of Longitude to reward improvements to general navigation, presented on behalf of Jacob Rowe, fails in the House of Commons Edmond Halley appointed second Astronomer Royal
1728	Sea trials of John Bates's longitude scheme, reportedly official

1730	John Harrison produces a manuscript account of his proposed sea clock
1735	John Harrison completes his first sea clock, 'H1', and brings it to London
1736	H1 achieves promising results in an Admiralty sea trial between Britain and Lisbon
1737	First known communal meeting of the Commissioners of Longitude grants Harrison funds to build a second sea clock, 'H2'
1739	Harrison completes H2
1740–1759	Harrison works on his third sea clock, 'H3'
1741	Longitude and Latitude Act (14 Geo 2 c 39) allows the Commissioners to fund coastal surveys with rewards of up to £2000
1742	Commissioners award £500 to William Whiston for a magnetic survey of the British coast James Bradley appointed third Astronomer Royal
1742–1743	Jane Squire publishes her longitude scheme in two editions
1751–1755	Leonhard Euler and Tobias Mayer correspond about the lunar-distance method
1753	Discovery of Longitude at Sea Act (26 Geo 2 c 25) allocates an additional £2000 for 'Experiments' and creates additional Commissioners
1754–1755	Tobias Mayer sends improved lunar tables to London
1755–1759	Harrison works on his sea watch, 'H4'
1756	Commissioners order sea trials of Tobias Mayer's lunar tables, method and observing instrument Earliest known use of 'the Board of Longitude' to denote a standing body
1757	John Bird and Captain John Campbell develop the marine sextant, trialled at sea in 1758–1759
1758–1765	Christopher Irwin pursues a reward for his marine chair
1760	Commissioners begin meeting communally at least annually John Harrison exhibits H4 to the Commissioners
1761	Nevil Maskelyne trials the lunar-distance method on a voyage to St Helena, using Mayer's tables
1761–1762	H4 sent on a sea trial to Jamaica with William Harrison
1762	Nathaniel Bliss appointed fourth Astronomer Royal Discovery of Longitude at Sea Act (2 Geo 3 c 18) allocates a further £2000 for bringing proposals to trial Board requests funds for a secretary and reimbursing longer journeys to attend meetings (agreed in 1763)

1763	Discovery of Longitude at Sea Act (3 Geo 3 c 14) makes £5000 available to Harrison upon disclosure of the details of H4, and grants him exclusive rights to any timekeeper reward for four years Publication of *An Account of the Proceedings*, anonymous but probably written for Harrison by James Short or Taylor White
1763–1764	Sea trials to Barbados of H4, Mayer's method and tables for lunar distances, and Irwin's marine chair
1765	Nevil Maskelyne appointed fifth Astronomer Royal Board agrees to print Mayer's tables and produce an annual nautical ephemeris or almanac Discovery of Longitude at Sea Act (5 Geo 3 c 11) allocates a further £2000 for bringing proposals to trial Discovery of Longitude at Sea Act (5 Geo 3 c 20) grants £10,000 to Harrison upon disclosure of the details of H4, with a further £10,000 payable when further timekeepers are made; gives rewards of £300 to Leonhard Euler and £3000 to Tobias Mayer's widow; authorises annual production of the *Nautical Almanac*; and appoints further Commissioners Harrison discloses details of H4 Publication of *A Narrative of the Proceedings*, anonymous but probably written for Harrison by Short
1766	Board publishes the first edition of the *Tables Requisite* Maskelyne tests H4 at the Royal Observatory and publishes *An Account of the Going*; Harrison publishes a response, *Remarks on a Pamphlet* Board agrees to reward John Bird (£500) for disclosing details of his methods of making precision instruments
1767	Board publishes: the first annual volume of the *Nautical Almanac*; *The Principles of Mr Harrison's Timekeeper*; Bird, *The Method of Dividing Astronomical Instruments*
1768	Board publishes Bird, *The Method of Constructing Mural Quadrants* Board recommends that Royal Navy masters are trained in the lunar-distance method using the *Nautical Almanac*
1768–1771	James Cook trials the lunar-distance method on his first Pacific voyage
1770	Discovery of Longitude at Sea Act (10 Geo 3 c 34) allocates £5000 for bringing ideas to trial, improving lunar tables and improving general navigation Harrison completes his second sea watch, 'H5'

	Larcum Kendall presents 'K1', a copy of H4, at a cost of £450 and is awarded an additional £50
1771	Board appoints William Wales and William Bayly as astronomers for James Cook's second Pacific voyage (1772–1775)
1772	William Harrison approaches George III through Stephen Demainbray, superintendent of the king's observatory at Kew, resulting in a trial there of H5
1772–1775	James Cook trials K1 and three timekeepers by John Arnold on his second Pacific voyage
1773	Board appoints Israel Lyons as astronomer on Constantine Phipps's Expedition towards the North Pole
	Following John Harrison's petition to Parliament, the Supply, etc. Act (13 Geo 3 c 77) grants him £8750 'as a further Reward and Encouragement'
1774	Discovery of Longitude at Sea Act (14 Geo 3 c 66) repeals all former Longitude Acts except clauses regarding appointments and publications, and establishes new rewards of up to £10,000 for longitude and navigation
1775	Board rewards Jesse Ramsden (£300) for his circular dividing engine
	Harrison publishes *A Description Concerning Such Mechanism*
1776	Board appoints James Cook, James King and William Bayly as astronomers on Cook's third Pacific voyage (1776–1780)
1777	Finding of the Longitude at Sea Act (17 Geo 3 c 48) authorises a further £5000 for funding experiments and issuing smaller rewards
	Board publishes Ramsden, *Description of an Engine for Dividing Mathematical Instruments*
	Thomas Mudge awarded £500 for his marine timekeepers
1778	Joseph Banks becomes president of the Royal Society
	Ramsden rewarded for his straight-line dividing engine
1780	Finding of the Longitude at Sea Act (20 Geo 3 c 61) authorises a further £5000 for funding experiments and issuing smaller rewards
1781	Finding of the Longitude at Sea Act (21 Geo 3 c 52) authorises a further £5000 for funding experiments and issuing smaller rewards
1785	Board repeats recommendation that masters of naval vessels be trained in the lunar-distance method using the *Nautical Almanac*

1790	Discovery of Longitude at Sea Act (30 Geo 3 c 14) authorises a further £5000 for funding experiments and issuing smaller rewards, and appoints the Secretaries of the Admiralty as Commissioners
1791	Board sends William Gooch as astronomer to join George Vancouver's expedition to the Pacific coast of North America
1793	Thomas Mudge awarded £2500, following an appeal to Parliament by his son Computation of the *Nautical Almanac* is temporarily suspended Following Gooch's death, the Board sends John Crosley as replacement astronomer for Vancouver's expedition (although Crosley remains on the *Providence* after failing to meet up with Vancouver's ships)
1796	Longitude at Sea Act (36 Geo 3 c 107) authorises a further £5000 for funding experiments and issuing smaller rewards
1801	Board appoints John Crosley as astronomer to Matthew Flinders's survey of New Holland (Australia)
1802	Board appoints James Inman as astronomer to Flinders's survey following Crosley's return due to ill health
1802–1803	Board authorises legal prosecutions of pirated versions of the *Nautical Almanac*
1803	Discovery of Longitude at Sea, etc. Act (43 Geo 3 c 118) institutes new rules for publishing the *Nautical Almanac* and authorises a further £5000 for funding experiments and issuing smaller rewards
1805	£3000 each awarded to the chronometer makers Thomas Earnshaw and John Arnold (posthumously)
1806	Discovery of Longitude at Sea, etc. Act (46 Geo 3 c 77) authorises a further £10,000 for funding experiments and issuing smaller rewards
1811	John Pond appointed sixth Astronomer Royal Board appoints John Murray as publisher of *Nautical Almanac*
1814	£1200 awarded to Joseph de Mendoza y Ríos for improved longitude tables
1815	Discovery of Longitude at Sea Act (55 Geo 3 c 75) authorises a further £10,000 for funding experiments and issuing smaller rewards, and changes the rules for certifying rewards
1818	Discovery of Longitude at Sea, etc. Act (58 Geo 3 c 20): reorganises the Board, creating three salaried Resident Commissioners gives it responsibility for rewards relating

	to navigation towards the North Pole and efforts to find the Northwest Passage; establishes a new range of rewards; creates salaried posts of Superintendent of the Nautical Almanac and Superintendent of Timekeepers
1820	The Board awards £5000 to the crews of the *Hecla* and *Griper* for reaching 100° W within the Arctic Circle Thomas Young, Secretary of the Board and Superintendent of the Nautical Almanac, produces a report on lunar tables The Board agrees to establish an observatory at the Cape of Good Hope
1821	Discovery of Longitude at Sea, etc. Act (1 & 2 Geo 4 c 2) establishes new rules relating to the reward for the Northern Passage
1826	£1000 awarded to Captain Edward Sabine for pendulum experiments
1828	Nautical Almanack Act (9 Geo 4 c 66) repeals all former Longitude Acts, dissolves the Board of Longitude and authorises the Admiralty to continue publication of the *Nautical Almanac* Admiralty creates a Resident Committee of Scientific Advice

Abbreviations

attr.	attributed
BL	British Library
BoL	Board of Longitude
CUL	Cambridge University Library
EIC	East India Company
FRS	Fellow of the Royal Society
LMA	London Metropolitan Archives
MP	Member of Parliament
NMM	National Maritime Museum, Greenwich
n.p.	no place
RN	Royal Navy
RS	Royal Society, London
TNA	The National Archives

A Note on Dating

Old Style dates have been used for dates to 1752, although the year has been taken as starting on 1 January; for example, 16 January 1741/2 is rendered as 16 January 1742.

1 Introduction

In July 1714, queen Anne gave royal assent to 'An Act for providing a Publick Reward for such Person or Persons as shall discover the Longitude at Sea'. This first Longitude Act established what at the time were life-changing rewards of up to £20,000 for a method for finding a ship's longitude when out of sight of land. Crucially, it also nominated a group of experts to assess and agree upon rewards for successful schemes. It is the work of these Commissioners of Longitude, later known as the Board of Longitude, that this book seeks to document. While parts of the story are well known, the aim here is to consider in addition those aspects that have to date been incompletely explored in order to present a more rounded account of what can be considered as one of the first state bodies to support improving initiatives in knowledge and practice.

Background and Sources

The Commissioners of Longitude were appointed from 1714 to encourage practical solutions to a navigational problem long known to seafarers. In principle, a ship's position can be found by determining its geographical coordinates in terms of latitude and longitude. Latitude, the north–south position, could be calculated relatively easily from the observed altitude of celestial bodies such as the Sun or Pole Star above the horizon using instruments including the mariner's astrolabe, mariner's quadrant, cross-staff and backstaff. However, longitude (the east–west position) could not be determined in such a straightforward manner. Efforts to do so when at sea, many of which relied on the fact that a difference in longitude between two places can be derived from the difference in their local times, were further complicated by the ship's movements and weather, hindering astronomical observations and affecting the smooth running of technologies such as clockwork. The lack of a reliable method of determining longitude did not completely compromise navigation, nor was it the only problem facing seafarers on long voyages, but it did contribute to the difficulties encountered. This was particularly true in areas such as the Pacific and Indian Oceans, into which increasing numbers of European ships sailed from the sixteenth century.[1]

Concerns about the lack of a better method for estimating longitude while at sea became more prominent in Britain during the nation's expansion of global trade and empire, as national economic and security concerns dovetailed with the commercial interests of powerful individuals and institutions such as the trading companies. The Longitude Act of 1714 emerged from a confluence of commercial and political interests with the determined public lobbying of reward-seekers William Whiston and Humphry Ditton, and from consultations with mathematicians including Isaac Newton and Edmond Halley. The Act nominated a group of Commissioners of Longitude – influential political, naval and learned figures who were deemed acceptable judges for dispensing the new funds and rewards. Operating initially as a group of independent experts, the Commissioners are known to have held meetings from 1737. It was only from the 1760s, however, that they became a more formal standing body, known as the Board of Longitude – a transformation in which discussions and debates with Yorkshire-born clockmaker John Harrison played a key role. From the 1770s, the Board's activities began to expand. In the nineteenth century it became involved in geodetic work, the foundation of a new observatory, the search for the Northwest Passage and other projects. In 1828, however, the Board was dissolved by Act of Parliament, its activities taken over by other bodies.

The provisions of the Act of 1714 were in several respects designed not to encourage innovation but rather to exert the power of judgement over longitude schemes in order to restrict them to those that were seen as plausible and viable. Over time, the Commissioners also took on an increasing number of responsibilities for practical undertakings, many of which had not been anticipated in 1714. They were partly, and in some cases wholly, responsible for some of the best-known technological, navigational and astronomical endeavours of the time. For example, they oversaw the creation and improvement of the mariner's sextant, a turning point in navigation. The Board was also an important source of funding and support for what eventually became the marine chronometer, and encouraged the adoption of new technologies for dividing the degree scales on observing instruments. It produced useful and innovative publications, including the *Nautical Almanac*, which is still published annually. It was an outfitter for well-known British voyages of science and empire, including those led by Captain James Cook and those that explored the Arctic and sought the Northwest Passage. It further contributed to international endeavours such as the trigonometric determination of longitude between different observatories and cities, and was central to the establishment

[1] Dunn and Higgitt, *Finding Longitude*, pp. 12–33.

of the Royal Observatory at the Cape of Good Hope. As a result of these activities and because of overlaps in membership, the Board was a frequent collaborator with other British institutions, including the Royal Society, the Royal Observatory in Greenwich and the Admiralty, as well as with trading companies such as the East India Company. Despite the periodic outbreak of war in Europe, it worked with foreign experts as well, including Tobias Mayer and Jérôme Lalande, and sometimes with foreign institutions such as the French Académie des Sciences and Bureau des Longitudes.

Offering a fuller account of the Board of Longitude depends on analysing the surviving archive. The majority of it can be found in the sixty-eight volumes now held at Cambridge University Library, which document in rich detail many, but not all, of the Board's wide-ranging activities. The current state and arrangement of this material are mainly due to George Airy, a member of the Board when it was abolished in 1828 and Astronomer Royal from 1835. Following its dissolution, the Board's papers were divided up between the Admiralty and the Royal Society. In 1840, Airy arranged for the papers to be brought to the Royal Observatory and by 1858 he was able to report to the Board of Visitors that:

> The Papers of the Board of Longitude are now finally stitched into books, in the arrangement in which we usually send our manuscripts to the book-binder. They will probably form one of the most curious collections of the results of scientific enterprise, both normal and abnormal, which exists.[2]

In doing so, Airy reordered the papers under subject headings that reflected his own classification of the material. In part, then, this project has sought to undo the taxonomy imposed by Airy's commendable act of preservation.

Two further points are worth noting about the surviving institutional archive held in Cambridge. First, no papers survive from before the first formally recorded meeting of Commissioners on 30 June 1737. Moreover, the papers documenting activity from before the appointment of a secretary in 1763 were put together at the creation of that post and were mostly minutes. Second, and as a consequence of this, most of the surviving papers deal with the Board's activities from the 1770s and appear to be most complete from the 1780s onwards. For this later period, they include rich documentation of the Board's decision-making, financial and other procedures and of its incoming correspondence from those hoping to attract the support of the Commissioners – many of them relatively unknown figures.[3] The aim of this book, then, is to give a more complete picture of the activities of the Commissioners in both of these periods, drawing upon their surviving

[2] Airy, 'Report of the Astronomer Royal ..., 1858', p. G5.
[3] For a fuller overview, see Schaffer, 'Papers of the Board of Longitude'.

archives as well as complementary archival and published evidence. While dealings with John Harrison dominated much of the Commissioners' time between 1737 and the early 1770s, other business was also being tackled in this period. From the 1770s onwards, the Board of Longitude was engaged in a wide range of activities including, but not limited to, publishing the *Nautical Almanac*, testing marine timekeepers and supporting voyages of exploration.

The Historiography of the Board of Longitude

Lacking a definitive history of the Board of Longitude, scholars outside the history of science have generally given it little attention. Research in the imperial and maritime history of the long eighteenth century, for example, has traditionally centred on operations, administrative infrastructure, the role of sea power in creating and maintaining empire, or the 'discoveries' of great men. Nicholas Rodger's *The Command of the Ocean* makes a single reference to the Longitude Act and does not otherwise discuss the Board's work, while major texts including Vincent Harlow's *The Founding of the Second British Empire* and Paul Kennedy's *The Rise and Fall of British Naval Mastery* adopt approaches in which state-sponsored technological enquiry plays little part.[4] There have been other studies of institutional aspects of the Royal Navy in the last decade or so, but more recent work in maritime and naval history has focused on the personal, affective and experiential or on the complex relationships between Britain and its empire, navy and nation.[5] Likewise, there remains a keen interest in science and empire but with the emphasis on activities in the field rather than on institutions in Britain. Even the eighteenth-century volume of *The Oxford History of the British Empire* makes only brief references to the Board of Longitude in two chapters, while John Darwin's *Unfinished Empire* makes no mention at all.[6]

Works offering more substantial consideration of the Board's work have tended to focus on specific individuals. John Beaglehole's publications on Captain Cook and his Pacific voyages accommodate governmental interest in science within the context of a maritime and commercial empire. Placing unprecedented emphasis on maritime skills and technologies, Beaglehole set new standards for the study of hydrography and marine surveying.[7] His approach has influenced later scholars, including those looking at science,

[4] Rodger, *Command of the Ocean*, p. 172; Harlow, *The Founding of the Second British Empire*; Kennedy, *The Rise and Fall*.
[5] Knight and Wilcox, *Sustaining the Fleet*; Cole, *Arming the Royal Navy*. For recent trends, see Colville and Davey (eds.), *A New Naval History*.
[6] Marshall (ed.), *The Oxford History of the British Empire, Vol. 2*; Darwin, *Unfinished Empire*.
[7] Beaglehole (ed.), *The Journals of Captain James Cook*; Beaglehole, *The Life of Captain James Cook*.

trade and imperial policy in the Pacific.[8] All focus on the Board's work in relation to Cook's and other voyages but tend not to explore its wider governmental, social and intellectual context.

The Commissioners and those seeking their support have received most attention in histories of science and navigation, for example Larry Stewart's chapter on 'The Longitudinarians'.[9] Indeed, the focus on projectors and the reception of their ideas that Stewart foregrounded has produced fruitful work in recent years, in particular through discussions of satirical responses to some of the longitude schemes put forward in the eighteenth century.[10] However, few modern publications focus on the institutional history of the Board of Longitude and those that do are comparatively brief. Notable partial inroads into the archives include Eric Forbes's index, Peter Johnson's useful overview of the Board's work and Derek Howse's initial foray into its finances, as well as his work on Nevil Maskelyne and the history of the Royal Observatory.[11] More commonly, aspects of the Board's history have been mentioned briefly in relation to specific aspects of the history of science and technology, such as instrument-making and nineteenth-century scientific reform.[12] A notable addition is Eric Forbes's biography of Hanoverian astronomer Tobias Mayer, who received a posthumous longitude reward. Forbes draws on an impressive range of archival sources, including many in Germany, to reveal the mediated discussions taking place behind the scenes in the 1750s and 1760s.[13]

Histories of navigation, from Eva Taylor's *The Haven-Finding Art* to more recent works, give some prominence to the longitude story and the consequences of the Act of 1714. Several make perceptive comments but many also repeat common assumptions about the nature and trajectory of 'the Board', about specific episodes in its history and sometimes about the search for longitude more broadly. None, for example, has fully recognised the extent of the independent longitude-related activities of the Commissioners, particularly before the first known communal meeting in 1737, or that these activities belong to the same narrative as that of the later Board. Similarly, few authors have explored the final years of the Board of Longitude and its transformation in 1828. Key factors in modern interpretations have included the nature of

[8] For example, Mackay, *In the Wake of Cook*; Fisher and Johnston (eds.) *From Maps to Metaphors*; Williams, *The Great South Sea*; Frost, *The Global Reach*.

[9] Stewart, *The Rise of Public Science*, pp. 183–211.

[10] Barrett, *Looking for Longitude*; Freiburg, 'Our Never Failing Guide'; Lynall, 'Scriblerian Projections'; Rogers, 'Longitude Forged'; Rogers, *Documenting Eighteenth Century Satire*.

[11] Forbes, 'Index of the Board of Longitude Papers'; Howse, *Nevil Maskelyne*; Howse, *Greenwich Time*; Howse, 'Britain's Board of Longitude'; Johnson, 'The Board of Longitude'.

[12] Miller, 'The Royal Society'; McConnell, *Jesse Ramsden*; Mörzer Bruyns, *Sextants at Greenwich*; Stimson, 'Some Board of Longitude'.

[13] Forbes, *Tobias Mayer*; see also Forbes, 'Tobias Mayer's Claim'; Forbes, *The Birth of Navigational Science*.

the later Board shaping expectations for the Commissioners' earlier decades, limited consultation of the surviving archives, and a focus on the years of exchanges with John Harrison (discussed in the next paragraphs). Another notable feature of works offering a long history of navigation is that the development of lunar distance and timekeeper methods from the 1760s marks a crucial stage that often forms an organising element in the narrative. For Taylor, indeed, it becomes the culmination.[14] A further consequence of this, at least in anglophone literature, has been for the British narrative to overshadow the long history of longitude in other seafaring nations.[15]

It is the Harrison story in particular that has come to dominate narratives about the Board of Longitude. A renewal of interest in the clockmaker began with Rupert Gould's restoration of his sea clocks and resulting publications after the First World War and continued through Humphrey Quill's considered biography.[16] The most dramatic transformation, however, resulted from the Longitude Symposium held at Harvard University in 1993. This inspired Dava Sobel's unexpected publishing sensation *Longitude: The True Story of the Lone Genius Who Solved the Greatest Scientific Problem of His Time*, as well as a much-cited scholarly collection of papers from the conference.[17] While Sobel's book has elicited some critical comment, Ulrike Zimmermann has noted that the effect of its tremendous success, aided by a well-received television film, has been to lift Harrison to heroic status, changing markedly the perceptions that would have been more usual in the century or so after his death.[18] In 2002 a public vote placed the clockmaker at number 39 in a BBC poll of the 100 greatest Britons, and in 2006 a memorial was laid in Westminster Abbey.[19] Harrison's story has since become something of a staple: in 2010, Adam Hart-Davis presented a laudatory BBC documentary as part of a series on the history of the world, while in 2013 Brian Cox's *Science Britannica* offered a version of the story as an example of successful targeted research.[20] These and other repetitions of the narrative have served to reinforce the perception that Harrison alone 'solved' the longitude problem and did

14 Taylor, *The Haven-Finding Art*, pp. 253–262; see also Bennett, *Navigation*, pp. 73–85; Dunn, *Navigational Instruments*, pp. 34–46.

15 The work of de Grijs in exploring the earlier history is a recent exception: de Grijs, *Time and Time Again*; de Grijs, 'European Longitude Prizes. I'; de Grijs, 'European Longitude Prizes. 2'; de Grijs, 'European Longitude Prizes. 3'; de Grijs, 'European Longitude Prizes. 4'. Dunn and Higgitt (eds.), *Navigational Enterprises* also aims to address the issue.

16 Gould, *The Marine Chronometer*; Quill, *John Harrison*.

17 Sobel, *Longitude*; Andrewes (ed.), *The Quest for Longitude*.

18 Zimmermann, 'John Harrison'; Higgitt, 'Revisiting and Revising', pp. 38–46; *Longitude* (television series). For discussions of Sobel's work by academic historians, see Gascoigne, 'Getting a Fix'; Miller, 'The "Sobel Effect"'; Schaffer, 'Our Trusty Friend'; Schaffer, 'The Disappearance of Useful Sciences'.

19 *Great Britons* (television series); Taylor and Wolfendale, 'John Harrison'.

20 'The Clock That Changed the World'; *Science Britannica*, episode 3, 'Clear Blue Skies'.

so only by overcoming a malicious Board of Longitude. This focus has also served in part to obscure both the Commissioners' activities before the 1760s and the importance and diversity of the Board's work after its dealings with Harrison had concluded.

A second important context for the public history of longitude, including that of the Board, has centred on the National Maritime Museum, which has displayed related objects and instruments, including notably Harrison's and Larcum Kendall's marine timekeepers, since its public opening in 1937. The story gained prominence as the museum took over the Royal Observatory site from the late 1950s, particularly after the appointment of former lieutenant-commander and navigational historian David W. Waters in 1960. Under Waters's direction, the newly created Navigation Department mounted two exhibitions directly related to the longitude story: *4 Steps to Longitude* (1962) and *The Nautical Almanac Bicentenary* (1967).[21] Harrison's timekeepers framed and featured prominently in the former and have been central to displays and publications since then.[22] They remain an important draw for visitors to the observatory.

To mark the 300th anniversary of the first Longitude Act, the National Maritime Museum mounted two exhibitions in 2014. *Longitude Punk'd* at the Royal Observatory took a light-hearted look though a steampunk lens at the creativity inspired by the quest for longitude.[23] At the National Maritime Museum, *Ships, Clocks & Stars* and the book published to accompany it (written by its lead curators) drew from the ongoing research project to reshape the narrative, placing Harrison and his timekeepers not as the culmination of the story but as a stage along the way.[24] Harrison remained an important figure in this account but there was a deliberate attempt to emphasise the wide range of actors involved, including the Commissioners of Longitude.[25]

The same year also saw the launch of the Longitude Prize 2014 (now renamed the Longitude Prize), a contemporary challenge prize that chose as its object the search for ways of tackling global antibiotic resistance.[26] Behind the naming was a belief in the longitude story's power 'to make the mind

[21] Dunn, 'Collecting and Interpreting', pp. 78–79. Each exhibition had a related publication: National Maritime Museum, *4 Steps to Longitude*; Sadler, *Man Is Not Lost*.

[22] Betts, *Harrison*; Betts, *Marine Chronometers at Greenwich*; Howse, *Greenwich Time*'; Higgitt, 'The Royal Observatory, Greenwich'; Lippincott, *A Guide to the Royal Observatory*.

[23] Atkinson, 'Steampunking Heritage'.

[24] Dunn and Higgitt, *Finding Longitude*; Higgitt, 'Challenging Tropes'; McAlpine, 'Ships, Clocks & Stars'; Zimmermann, 'John Harrison', pp. 125–126; Schaffer, 'Chronometers, Charts, Charisma'. *Ships, Clocks & Stars* toured to the Folger Shakespeare Library, Washington, DC (16 March–23 August 2015), Mystic Seaport, Connecticut (19 September 2015–28 March 2016) and the Australian National Maritime Museum, Sydney (5 May–30 October 2016).

[25] Barrett and Dunn, 'A Mechanic Art', discuss the role of portraits in this context.

[26] Rees, 'A Longitude Prize for the Twenty-First Century'.

race, the chest swell and the heart beat faster'.[27] This co-option of the narrative to address contemporary concerns reflects recent attempts to interpret the Longitude Act as 'the definitive example of a government-sponsored attempt to promote innovation' and 'a model for challenge-based open innovation in the twenty-first century'.[28] One aim of this book is to gain a better understanding of the significance of collective and public endowment and its implications for a model of technical change based on existing understanding of what is feasible and what is practical. Lacking this context, figures like John Harrison become nothing more than the lone genius, heroic outsider and emblematic 'man in his shed', whose brilliance can be harnessed with the right incentive.[29] As Rebekah Higgitt and James Wilsdon note,

> While the well-known version of the longitude story would seem to back claims about the efficacy of one-off inducement prizes … the Board was necessarily much more flexible in the range of funding mechanisms they used than the simple version of the story suggests.[30]

Rethinking the Board of Longitude

The chapters in this volume seek to recast the history of the Commissioners of Longitude and the Board that they became, focusing not just on their most notorious episode but on the full chronological narrative. Some of our revisions are indicated in the chapter outlines at the end of this introduction but it is first necessary to touch on two key areas, the first concerning terminology and the second the relationship between the Commissioners of Longitude and Parliament.

In twentieth- and twenty-first-century writings relating to the Longitude Act and its consequences, the rewards offered have generally been referred to as the 'Longitude Prize', a term now entrenched in academic and public narratives. However, this phrase was very rarely used at the time and only became common in historical accounts in the late twentieth century. Prizes in the eighteenth century were won in competition, through combat or by chance, as in a lottery. Thus, when virtuoso William Stukeley visited John Harrison at

[27] Highfield, 'Calling all Geniuses'; see also Longitude Prize, 'Origins of the Longitude Prize'. With the BBC as a partner in the prize, several television and radio programmes included segments on the historical narrative; for example, *Inside Science* (BBC Radio 4, 24 April 2014), *The One Show* (BBC One, 15 and 19 May 2014), *Horizon*, 'The £10 Million Challenge' (BBC Two, 22 May 2014) and *Coast*, 'Sea and the City' (BBC Four, 12 August 2014).

[28] Burton and Nicholas, 'Prizes, Patents', p. 35; Spencer, 'Open Innovation', p. 43. Khan, *Inventing Ideas*, pp. 72–74, argues that the Longitude Act failed to encourage invention as part of a broader argument about the failure of incentive prizes.

[29] For critiques of such narratives, see Charney, 'Lone Geniuses'; Jordanova, 'On Heroism'.

[30] Higgitt and Wilsdon, 'The Benefits of Hindsight', pp. 82–83.

Barrow-in-Furness in 1742 and recorded that this 'wonderful genius' might 'carry the prize of longitude found out', it was without specific reference to the provisions of the Act of 1714.[31] By contrast, something given as remuneration or compensation for information or for work done was a *reward*.[32] All Acts dealing with longitude in the eighteenth and early nineteenth centuries consistently used the term 'reward'. Instances of the word 'prize' were so unusual that one might even list them. In 1768 and 1774, a couple of references to 'prize' appeared in editions of the *British Palladium* by army officer and mathematician Robert Heath. The latter, found in an attack on Astronomer Royal Nevil Maskelyne, the Board and the *Nautical Almanac*, is particularly revealing:

The *British Computers* make as puzzling a *Mystery* of their *mixed* and *borrowed* Calculations (and some *no Use at Sea*) as of the Longitude they seek. But we, on-board the *Navy*, make the same Use of the *Nautical Ephemeris* as we do of a *Pack of Cards* or the *Back-gammon Tables*; to pass an idle Hour or to kill Time! For, as we find none is paid for *chacing* the *Longitude-Prize* but *Longitude Schemers* and *Projectors* (for whose profit we are *annually* out of Pocket by being *compelled* to buy their Work,) we have long given over the *Chace* ourselves, without endeavouring to come up with what is not worth our *picking up*.[33]

Deploying a maritime analogy, Heath used prize in the sense of a ship taken during wartime, a familiar meaning in a century dominated by naval conflict.[34]

Maskelyne's only published use of prize in this context was during a dispute in the 1790s with Thomas Mudge senior and junior, who claimed that the Astronomer Royal's interest in winning cash from the Board for his astronomical tables biased him against watchmakers such as the senior Mudge (see Chapter 6). The lawyer Thomas Mudge junior alleged that Maskelyne's search for 'one of the specific rewards offered by the Act' meant 'he therefore, of course, cannot wish well to the mechanics who are candidate for the same prize'. Maskelyne reproduced Mudge's remarks, including the term 'prize', yet then insisted that mathematicians making astronomical tables and clockmakers improving timekeepers 'cannot be candidates for the same reward', reverting to the term used in successive Acts. The mathematician John Hellins, former computer at the Royal Observatory, likewise confirmed that

[31] William Stukeley, 'Memoirs of the Royal Society. Vol. II', 1750, Spalding Gentleman's Society, fols. 30–31 www.sgsoc.org/wp-content/uploads/2021/07/Stukeley-MRS-2-Page-Images.pdf [accessed 16 March 2024].

[32] 'reward, n', *OED Online* (Oxford: Oxford University Press, 2021) www.oed.com/view/Entry/165009 [accessed 6 May 2021].

[33] *The British Palladium ... for the Year 1774*, p. 4. The earlier reference is *The British Palladium ... for the Year 1768*, p. 72; see also Higgitt, 'Prize Fights'.

[34] 'prize, n.2', *OED Online* (Oxford: Oxford University Press, 2021) www.oed.com/view/Entry/151648 [accessed 6 May 2021].

Maskelyne's large income prevented bias towards timepieces' performance, 'even if their Makers and the Astronomer-Royal had been competitors for the same reward (which was not the case)'.[35]

French-language journalists who first reported the Act of 1714 used terms such as *récompense* to describe the rewards on offer. This was somewhat of a contrast with usage in scientific academies across Europe, which explicitly offered prizes (*prix, praemium*) for such schemes.[36] In a widely read French account of natural philosophy, pre-eminent academic mathematician Leonhard Euler, awarded £300 for his lunar theory having contributed to navigation tables, described alternative methods for longitude at sea. He mentioned Harrison had been granted sums for marine timekeepers: 'l'inventeur reçut un partie du prix destiné à la découverte de la longitude'. Writing in 1761, Euler had no further news of Harrison's scheme and guessed that it had failed. In 1795, Euler's text was translated by a Scottish divine, Henry Hunter, who rendered the phrase as 'the parliamentary prize proposed for the discovery of longitude', adding that meanwhile Harrison's 'marine time-piece has been tried with success'.[37] In sum, reference to a longitude prize was effectively confined to a few satirical and polemical uses, or else translation from foreign accounts. In what follows, therefore, we will generally use reward or rewards to refer to the monies available under the terms of the Acts.

Two other terms will be used in a chronologically specific way. As Chapter 5 explains, 'Board of Longitude' was a phrase not generally employed until the 1760s, when it first became appropriate to describe the Commissioners as a committee with bureaucratic trappings that met regularly. For the earlier chapters, therefore, we talk of the Commissioners of Longitude or simply the Commissioners to highlight what is a significant reappraisal of the immediate consequences of the Act of 1714. Likewise, terms such as 'marine timekeeper' or simply 'timekeeper' have been used for all such devices prior to 1780, when the term 'chronometer' first began to be adopted for the navigational instrument that came into more widespread use in the nineteenth century (see Chapter 6).

The relations between the Commissioners of Longitude and Parliament, mediated by no fewer than twenty-one Acts passed between 1714 and 1828, also require clarification. The first of these Acts nominated the Commissioners as experts fit to assess schemes and proposals. They were able significantly to interpret their roles under the terms of the relevant legislation, but criticisms of their resulting conduct were sometimes made in public and private, and

[35] Mudge, junior, *Narrative*, pp. 1, 4; Maskelyne, *An Answer*, pp. 54, 126; John Hellins to Francis Maseres, 25 September 1792, CUL RGO 14/23, p. 179r.

[36] *Journal Littéraire* 4, part 1 (May–June 1714), p. 235; Fauque, 'Testing Longitude Methods', p. 160; Calinger, *Leonhard Euler*, p. 30.

[37] Euler, *Lettres*, III, p. 45; Hunter (trans.), *Letters of Euler*, II, p. 49.

challenges to their judgements were made to Parliament and government lawyers. The latter approach brought success in the case of Thomas Mudge but failed for Thomas Earnshaw (see Chapter 6).

Parliament received vast numbers of petitions about diverse subjects in this period, estimated at 38,000 between the Restoration and the end of the eighteenth century, and especially from the 1760s. Most were launched by ad hoc groups, for whom social status significantly influenced impact, and later in the century they focused more explicitly on the redress of grievances against official bodies' conduct. Though petitioners did not always expect their demands would be met by legislators, they often brought issues of justice and political economy to public debate.[38] John Harrison famously used a direct appeal to Parliament to circumvent what had become frustrating discussions with the Commissioners, allowing him to receive most of the remaining payment he considered his due. Significantly, Harrison's published statements linked his treatment to the broadest issues of property rights and parliamentary powers. He claimed he had long devoted his efforts 'on the Faith of an Act of Parliament', that of 1714. He rejected the 'Doctrine … that the Commissioners were invested with a Discretionary Power of ordering other Trials and the fulfilling of other conditions than those specially annexed by Act of Parliament to the Reward' and insisted that before the Act of 1765, which removed one of the lines of argument he and his supporters were pursuing, he 'had as full and perfect a *Right* to the Reward of 20,000l. as any Free-holder in Britain has to his Estate'.[39]

Eighteenth-century British parliaments were widely understood as legislatively supreme, especially on such issues as subjects' property and rights. The history of the Commissioners' conduct should thus be viewed within common patterns of parliamentary legislation in Hanoverian Britain. The long sequence of longitude statutes existed within a huge overall increase in the number of Acts passed in the period, more than 14,000 between the Glorious Revolution and the end of the eighteenth century. Most statutes were articulations not of positive or general principles but rather of local and particular provisions. They were often uncoordinated and unplanned and frequently made in response to active lobbying. At least a fifth repealed, extended or amended existing laws, as was true of most of the Longitude Acts.[40] Hanoverian parliaments preferred lengthy series of specific Acts to the articulation of singular and clear policies. Historians concur that law-making in this period did not and could not end with an Act's passage and that provisions were often subtly altered and reconstrued.

[38] Hoppit, *Britain's Political Economies*, pp. 153, 159; Innes, 'Legislation and Public Participation', pp. 114–115.

[39] Harrison, *Remarks*, 1st ed., pp. 20–21.

[40] Langford, *Public Life*, p. 156; Hoppit, *Britain's Political Economies*, p. 43; Loft, 'A Tapestry of Laws, pp. 276–278.

Statutory meanings were frequently changed through shifts of terms' definitions and legal statements could be and were adjusted by interest groups and administrative institutions, however explicit they had appeared.[41] Complaints against this mutability and multiplicity of legislative powers were also common. In 1762, Tobias Smollett's *Critical Review* praised the fixed 'method followed in geometry and natural philosophy' in contrast to the unfortunate shifts in British legal conduct: it seemed that 'a party shall be deprived of his right' because of changeable legal authority.[42] Historians of eighteenth-century property law have influentially argued that this applied especially in the establishment of large numbers of commissioners, 'the new governors of Georgian England', to whom discretionary authority was transferred over wide ranges of public affairs, notably in trade and transport.[43]

Within this broader context of legislation and revision, many of the Longitude Acts were designed to renew and extend funding and warrant for the activities of the Commissioners of Longitude. They were periodically required to report on their spending to justify and obtain further funds, during periods of economic reform and well-publicised campaigns against 'Old Corruption'. This generally occurred through new Acts, which typically reported the exhaustion of the Commissioners' existing reserves and approved further expenditure of up to £5000 or even £10,000 (see Chapter 8 and Appendix 2). In May 1781, for example, Parliament received accounts from the Board of Longitude outlining expenditures on fifteen different beneficiaries. These included the costs of computation and publication of the *Nautical Almanac*, a reward to the clockmaker John Arnold for his improvement of watches and a reward to James Cook's widow, Elizabeth, in recognition of her late husband's work on astronomical observations during his last voyage to the Pacific.[44]

Parliament on occasion questioned these accounts or demanded their production, especially as the Board's expenditures continued to rise in the decades after the launch of the *Nautical Almanac*. One financial critic was Samuel Horsley, a bishop and mathematical writer who was a member of the parliamentary subcommittee that helped back Mudge's petition against the Board (see Chapter 6). Horsley was a convinced enemy of Joseph Banks, the leading Commissioner of Longitude and the Royal Society's president, after the dissensions at the Royal Society during the 1780s set some mathematical fellows against the leadership. In May 1796, the bishop spoke in the House of Lords to delay an Act that would have authorised a further £5000 for the Board. He pithily recalled the Commissioners' repeated requests for cash, a total he

[41] Langford, *Public Life*, pp. 162–163; Hoppit, *Britain's Political Economies*, p. 101; Loft, 'Tapestry of Laws', pp. 279, 305.
[42] 'The Frederician Code', *The Critical Review*, 13 (January 1762), pp. 1–12 (pp. 6, 3).
[43] Langford, *Public Life*, p. 208; Lemmings, *Law and Government*, pp. 142–148.
[44] *Journals of the House of Commons*, 38, pp. 498–500 (30 May 1781).

estimated as £38,000, and stated that the *Nautical Almanac* and related publications ought to bc self-funding 'as they were bought throughout Europe'. Horsley noted that 'the reward to be given for discoveries' was not included in the new financial demand, so his 'imagination therefore had not been able to compass the probable causes of the expenditure of the Board'.[45] Although the Act of 1796 ultimately passed, Horsley's intervention was but one indication of how complex relations were between the Commissioners and Parliament. There is evidence throughout this book of ingenious manoeuvres, political and public, by both allies and critics of the conduct of the Commissioners. The Commissioners could, on rare occasions, limit challenges to their authority through fresh legislation, steered carefully through the necessary procedures, to clarify or change the framework of their conduct. This happened in both 1765 and 1774, when assessment criteria were revised, likely to avoid some of the problems that had arisen with John Harrison (see Chapter 6).[46]

Claimants' rival interpretations of the Act of 1818, which was the mechanism for a radical shake-up of the Board by Banks and his allies (see Chapter 11), show how flexibly the terms of the Acts could be applied. In 1820, for example, Lieutenant of Marines Edward Naylor unsuccessfully demanded that the Board return papers related to his calculation of magnetic variation, as detailed in Chapter 9. In defence of what he saw as his 'property', Naylor consulted the Act of 1818 and insisted that no clause therein sanctioned the retention of documents, 'the Board not having by any reward to me made them their own'. Thomas Young, the Board's Secretary, refused the demand but noted that Naylor would succeed if he asked for the documents' return 'as a favour'.[47] Discretionary interpretation like this was pervasive.

In early 1822, whaler and Arctic naturalist William Scoresby requested magnetic instruments and a chronometer from the Board for a northern voyage. According to Scoresby, Admiralty Secretary and Commissioner of Longitude John Barrow had hinted that provisions of the Act of 1818 could be manipulated to offer £2000 for circumnavigating Spitzbergen. Nothing came of this and, although the Board's Instruments Committee approved some loans, Barrow wrote from the Admiralty that 'we have no authority to dispose of His Majesty's property to private ships and I believe the Board of Longitude has as little. Mr Scoresby must therefore do the best he can with his private means.' Scoresby furiously riposted that 'I do think if I understand the meaning of the act empowering the Board of Long[itude] to dispense 500£ a year in the promotion of geographical knowledge &c, that it is part of the intention of this act

45 *Parliamentary Register*, 45, p. 341 (13 May 1796). For Horsley against Banks, see Heilbron, 'A Mathematicians' Mutiny', pp. 84–88; Homes, 'Friend and Foe', p. 247.

46 Barrett, '"Explaining" Themselves', demonstrates how carefully the Commissioners, notably William Wildman, Viscount Barrington, steered the 1765 Act through Parliament.

47 Edward Naylor to Thomas Young, 15 February 1820, CUL RGO 14/12, p. 378r.

to promote geographical nautical knowledge as well to his Majesty's subjects and His Majesty's servants'. This view of the Act's meaning was repeated in Scoresby's successive petitions to the Board. The acceptance of interpretations of the Act often hinged on the connections linking principal Commissioners with individuals including mariners, ministers and naval officials, and institutions including Parliament, the Board of Trade and the Admiralty. These variable legal and informal linkages generated resources as well as challenges in the later history of the Board up to its disappearance in 1828.

Content Summary

The chapters that follow reinterpret the history of the Commissioners as well as that of the British search for longitude. In doing so, they shed light on many other topics arising from longitude's intersection with diverse interests and the expansion of the Board's activities in its later decades. We have also highlighted what we hope are productive avenues that future scholars might choose to follow. The account is told in two strands, and complemented by three appendices (including the text of the first Longitude Act and analyses of the Board's financial activities), a timeline and a glossary to help the reader. The first strand, comprising Chapters 2–5, 8, 11 and the Epilogue, offers a chronological account from the first Longitude Act and the early work of the Commissioners through the development of the Board of Longitude as a standing body from the 1760s to its dissolution in 1828. The second strand focuses on specific aspects of the Board's work from the 1760s onwards, arranged to align with the book's overall chronology.

Opening the more chronological strand of the book's narrative, the next chapter explores the long- and short-term roots of the Act of 1714, in terms of both the development of techniques and instruments for navigation and related practices and the judging and funding of longitude proposals in sixteenth- and seventeenth-century Europe. It thus exposes the degree of continuity with earlier precedents in the events and discussions leading to the new British rewards. These saw the self-interested lobbying of two projectors gain momentum through a confluence of national and political interests, before becoming enshrined in law as rewards open to all comers. The Act of 1714 and its immediate aftermath are scrutinised in the third chapter, which shows the extent to which the wording of the Act drew on precedents from the previous century and argues that the Act was never intended to create a 'Board of Longitude' as a formal, standing committee. Rather, it nominated individuals – by name or by virtue of their official role – fit to judge potential ideas. It is shown that the Act did indeed foster activity and discussion around longitude matters over the next two decades, albeit with a remarkable continuity in the types of schemes being discussed.

Chapter 4 covers the two decades from the first minuted meeting of the Commissioners of Longitude in 1737. During this time, small groups of Commissioners were called together sporadically for ad hoc meetings, principally to agree funding for specific projectors, notably clockmaker John Harrison and longitude veteran William Whiston. Yet it was also a period in which public opinion sometimes reverted to mockery of those seeking the seemingly impossible longitude dream, as decades began to pass without the allocation of a longitude reward. However, the chapter emphasises the value of understanding schemes and projectors, including women, whom some more modern authors have treated with derision.[48] Focusing on the period from the early 1760s to the resolution of the Harrison affair in 1773, the fifth chapter argues that this was the period in which the 'Board of Longitude' came into being, largely in response to disputes surrounding the sea trials of Harrison's fourth marine timekeeper ('H4'). These debates, which focused on questions of adequate testing and the judging of trials, disclosure and replicability, saw the Board harden its stance through the use of legislation to ensure resolution.

Chapter 8 surveys the workings of the Board in a period of increased expenditure and bureaucracy following the termination of dealings with Harrison. Managed through a permanent secretary, the Board more resembled an office of state, although personal and patronage relations still played vital roles in its conduct. The final decade of the Board's formal existence following the Napoleonic Wars is reinterpreted in the closing chapter, which explains how an Act of 1818 brought the Board more tightly under Admiralty control, and how that of 1828 effectively shifted its work inside the Admiralty. The epilogue considers the decades after the dissolution of the Board of Longitude and the ways in which these functions were subsumed by and distributed between other bodies.

Beginning the second strand of the volume, which focuses on specific aspects of the Board's work from the 1760s onwards, Chapter 6 looks at the Commissioners' relationships with watchmakers in the decades after the end of dealings with Harrison. The Board continued to field proposals and increasingly relied on land trials at the Royal Observatory, alongside a small number of long-distance voyages, for testing the nascent technology. However, the Commissioners became embroiled in two debates that further shaped their horological dealings and saw their authority contested in Parliament. The creation and management of the *Nautical Almanac*, one of the Board's most important concerns, is explored in the seventh chapter. Astronomer Royal Nevil Maskelyne managed the new publication, coordinating the work of a group of computers and comparers, including a notable mother-and-daughter pair. The organisation and increasing costs of this endeavour came to dominate the

48 Gingerich, 'Cranks and Opportunists'; contrast Barrett, *Looking for Longitude*, pp. 136–142.

Board's subsequent affairs. After Maskelyne's death, work seems to have suffered, a situation that the Act of 1818 sought to correct by placing the *Nautical Almanac* under Thomas Young's management, although controversy dogged the publication for the next decade.

Chapter 9 surveys the intensification of administrative labour and the management of correspondence and lobbying, by exploring the ways in which the Board handled the myriad schemes presented by mathematicians, mariners, inventors and entrepreneurs. Labels of impracticality, eccentricity and derangement have long been assigned to many such schemes, notably in the classification imposed by Airy when reorganising the Board's archives. The chapter instead considers how schemes were assessed and managed, with the Board notably distinguishing between projects reckoned impossible or unsound, and those that it judged irrelevant or beyond its scope. One of the most significant of the Board's enterprises in its final years, management of a new observatory proposed in 1820 for the Cape of Good Hope, is surveyed in Chapter 10. Although funds came from Navy estimates, long-distance management proved difficult, revealing both the opportunities and challenges in issues of scientific and geographical management in an epoch of empire and reform.

Conclusion

This volume represents a first attempt at more completely describing the work of the Commissioners of Longitude and of the Board that they became. We must acknowledge, of course, that many questions remain. In addition, there is much that can still be done to explore more fully the surviving papers of the Board of Longitude, not to mention the many additional archives and publications that shed further light on this fascinating body.[49] Of those questions that remain only partly resolved, a major one concerns the passage of the Longitude Act 1714. As Chapter 2 discusses, Mobbs and Unwin have shed valuable light on its journey through Parliament, in particular its connection to other legislation in the period.[50] What remain less clear are some of the more immediate drivers, particularly the role of William Whiston and Humphry Ditton in the process. Likewise, while Chapters 3 and 4 draw on a range of sources including published treatises and periodicals to uncover some of the work of the Commissioners of Longitude before 1737, there is certainly more to discover in unpublished correspondence and other sources.

Several areas for further work in what remains of the Board of Longitude archive, all touched on briefly in this volume, are also apparent. First, this

[49] Fulford, 'The Role of Patronage', pp. 471–474, provides one example of other archives in which further information is held.

[50] Mobbs and Unwin, 'The Longitude Act of 1714'.

volume has not gone into great detail about the work of the Board, and in particular of Nevil Maskelyne, with respect to the astronomers sent on voyages of exploration. As Chapter 6 explains, many of these voyages have rich secondary literatures, and some initial work has been published elsewhere to bring together an overview from the Board's perspective.[51] Nevertheless, a full account remains to be made. This could extend to a further area, the Board's involvement in pendulum and survey work in the early nineteenth century, which has to date been touched on relatively lightly.[52]

It would also be valuable to investigate the Board's final decade, in which it appointed committees that looked into matters considerably beyond its original remit, including the challenge of measuring a ship's tonnage (in 1821) and, in a joint committee with the Royal Society, the production of high-quality optical glass.[53] There also remains enormous potential in undertaking a more systematic study of the incoming schemes to the Board, including an analysis of how this changed over time, and of their proposers, many of whom are otherwise unknown or only sporadically recorded elsewhere. The surviving archive provides in particular fascinating evidence of the inventiveness evident among the 'middling sort', ranging from ambitious artisans and natural philosophers to authors of a vast range of mathematical and mechanical schemes.[54] Lastly, and strongly related to these questions of proposals and schemes proposed to the Board by a host of protagonists, there is scope to examine more closely the role of patronage and related practices such as subscription in eighteenth- and early nineteenth-century science through the activities of the Commissioners and their correspondents. Several authors have noted the continued importance of patronage throughout the period and long into the nineteenth century, even as government became increasingly bureaucratised within an expanding military-fiscal state.[55] This was an important context for the activities of the Board, which straddled the intersection between institutions of state and networks of

[51] Initial work can be found in Higgitt, 'Equipping Expeditionary Astronomers'; Eóin Phillips, 'Making Time Fit', pp. 114–163; Phillips, 'Instrumenting Order'.

[52] Higgitt and Dunn, 'The Bureau and the Board', pp. 210–212; Waring, 'Thomas Young', pp. 103–155.

[53] For the Glass Committee, see Frank James, 'Michael Faraday's Work'; Jackson, *Spectrum of Belief*, pp. 124, 145–150.

[54] Publications discussing specific schemes include Bryden, 'Georgian Instruments Patents'; Bryden, 'William Ross'; Dunn, 'Scoping Longitude'; Dunn, 'The Impudent Mr Jennings'; Dunn and Higgitt, *Finding Longitude*, pp. 177–185; Forbes, 'Schultz's Proposal'; May, 'The Gentleman of Jamaica'; Schaffer, 'Swedenborg's Lunars'. Bryden, *Innovation in the Design of Scientific Instruments*, identifies recipients of Society of Arts premiums who also approached the Board of Longitude.

[55] Fulford, 'The Role of Patronage'; Gascoigne, *Science in the Service*, pp. 8–10, 16–19, 22, 199–202; Gascoigne, *Science and the State*, pp. 66–67, 84–92; Gillan, 'Lord Bute'; Shapin, 'Science and the Public', pp. 1003–1004; Stewart, *The Rise of Public Science*, pp. 151–160; Wills, 'The Diary of Charles Blagden', pp. 17–18.

individuals. While the significance of patronage to its work is touched upon in some of the chapters that follow, it remains to be explored in depth.

Drawing on the Board of Longitude archive and material elsewhere that supplements this core material, the intention of this volume is to document how the Commissioners of Longitude functioned, what they did, and how they perceived and presented their roles in official documentation and to the wider world. Based on an unprecedented analysis of the surviving Board of Longitude archives, the story told here is principally centred on the perspectives of the British Empire and especially of its metropolitan institutions. This approach is certainly not adopted with the intent of presenting that as the default viewpoint, or of ignoring the impacts of imperial activities or overlooking the manifold involvements in navigation and longitude by practitioners beyond and outside Europe. An aim, rather, is to encourage work on the global and colonial dimensions and impacts of these histories. The book offers a preliminary account of the narrative available from the surviving archive and it is to be hoped it will help other researchers navigate that rich material. We look forward to the work that other historians will carry out to explore in more detail the questions we have raised.

2 Sailing the Oceans and Seeking Longitude before 1714

Alexi Baker

The Longitude Act of 1714 was unusual in a number of ways and in the long term spurred important developments in navigation and science. However, the tenets of the legislation, as well as the dynamics of the search for longitude in the decades immediately after the Act's introduction, exhibited significant continuities with the past. The schemes pursued and those given most credence were generally the same as before, and reward-seekers would largely continue to approach the same individual and institutional authorities as they had previously. A specific confluence of commercial and political interests and self-interested lobbying spawned the new rewards. However, these had all emerged from and were shaped by more than two centuries of efforts and discoveries in navigation, astronomy, natural history, timekeeping and attempts to determine longitude on land and at sea.

Longitude in Europe before 1714

A number of developments in Europe in the early modern period underpinned later attempts to address the longitude problem. Success in measuring longitude on land through astronomical observations encouraged the search for a method that would work at sea, even though techniques such as delicate observations of satellite positions could not easily be transferred directly from one arena to the other. This latter point was not, however, something that every early modern reward-seeker or commentator appreciated. Marine navigation also benefited from innovations in technique and technology that contributed to developments in the Georgian era, and in some cases continued to coexist with them.

For the determination of position, the tools of the early modern European navigator included the magnetic compass and the log and line. These gave the ship's heading and speed or distance travelled, the data essential for determining current location by a method known as dead reckoning. At its most basic, dead reckoning involved plotting the ship's course from position to position using the direction and distance travelled. For longer voyages, recognising that the ship was travelling on the surface of a sphere, the mariner needed printed tables to convert the movements more accurately into changes of position north–south (difference of latitude) and east–west (generally called 'departure'). The ship's position was obtained from its previous one by applying the difference of latitude and the departure and, with adjustments for the effects of wind and currents where known, plotting the new position on the chart and/or noting it in the written log. During the sixteenth and seventeenth centuries, dead reckoning was supplemented by new techniques such as measuring latitude directly from observations of the Sun or Pole Star using a cross-staff, quadrant, astrolabe or backstaff. Direct measurement of latitude therefore allowed the navigator to check and correct the dead-reckoning determination.[1] Longitude could not be measured in such a direct manner, however, so there was no way of assessing the result given by dead reckoning.

These tools and techniques had limitations. For example, dead reckoning could not easily take account of factors such as ocean currents, and errors tended to build up over longer distances. Nevertheless, European ships were able to use dead reckoning (which has remained a staple of navigation) to cross large expanses of ocean reliably.[2] This was aided by individual navigators' experience and tacit and local knowledge. Moreover, accurate estimates of a ship's longitude were not equally important everywhere. Near coastlines, which were notoriously dangerous, local knowledge and keeping a good lookout were of greater importance. On long-haul routes, particularly the lucrative trade routes to Asia, knowing the longitude gained greater significance.[3] As Samuel Pepys noted, 'the East Indies masters are the most knowing men in their navigations, as being from the consideration of their rich cargoes, and the length of their sailing, more careful than others'.[4] As a result, the East India Company would prove particularly receptive to new longitude technologies.

In addition to innovations in latitude-finding and navigation generally, the sixteenth and seventeenth centuries saw the formulation of a range of approaches to longitude-finding that would remain of interest into the nineteenth century. Proposals outlining in principle the use of lunar observations, mechanical timekeepers and magnetic variation to better estimate longitude emerged during the sixteenth century, as did suggestions for related technologies. The idea of using Jupiter's moons accompanied the

[1] Bennett, *Navigation*, pp. 42–53, 67–73; Hicks, *Voyage to Jamestown*, pp. 50–54, 103–124, 142–146; Schotte, *Sailing School*, pp. 31–41.

[2] Carter and Carter, 'The Age of Sail', and Peck, 'Theory Versus Practical Application', make forceful arguments for the effectiveness of dead reckoning; see also Peck, 'The Controversial Skill'. Huxtable and Jackson, 'Journey to Work', show how effective it could be for transatlantic voyages in the 1760s.

[3] Cook, 'Establishing the Sea-Routes'; Landes, 'Finding the Point', p. 25; de Grijs, 'European Longitude Prizes. 2'.

[4] Chappell (ed.), *The Tangier Papers of Samuel Pepys*, pp. 127–128.

development of the telescope in the early seventeenth century. Inventors and clockmakers from Galileo Galilei to Christiaan Huygens also took important steps towards the development of successful marine timekeepers during the 1600s.

Since longitude is a measure in the direction of the Earth's rotation, the longitude difference between two places can be thought of as the difference between their local times (given by the Sun's position). Thus, one hour of time difference is equivalent to a 15-degree difference of longitude. Many longitude schemes exploited this fact by seeking a way to determine the time where the ship was and, simultaneously, at a reference point with a known geographical position.[5] Astronomical methods tried to use the observed motions of celestial bodies as a timekeeper to determine longitude in this manner. European explorers including Columbus and Vespucci had initially tried to find longitude by observing lunar eclipses and occultations, but the viewing technologies and predictive astronomical tables available were not yet sufficiently accurate.[6] Early modern inventors proposed new and improved astronomical instruments to this end, as well as viewing platforms to stabilise the shipboard observer. For example, the French Protestant mathematician Jacques Besson mentioned a number of innovations in *Le Cosmolabe*, first published in 1567. This included a stabilised chair, depicted within a gimballed globe mounted on perpendicular axes on the poop deck of a ship. Such innovations were often called 'marine chairs' or 'marine observatories' in English and continued to be proposed to the Commissioners of Longitude into the nineteenth century.

In 1514, Johann Werner of Nuremburg gave possibly the first printed account of how terrestrial longitude might be found from lunar distances. This involved measuring the angular distance between the Moon and the Sun or another star to determine the time at the reference meridian, to be compared with local time to find the longitude difference.[7] Werner's ideas were expanded upon and disseminated by authors such as Peter Apian and in England by William Cuningham.[8] Early in the following century, there were several attempts to perform the method at sea, as well as further refinements.[9] For instance, Jean-Baptiste Morin proposed an improved method that accounted for lunar

[5] Dunn and Higgitt, *Finding Longitude*, pp. 20–21. For local time and its determination, see Andrewes, 'Finding Local Time'; Cotter, *A History of Nautical Astronomy*, pp. 32–48.

[6] De Grijs, *Time and Time Again*, pp. 2-44–2-47; de Grijs, 'A (Not So) Brief History', pp. 498–500; Howse, *Greenwich Time*, pp. 182–183; Morison, *Admiral of the Ocean Sea*, pp. 183–196; Randles, 'Portuguese and Spanish Attempts', pp. 235–236; Turner, 'Longitude Finding', pp. 407–408.

[7] Cotter, *A History of Nautical Astronomy*, pp. 195–198; Howse, *Greenwich Time*, pp. 183–185. De Grijs, 'A (Not So) Brief History', provides an overview of the long history of lunar methods.

[8] Howse, *Greenwich Time*, pp. 20–23; Turner, 'Longitude Finding', pp. 410–411.

[9] Dunn and Higgitt, *Finding Longitude*, p. 53.

parallax (see later in this chapter).[10] Morin was, however, critical of the later lunar theory of Leonard Duliris of Guienne, which employed the Moon's position in the zodiac and its height above the horizon. In England, the astronomer Thomas Streete reportedly devised a method in 1664 but did not publish details for lack of financial incentives. Later reward-seekers would, however, make use of Streete's published lunar tables for their own proposals.[11] In 1674, a Frenchman now known only as Le Sieur de St Pierre put forward a lunar-distance scheme that attracted the attention of Charles II, thanks to the advocacy of Louise de Kérouaille, Duchess of Portsmouth. As discussed below, the proposal fell by the wayside but contributed to the foundation of the Royal Observatory and the wording of the Longitude Act of 1714.[12]

A second astronomical method used eclipses and conjunctions of Jupiter's moons to find the reference time. Galileo had discovered the moons in 1610 using the recently invented telescope. He suggested that their reliable motions could provide a way of determining a ship's longitude, but it proved difficult to put into practice due to the challenges of using a telescope on a pitching and rolling ship, not to mention the impossibility of doing so in poor weather. From the late seventeenth century, however, such observations were successfully used to determine longitudes on land, thanks to the work of astronomers including John Flamsteed and Giovanni Domenico (Jean-Dominique) Cassini. Despite the practical observing challenges, longitude-seekers would continue to pursue a shipboard method based on the Jovian moons, often suggesting marine chairs or similar devices to facilitate the observations.[13]

A long-understood alternative to using such astronomical observations was to carry the reference time using an artificial timekeeper, to be compared to local time determined astronomically. This idea was suggested in print by Gemma Frisius in *De principiis astronomiae cosmographicae* (1530), but seems to have been proposed some years earlier by Fernando Colombo.[14] Frisius extended the proposal to shipboard use in his 1553 edition, suggesting the use of a large hourglass or water clock.[15] It would be more than two centuries before clock- and watchmakers began to achieve the reliability in timekeeping at sea necessary for determining longitude. However, during the intervening years authors across Europe suggested many new designs for maritime timekeepers, in part spurred on by seventeenth-century advances in precision timekeeping on land.

[10] Howse, *Greenwich Time*, p. 29; Marguet, *Histoire générale*, pp. 17–21.
[11] Turner, 'In the Wake of the Act', pp. 117–119. [12] Howse, *Greenwich Time*, pp. 33–43.
[13] De Grijs, 'European Longitude Prizes. I', pp. 481–484; De Grijs, 'European Longitude Prizes. 2'; Dunn, 'Scoping Longitude', p. 143; Turner, 'Longitude Finding', p. 409; Van Helden, 'Longitude and the Satellites'.
[14] Figueiredo and Boistel, 'Monteiro da Rocha', p. 232 n. 48.
[15] Andrewes (ed.), *The Quest for Longitude*, p. 2; Howse, *Greenwich Time*, pp. 24–25; Turner, 'Longitude Finding', pp. 411–412.

The maritime context was more challenging than the terrestrial because the ship's movements and changes in temperature and humidity interfered with the workings of timekeepers. This was particularly true for pendulum clocks, which marked a significant leap in terms of precision on land but could not be deployed at sea. Christiaan Huygens designed the first working pendulum clock in 1656–1657, building on Galileo's discovery of the isochronism of pendulums. Huygens also sought to put the innovation to maritime use, constructing and trialling longitude timekeepers from the 1660s, initially in collaboration with Scottish nobleman Alexander Bruce. In 1668, one of these prototypes allowed the estimation of longitude with an error of only about 60 miles, and continued to run through storms and a sea battle.[16] In 1674–1675, Huygens turned his attention to watches rather than pendulums, employing a balance spring, an innovation that may have been pioneered by Robert Hooke.[17] Timekeepers remained common among the longitude schemes of later decades, often as clocks rather than as watches. In May 1712, for instance, the naturalist and theologian John Hutchinson went so far as to petition the House of Commons for legal protection for his timekeeper design, since 'in the Opinion of the greatest Mathematicians, and the best Judges of such Movements, it is capable of moving more equally, and less liable to be disturbed by external Motion; so that it may be of great Use to the Publick for finding the Longitude'.[18] The proposal instigated the introduction of a Time Movement Bill, but faced stiff opposition from the Worshipful Company of Clockmakers. In the end, the bill was counted out the following month by the prorogation of Parliament.[19]

Magnetic schemes, which did not depend on the longitude–time relationship, also first arose from the late sixteenth century and would be entertained throughout the lifetime of the British Commissioners of Longitude. These schemes relied on the mapping of either magnetic variation or magnetic dip. By accumulating this global magnetic data, lines of equal variation or dip could be drawn onto a chart that also included lines of latitude and longitude. At sea, a local magnetic observation and a latitude determination from astronomical observations would in principle allow the navigator to plot the ship's position on such a chart.[20]

[16] De Grijs, *Time and Time Again*, pp. 4-28–4-45, 5-28–5-49; De Grijs, 'European Longitude Prizes. I'; Dunn and Higgitt, *Finding Longitude*, pp. 58–63; Howse, *Greenwich Time*, p. 29; Leopold, 'The Longitude Timekeepers'; Jardine, 'Accidental Anglo–Dutch Collaborations'; Jardine, *Going Dutch*, pp. 263–290; Jardine, *Temptation in the Archives*, pp. 33–44, 122–128; Olmsted, 'The Voyage of Jean Richer'; Van der Kraan, 'The Dutch East India Company'.

[17] De Grijs, *Time and Time Again*, pp. 5-11–5-20; Jardine, 'Scientists, Sea Trials'.

[18] *Journals of the House of Commons*, 17, p. 223.

[19] Atkins and Overall, *Some Account*, pp. 248–250; Mobbs and Unwin, 'The Longitude Act', pp. 154–156.

[20] De Jong, 'Navigating through Technology', pp. 264–272; Dunn and Higgitt, *Finding Longitude*, pp. 43–45; Jonkers, *Earth's Magnetism*; Jonkers, 'Finding Longitude at Sea'; Turner, 'Longitude Finding', pp. 408–409; de Grijs, 'European Longitude Prizes. I', pp. 474–479, 486–487; de Grijs, 'European Longitude Prizes. 2'. For examples of magnetic variation applied to navigation, see Bennett, 'Adventures with Instruments', p. 305; Davids, 'Finding Longitude at Sea'; Schotte, 'Expert Records', pp. 291–292.

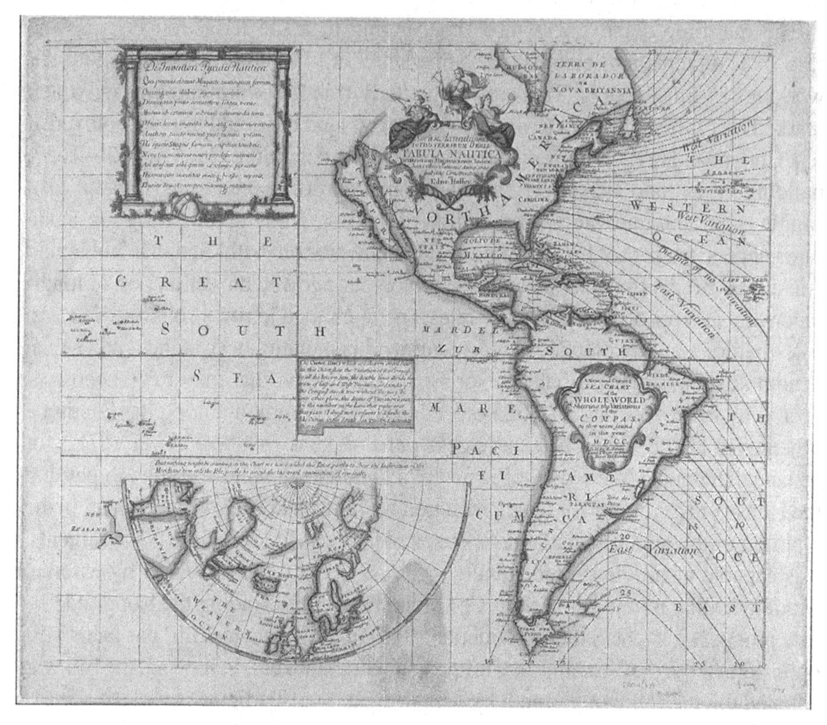

Figure 2.1 Edmond Halley, *A New and Correct Sea Chart of the Whole World Shewing the Variations of the Compass* (London, 1702), NMM G201:1/1. © National Maritime Museum, Greenwich, London

Voyagers such as Columbus had already noted magnetic variation in their compass readings in the fifteenth century. In 1581, English instrument maker Robert Norman described the effect and also proposed an instrument for measuring magnetic dip in *The newe Attractive*. Contemporaries including Thomas Blundeville, Edward Wright and Henry Briggs suggested that the phenomenon might also be used to find longitude, although William Gilbert's suggestion that variation was due only to local effects seemed to undermine the idea. In France in 1603, however, the Huguenot minister Guillaume de Nautonnier published a three-part work on finding longitude through magnetic variation, which helped gain him the position of royal geographer and a pension of 1200 livres from Henri IV. His proposals were soon being discussed in England by authors such as Anthony Linton and Edward Wright. By the mid-1630s, the work of navigational mathematicians including Henry Gellibrand posed further practical challenges through the identification of secular variation, yet also offered hope

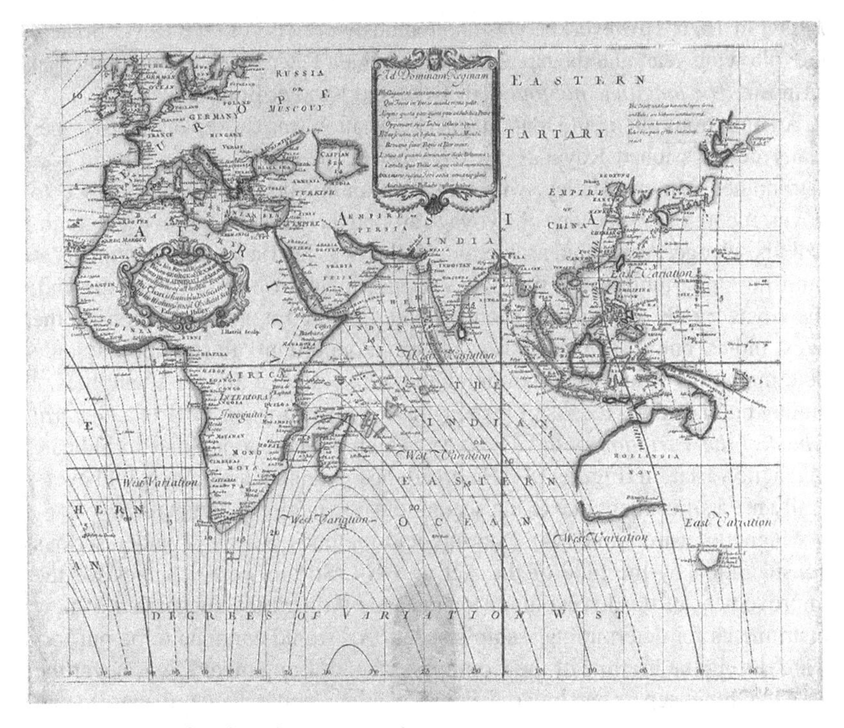

Figure 2.1 (cont.)

in the suggestion that some general law governing this phenomenon would be found.[21]

Magnetic schemes gained some prominence in England in the 1670s through the work of mathematical practitioner Henry Bond, who had been investigating magnetism since the 1630s and gained some support from the Dutch East India Company (VOC) and the Royal Society for his proposals. In 1674, Charles II appointed commissioners, including Robert Hooke, William Lord Brouncker, John Pell, Samuel Morland, Seth Ward and Silius Titus, to consider Bond's magnetic method for finding longitude. Hooke somewhat suspiciously produced his own magnetic scheme at this time, although it did not gain traction. The commissioners met and corresponded about Bond's proposal at least a few times that summer, and their deliberations attracted the interest of the king himself. By December, however, most had turned their attention to Le Sieur de St Pierre's lunar method. They were reportedly never fully convinced by Bond's theories but did recommend in March 1675 that he be rewarded. The mathematician secured a royal annuity of £50 and went on to publish *Longitude*

[21] Turner, 'In the Wake of the Act', p. 118.

Found in 1676. However, he was anonymously criticised to the Royal Society the following year, and the naval timber merchant Peter Blackborrow published a riposte, *The longitude not found*, in 1678, shortly after Bond's death.[22]

An ambitious attempt to map magnetic variation was carried out on two specially commissioned Royal Navy voyages in 1698–1700. Their captain was astronomer Edmond Halley, who with Benjamin Middleton had proposed to government in 1693 the idea of a voyage around the world to improve geographical knowledge. Appointed as captain, Halley was instructed by Parliament 'to omit no Opportunity of noting the Variation of the Compass'.[23] In the end, the voyages only took in the Atlantic, including the English plantations in the West Indies, but also gave the astronomer an opportunity to attempt longitude determinations from observations of the Moon and of Jupiter's satellites.[24] Halley published the results of his magnetic observations in 1701 in a *General Chart of the Variation of the Compass*, which used isogonic lines to indicate magnetic variation (Figure 2.1). He noted, however, that his chart would eventually need adjustment due to the secular change in global magnetism.[25]

Magnetic, astronomical and horological methods were still under serious consideration by the time of the Act of 1714. So too were schemes for the improvement of dead reckoning, for instance through the suggestion of new instruments for determining a ship's speed. All would continue to be put forward during the lifetime of the Commissioners of Longitude. It was never the case that timekeepers and lunar distances were the only schemes discussed and encouraged in the British search for longitude, even if they did come to dominate the attention and expenditures of the Commissioners.

In addition to the establishment of these enduring approaches to the determination of longitude, the sixteenth and seventeenth centuries saw the creation of a number of rewards for effective schemes across Europe. These precedents helped plant the idea of establishing a British reward and in some cases informed the terms of the Act of 1714. Some of the best-known rewards included those announced by Philip II of Spain in 1567 and thirty-one years later by Philip III. The latter comprised a lump sum of 6000 ducats, an annuity of 2000 ducats for life and up to 1000 ducats towards expenses. No one ever won the largest Spanish sum but significant amounts were issued in funding and lesser rewards, as would later occur with the British rewards. These smaller sums were mainly awarded for magnetic schemes and compass innovations during

[22] Bryden, 'Magnetic Inclinatory Needles'; Jonkers, *Earth's Magnetism*, pp. 85–89; Taylor, 'Old Henry Bond'; Warner, 'Terrestrial Magnetism'; Willmoth, *Sir Jonas Moore*, p. 178.

[23] *Journals of the House of Commons*, 13, p. 87.

[24] Bennett, 'Catadioptrics and Commerce', p. 252; Cook, 'Edmond Halley and the Magnetic Field', pp. 480–485; Cook, *Edmond Halley*, pp. 267–268.

[25] Cook, *Edmond Halley*, pp. 256–291; Thrower, 'Cartography', pp. 56–59; Murray and Bellhouse, 'How Was Edmond Halley's Map'.

the early seventeenth century.[26] There were reports of rewards being advertised by Portugal and Venice as well, and of the States General of Holland offering 30,000 florins.[27] Such rewards drew interested parties possessing all manner of motivations and degrees of knowledge, as would also be the case during the eighteenth century. Galileo tried but failed to win the second Spanish reward between 1616 and 1632 with a method based on observations of Jupiter's moons, and was still involved in protracted negotiations with the Dutch Republic over his tables when he died in 1642.[28]

European nations and learned bodies also considered specific longitude schemes on an ad hoc basis during these centuries. For example, Cardinal Richelieu set up a commission comprising an admiral and five scholars to examine Jean-Baptiste Morin's lunar-distance method in France in 1634. The Commissioners judged that, while his concept was feasible, the accompanying tables had too many imperfections. Morin also failed to win the Dutch reward but was eventually granted a French pension of 2000 livres. In 1668 André Reusner of Neystett sought up to 160,000 livres from Louis XIV for an 'odometer' to measure distance travelled on land or at sea. Seven worthies including the finance minister and the lieutenant general of the navy examined and dismissed the invention.[29] Likewise, as discussed in Chapter 3, the English precedents that had the greatest influence on the Act of 1714 were Charles II's appointment of commissioners to judge two schemes in 1674 and the establishment of Thomas Axe's private reward in 1691.

Several institutions that would prove central to Georgian longitude and navigation were founded in England and France from 1660. These included the Royal Society (1660), the Académie Royale des Sciences (1666), the Observatoire de Paris (1667), the Royal Mathematical School at Christ's Hospital (1673) and the Royal Observatory, Greenwich (1675). The Royal Society, for which the longitude problem was recognised as a pressing challenge from its outset, would in particular be a frequent collaborator of the Commissioners of Longitude appointed in 1714. It was the foundation of the Royal Observatory, however, that was of most importance to later events, since its Astronomer Royal was the most active of the Commissioners for much of their history.

The creation of the Royal Observatory owed much to competition from France. The Royal Society had been planning an observatory in Chelsea

26 De Grijs, 'European Longitude Prizes. I'.

27 De Grijs, 'European Longitude Prizes. 3', casts plausible doubt that there was a Venetian prize, the existence of which has been assumed from a later source.

28 De Grijs, *Time and Time Again*, pp. 3-19–3-25; de Grijs, 'European Longitude Prizes. 2'; Jonkers, 'Rewards and Prizes'; Van Helden, 'Longitude and the Satellites', pp. 89–92.

29 Howse, *Greenwich Time*, pp. 29–31; Olmsted, 'The Voyage of Jean Richer', pp. 617–618; see also Jullien (ed), *Le calcul des longitudes*; Davids, 'Dutch and Spanish Global Networks'; Turner, 'In the Wake of the Act', pp. 120–122.

Figure 2.2 E. Kinhall, 'A View of the Observatory in Greenwich Park' and 'A View of Greenwich, Deptford and London, taken from Flamsteads Hill in Greenwich Park', 1750, NMM PAD2182 © National Maritime Museum, Greenwich, London

when, in 1674, Charles II appointed commissioners to examine Le Sieur de St Pierre's proposal for finding longitude by lunar distances. Jonas Moore, who was intending to pay for the Chelsea observatory and wanted it staffed by young astronomer John Flamsteed, skilfully leveraged the commissioners' work to involve the king in the foundation of a national observatory. Moore also helped Flamsteed become assistant to the longitude commissioners. This allowed Flamsteed to tell Charles of French successes in surveying, aided by the Observatoire de Paris, and to highlight the fact that any attempt to perfect the lunar-distance method would need far better astronomical data. These considerations encouraged the king to establish his own royal observatory, for which he issued a warrant on 4 March 1675. Flamsteed became its first 'astronomical observator' and was directed 'forthwith to apply himself with the most exact Care and Diligence to the rectifying the Tables of the Motions of the Heavens, and the places of the fixed Stars, so as to find out the so much desired Longitude of Places for perfecting the art of Navigation'.[30] Following Christopher Wren's suggestion, the new observatory was built on royal

[30] Royal Warrant, quoted in Howse, *Greenwich Time*, p. 42.

property in Greenwich Park (Figure 2.2), with the first observations made there on 16 September 1676.[31]

Thus, by the early eighteenth century, much was in place in Britain and abroad that would inspire and shape the Act of 1714 and ensuing events. These included institutions such as the Royal Observatory, precedents in legislation and more than two centuries of attempts to find practical ways of making the well-understood principles of longitude determination work at sea. There were other continuities too. For example, there was a surprising degree of international dialogue and collaboration over longitude matters, despite frequent wars. This appears to have been fuelled at least in part by the self-interest of individual reward-seekers and other innovators, as well as by the broader international dialogue over what would today be called science and technology.

Public perceptions of the 'projectors' who proposed longitude projects in Europe also followed familiar dynamics over the centuries. As Barrett notes, projecting was a matter of some concern by the early eighteenth century, representing 'the imposition on the public of new, naïve, ridiculous, impossible or even malicious schemes, the perversion of established institutions and knowledge by untested inventions'.[32] In other words, projectors putting forward such scientific, technological or financial schemes often inspired suspicion or ridicule (even if the term 'projector' was not always used pejoratively). This had long been the case in the search for successful longitude schemes. Authors including Miguel de Cervantes, for example, mocked longitude-seekers in this vein in the wake of the Spanish rewards.[33] Such negative assumptions continued through the eighteenth century and were amplified in the case of longitude by a sense that there might not *be* a solution to such a longstanding problem – as was equally the case with efforts to prolong human life or create the Philosopher's Stone or perpetual motion machines (see Chapter 9).

In practice, longitude projectors included all manner of people with a wide variety of motivations. Some never truly intended to vie for longitude funding or rewards but tried to leverage the mechanisms and the high profile of longitude in Britain to secure charity, patronage or public attention. Even the best-informed and innovative of longitude projectors were usually spurred on by a combination of care for the common good and a desire for personal profit and compensation. This was as true for Galileo pitching his Jovian moon method to different nations as for the two projectors who set the ball rolling in Britain in 1713.

[31] Howse, *Greenwich Time*, pp. 43–45.
[32] Barrett, *Looking for Longitude*, p. 107; see also Keller and McCormick, 'Towards a History'.
[33] Dunn and Higgitt, *Finding Longitude*, p. 36.

Towards a British Longitude Reward

The immediate build-up to the establishment of British longitude rewards in 1714 brought together personal, commercial and political interests. It was initiated in part by two reward-seekers, William Whiston and Humphry Ditton, who had some authority in mathematical circles as well as some influential connections. However, they were also hindered by contentious religious and political beliefs, not to mention public suspicion towards projectors.

In 1713 Whiston and Ditton hinted in letters and advertisements in newspapers at details of a new scheme for finding longitude at sea and began lobbying for financial compensation. Whiston was a controversial theologian and mathematician who had largely taken up public lecturing in the wake of his expulsion for heterodoxy from a Cambridge professorship three years earlier, and was favoured by the Whig party. Ditton, formerly a dissenting minister, was master of the Royal Mathematical School at Christ's Hospital.[34]

In 1713, the two men were intentionally vague about their scheme. It would turn out to involve firing mortars from ships moored at known locations along key routes at a predetermined time each day. Navigators could measure their distance and bearing from those known positions using a magnetic compass and by measuring either the elevation of the mortar's flare or the time difference between seeing and hearing the explosion. A grand fireworks display on the Thames on 7 July 1713, in celebration of the Tory-negotiated end to the War of the Spanish Succession, had apparently planted the idea in Whiston's mind.[35] Just a week later, he and Ditton published a letter in *The Guardian*. At that time, the newspaper was edited by Richard Steele, another Whig who had been intrigued by the recent pyrotechnic displays. One of the publication's contributors, Joseph Addison, had previously also collaborated with Steele on *Tatler* and *The Spectator*, and would later as a Member of Parliament be one of the authors of the Act of 1714.[36] Whigs, who had ceded political ground to the Tories since 1710 due to the unpopularity of the War of the Spanish Succession, would constitute a number of the participants in the parliamentary longitude events.

Steele penned a laudatory introduction for the *Guardian* letter, praising Whiston's astronomical publications and saying that his and Ditton's communication deserved:

> the utmost Attention of the Publick, and in particular of such who are Lovers of Mankind. It is on no less a Subject, than that of discovering the *Longitude*, and deserves a much higher Name than that of a Project, if our Language afforded any such Term.[37]

[34] Farrell, *William Whiston*, pp. 116–183; Forbes, Murdin and Wilmoth (eds.), *The Correspondence of John Flamsteed*, III, pp. 712–713, indicates that Ditton died in 1714 rather than 1715.
[35] Werrett, *Fireworks*, pp. 96–98. [36] Winton, 'Steele, Sir Richard'.
[37] *The Guardian*, 14 July 1713.

In their letter, Whiston and Ditton discussed existing obstacles to navigation and criticised longitude methods based on Jupiter's moons, Earth's moon or a pendulum timekeeper (although Whiston would later work on magnetic and Jovian schemes):

'Tis well known, Sir, to your self, and to the Learned, and Trading, and Sailing World, that the great Defect of the Art of Navigation is, that a Ship at Sea has no certain Method, in either her Eastern or Western Voyages, or even in her less distant Sailing from the Coasts, to know her Longitude, or how much she is gone Eastward or Westward; as it can easily be known in any clear Day or Night, how much she is gone Northward or Southward: The several Methods by Lunar Eclipses, by those of *Jupiter's* Satellites, by the Appulses of the Moon to fixed Stars, and by the even Motions of Pendulum Clocks and Watches, upon how Solid Foundations soever they are built, still sailing in long Voyages at Sea, when they come to be practis'd; and leaving the poor Sailors frequently to the great Inaccuracy of a Log-line, or Dead Reckoning. This Defect is so great, and so many Ships have been lost by it, and this has been so long and so sensibly known by Trading Nations, that great Rewards are said to be publickly offer'd for its Supply.

The two explained that their own scheme was 'easily intelligible by all, and readily to be practised at Sea as well as at Land', and that it could more accurately find both latitude and longitude. They were certain that it deserved a reward due to its great public benefit, but assured readers that they did not 'desire actually to receive any benefit of that nature till Sir *Isaac Newton* himself, with such other proper Persons as shall be chosen to assist him, have given their Opinion in favour of this Discovery'.[38]

Whiston and Ditton's appeal seems to have attracted little interest at first, since on 8 December 1713 they reached out to the public in another of Steele's short-lived newspapers, *The Englishman*. Steele's preface emphasised once more the projectors' renown and the importance of longitude to the nation and the world, concluding that '[t]he memorable Case of *Columbus*, the Neglect of whom at the English Court, was the Occasion that the Indian World fell into Spanish Hands, might, one would think, be an Admonition to all who have Means and Authority, to overlook no Opportunities of this sort'. Whiston and Ditton's letter dismissed the ridicule and rumours to which their scheme had been subjected. They insisted that they had not yet revealed details because 'want of Publick Spirit in *Britain*' had denied them the guarantee of a reward should expert opinion and testing prove the scheme's worth. Adopting a strategy common among early modern projectors, they regretfully concluded that they would have to seek encouragement from other nations: 'we beg leave here to Desire all the Patrons and Well-wishers of Learning and Ingenuity of all Nations, to give us the best

[38] Ibid.

Information they can of the several Rewards which are anywhere set apart for this Discovery'.[39] The suggestion of taking national innovations abroad, and Steele's reference to Spanish power in the Americas, would presumably have had additional resonance given the recently concluded war and the launch of the South Sea Company.

Whiston and Ditton's scheme attracted scorn from the outset, as they mentioned in *The Englishman*. This was only partly due to objections to the vague nature of their proposal, or to distaste for their lobbying. It was more often a reaction against Whiston's 'heretical' Arian religious beliefs and his association with the Whigs, with Ditton largely escaping such opprobrium. Whiston was a particular target of the predominantly Tory 'Scriblerians' – satirists including Jonathan Swift, Alexander Pope, John Gay and John Arbuthnot. For example, in early 1714 the pamphlet *Will-with-a-Whisp; or, the Grand Ignis fatuus of London* attacked Whiston's religious and astronomical ideas and 'His boasted Longitude'.[40] As Werrett notes, it also referenced the pyrotechnic nature of the scheme and the common association of fiery spirits with false religious beliefs.[41] This attitude towards Whiston's longitude efforts would persist throughout his decades of work on navigation and mapping, and even after his death. It took many forms, from scathing comments and scatological songs to visual parodies, notably the 'longitude lunatic' drawing a Whistonesque scheme in William Hogarth's depiction of Bedlam in the final image in his print series *A Rake's Progress*.[42]

As 1713 turned to 1714, Whiston and Ditton continued their efforts to establish and obtain a reward. They campaigned not only through newspapers but also by advertising their scheme in their own publications and lectures. For example, Ditton attached their longitude advertisement to his *New Law of Fluids*.[43] Nevertheless, on 3 April 1714, the astronomer and mathematician Abraham Sharp suggested to Astronomer Royal John Flamsteed that their lobbying was going nowhere:

> I have but lately seen Whistons and Dittons proposeall concerning the longitude alltho it have been published near nine Months agoe 'tis probable they meet not with that encouragement they expected else methinks something further would ere this have been produc'd, sure Two such noted Mathematicians should not amuse the nation with such mistakes and fallacys … since that would be utterly to ruin their Credit which I presume they in a great measure maintain themselves by …[44]

39 *The Englishman*, 8–10 December 1713.
40 Anonymous, *Will-with-a-Wisp*; *Post Boy*, 29 April–1 May 1714.
41 Werrett, *Fireworks*, p. 98.
42 Barrett, *Looking for Longitude*; Lynall, 'Scriblerian Projections'; Rogers, *Documenting Eighteenth Century Satire*, pp. 46–54.
43 *Post Boy*, 9–12 January 1714.
44 Abraham Sharp to John Flamsteed, 3 April 1714, in Forbes, Murdin and Wilmoth (eds.), *The Correspondence of John Flamsteed*, III, pp. 695–697 (p. 696).

Pushing ahead, the two projectors had a one-page 'petition' printed on 29 April. There has been some confusion as to the nature of this document, which was previously thought to mark the start of the parliamentary consideration of a longitude reward. This was presumably due to a later account, that, 'encouraged by some Gentlemen to apply themselves to the House of Commons for a Reward', Whiston and Ditton mounted a formal petition. There is, however, no record that such a petition was formally submitted to or considered by Parliament.[45] The broadsheet repeated the projectors' claims about the importance of longitude for the safety and trade of the nation and their offer to prove their solution before appropriate judges. It also suggested clauses that would later be echoed in the Longitude Act:

> That the lowest Reward may be allotted to the discovering the same within one whole Degree of a great Circle, or 70 Measur'd Miles; a greater to the discovering it within one half; and a still greater to the discovering it within one Quarter of that Measure: And that withal, if it be thought fit, proper Rewards may be also allotted to such as shall afterward make any farther considerable Improvements for the perfecting so important a Discovery.[46]

The clearest turning point came when a petition was submitted to the House of Commons on 25 May by 'several Captains of her Majesty's Ships, Merchants of *London*, and Commanders of Merchant-men'. No copy of the text appears to survive but, given the timing and what is known of the contents, it may have been orchestrated by the two projectors or their associates.[47] The petition suggested that public 'Encouragement' could lead to a working longitude solution, 'for the Safety of Mens Lives, her Majesty's Navy, the Increase of Trade, and the Shipping of these Islands, and the lasting Honour of the *British* Nation'. The same day, the House of Commons referred the matter to a committee of fifty-one MPs (overwhelmingly Tories) deemed to have relevant knowledge or interests – a number of whom would soon be named Commissioners of Longitude – and to the officials of the Admiralty and Navy. Nineteen of the MPs nominated had previously sat on the committee called to consider the Time Movement Bill in 1712. The longitude committee also included more ministers and officers of the Crown

[45] Anonymous, *The History and Proceedings*, V, pp. 144–145; Mobbs and Unwin, 'The Longitude Act', pp. 158–159. There is no reference to the petition in *Journals of the House of Commons*.

[46] Whiston and Ditton, *A Petition about the Longitude*, dated 29 April 1714, copy at BL 516.m. 18.(52.); text reproduced in Turner, 'In the Wake of the Act', p. 128.

[47] Petitions to the House of Commons from before 1834 were destroyed when the old Houses of Parliament burned down. Those to the House of Lords were routinely destroyed before 1950. A small number survive in the Parliamentary Archives, but so far no one has been able to locate the 1714 petition among them. Mobbs and Unwin, 'The Longitude Act', p. 159, believe it unlikely that Whiston and Ditton were behind the petition, pointing to a petition of 30 April with similar wording that sought to preserve the navigation of the river Thames.

than was the case for any other legislation in the last parliament of queen Anne's reign, indicating active government support.[48]

As Mobbs and Unwin note, the passage of what became the Longitude Act was 'concurrent with other statutory measures designed to protect shipping and cargoes at a time of trade and commercial expansion'.[49] Owing to its alignment with other maritime interests and initiatives in the same period, such as the Stranded Ships Act, MPs from both political parties proved willing to offer support. In doing so, they expanded the proposal beyond Whiston and Ditton's personal interests. Many of the politicians involved had strong maritime and trading interests and some shared Whiston's Whig affiliation. Nevertheless, it seems that it was the strong cross-party consensus that ensured a swift passage through Parliament.[50]

The nominated committee questioned Whiston and Ditton on 4 June 1714 in the presence of a group of 'celebrated Mathematicians': Isaac Newton (president of the Royal Society), Edmond Halley (Savilian professor of geometry at Oxford), Samuel Clarke (clergyman and philosopher) and Roger Cotes (Plumian professor of astronomy and experimental philosophy at Cambridge). These four, who also appeared as witnesses, had reportedly encouraged Whiston and Ditton to petition Parliament and were already connected to each other and to Whiston in different ways through shared beliefs and experiences in religion and natural philosophy. Newton, Halley and Cotes would soon also number among the Commissioners of Longitude by virtue of the positions they held. What was novel at this stage was that the committee chose to draw so fully on such expert evidence.[51]

In his evidence to the committee, Newton famously opined that, 'for determining the Longitude at Sea, there have been several Projects, true in the Theory, but difficult to execute'. This was a crucial point that the resulting Act would acknowledge in notably similar wording: that 'in the Judgment of able Mathematicians and Navigators, several Methods have already been discovered, true in Theory, though very difficult in Practice'. In other words, the longitude problem was not one of abstract principle, which was already well understood, but of its application in maritime practice. The projects Newton identified were 'a Watch to keep Time exactly', 'the Eclipses of *Jupiter's Satellites*', 'the Place of the Moon' and 'Mr. *Ditton's*'. He noted that two watches would be necessary for the first approach, one for the second and third. He also added that Whiston and Ditton's scheme would require vessels to 'sail due East or West, without varying their Latitude' and that it would

[48] *Journals of the House of Commons*, 17, pp. 641–642; Mobbs and Unwin, 'The Longitude Act', pp. 160–161.
[49] Mobbs and Unwin, 'The Longitude Act', p. 153. [50] Ibid., pp. 165–169.
[51] Ibid., p. 162.

'keep' the longitude as a ship advanced rather than 'finding' it, although he allowed that it might be of some use near coasts.[52] This latter consideration would result in the Act of 1714 offering the first half of a reward once the required number of Commissioners agreed that the proposed method might guarantee the safety of ships close to shore. Each of the mathematicians testified that Whiston and Ditton's scheme could work in theory, but that they anticipated practical difficulties. Halley and Cotes thought that the idea needed to be proved through sea trials. Overall, the consensus seems to have been that, as Clarke opined, 'there could no Discredit arise to the Government, in promising a Reward in general, without respect to any particular Project'.[53]

While the writing and passage of the Act of 1714 proceeded with unusual speed, accounts of the committee meeting on 4 June suggest that the response was not unequivocally favourable. Whiston recorded that the chairman sought to exploit an opportunity of 'trying to drop the bill' and declared 'his own Opinion to be that, "Unless Sir Isaac Newton would say, that the Method now proposed was likely to be useful for the Discovery of the Longitude, he was against making a Bill in general for a Reward for such a Discovery"'.[54] Similarly, Newton later recalled that 'the chairman of the Committee of the House of Commons being a seaman represented that they did not want [i.e. lack] longitude and so far as I can observe seamen are generally of that opinion'.[55] The chairman in question was William Clayton, a tobacco merchant and shipowner who was the Tory MP for Liverpool and experienced at managing important bills through the Commons.[56] In addition to the possible role played by party politics – with Whiston and many proponents of a longitude reward being Whigs, and Clayton a Tory – there were mixed responses to longitude efforts among British seamen at large.

Clayton reported back to the House of Commons on 7 June, when it was agreed that MPs would consider the committee's report the following week.[57] Despite his recorded opposition to the establishment of a new reward, Clayton would soon be named as one of the first Commissioners of Longitude (although he only held the role briefly, dying in 1715). Three days later, the day before the committee's report and the possibility of a parliamentary bill were to be considered by the House of Commons, Whiston and Ditton put out a broadsheet setting out eleven reasons why a bill was needed. They suggested that

[52] Howse, *Greenwich Time*, pp. 55–57; *Journals of the House of Commons*, 17, pp. 677–678.
[53] *Journals of the House of Commons*, 17, pp. 677–678; Turner, 'In the Wake of the Act', p. 116.
[54] Whiston, *The Longitude Discovered*, Historical Preface (dated 1741), p. v; Westfall, *Never at Rest*, p. 835.
[55] Isaac Newton, draft letter about projects for determining longitude at sea, n.d. [c. 1721], CUL MS Add.3972, p. 40r.
[56] Higman, 'Locating the Caribbean', pp. 299–301; Mobbs and Unwin, 'The Longitude Act', p. 161.
[57] *Journals of the House of Commons*, 17, pp. 671, 677–678.

the proposed reward should be open to anyone (although they clearly hoped to claim it for themselves) and that Newton's testimony favoured their plan and the lunar-distance method. They described their own scheme as 'easy to be understood and practis'd by Ordinary Seamen, without the Necessity of any puzzling Calculations in Astronomy' and said that it would 'save the Nation great Sums of Money'. They also offered to let Halley 'go equal Shares with us in the Reward' if he would see to the 'Tryal, and Practice, and Improvement of this Method'.

Whiston and Ditton further claimed that, as well as allowing seamen to find their longitude, their proposed method would 'prevent the Loss of abundance of Ships and Lives of Men; as it would certainly have sav'd all Sir *Cloudsly Shovel's* Fleet, had it been then put in Practice'.[58] This suggestion appears to be the reason for the modern perception that the Act of 1714 was a direct response to the Isles of Scilly disaster of 1707, a tragedy in which four ships from a fleet commanded by Sir Cloudesley Shovell sank close to the Isles of Scilly, killing hundreds of seamen and also Shovell.[59] Rodger, for instance, writes that the celebrated admiral's death 'caused a profound shock, and led in due course to the 1714 Longitude Act'.[60] Twentieth-century authors tend, however, to have been less definitive about a causal link. Howse, for instance, simply states that the magnitude of the disaster 'made such an impression on the British public that they became more than ever receptive to any suggestion that might make navigation safer'.[61] To date, no evidence has been found of a specific connection being made between the Isles of Scilly wreck and the difficulty of determining longitude in the years following the disaster. Nor have calls, repeated or otherwise, for parliamentary action to address the longitude question been identified. The immediate reporting in newspapers was factual rather than speculative, as seems to have been typical at the time.[62] Although in correspondence with the Admiralty the captain of one of the surviving ships blamed the poor state of the compasses, public speculation largely focused on the weather as the cause.[63] Pamphlets lamenting Shovell's loss talked, for instance, of a 'dismal Fog', but made neither a link to the longitude problem nor a call for action.[64]

[58] Whiston and Ditton, *Reasons for a Bill*, dated 10 June 1714.

[59] For a detailed account, see McBride and Larn, *Admiral Shovell's Treasure*, which makes a direct link to the Longitude Act. May, 'The Last Voyage', analyses logbooks from surviving ships.

[60] Rodger, *Command of the Ocean*, p. 172. [61] Howse, *Greenwich Time*, p. 54.

[62] *Daily Courant*, 28 October 1707: 'To Day came in Her Majesty's Ships the Royal Anne [etc] … They bring the bad News of Sir Cloudsly Shovel's being lost in the Association on the Rocks of Scilly the 22d Instant about 8 at Night'; see also *Daily Courant*, 1 November 1707.

[63] William Jumper to the Admiralty, 29 October 1707, TNA ADM 1/1981; May, 'Naval Compasses in 1707'.

[64] Anonymous, *The Life and Glorious Actions*, p. 8.

Whiston and Ditton's broadsheet is the earliest publication found to date that identifies the disaster in relation to longitude and needs to be read carefully. Most of its numbered points set out the case for supporting their own signalling scheme and do not make a direct connection between the 1707 disaster and the fleet's ability to determine their longitude. The two projectors were instead making the specific claim that their own method would improve all aspects of navigation, particularly close to shore.

The House of Commons considered the longitude committee's report on 11 June and unanimously agreed to its suggestion that:

a Reward be settled by Parliament, upon such Person or Persons, as shall discover a more certain and practicable Method of ascertaining the Longitude, than any yet in Practice, and that the said Reward be proportioned to the Degree of Exactness to which the said Method shall reach.[65]

Events proceeded swiftly in order for the new bill to become law before the imminent prorogation of Parliament. A group of five Tories and four Whigs strongly representing naval and port interests managed the bill through this process, concurrently with the Stranded Ships Bill (which was largely managed by Clayton).[66] Leading Whig MP James Stanhope presented the Longitude Bill on 16 June (and would steer it through the Commons). It was read the following day, then read again and committed on 22 June. Amendments were made and the bill engrossed on 30 June and 1 July. Two days later, before the bill was passed by the House of Commons, four names were added to the list of proposed Commissioners: Thomas Herbert, Earl of Pembroke, former Lord High Admiral; Philip Bisse, Bishop of Hereford and FRS; the high Tory George Smalridge, Bishop of Bristol; and the Tory lawyer and peer Thomas Trevor.[67] The House of Lords passed the bill on 7 July and queen Anne gave her royal assent to the Longitude Act (Appendix 1) two days later – just three weeks before her death.[68]

Conclusion

The Longitude Act of 1714 stirred interest across Europe even before its official passage. Whiston and Ditton were advertising their published proposal before it had received royal assent. Isaac Newton had already received longitude proposals from Lorraine and from Paris. Likewise, the *Journal*

[65] Anonymous, *The History and Proceedings*, V, pp. 144–145.
[66] Mobbs and Unwin, 'The Longitude Act', pp. 165–168.
[67] *Journals of the House of Commons*, 17, pp. 715–716.
[68] Howse, *Greenwich Time*, p. 58; *Journals of the House of Commons*, 17, pp. 686–687, 692, 711–712, 715–716; *Journals of the House of Lords*, 19, p. 750; Mobbs and Unwin, 'The Longitude Act', p. 167, give a full parliamentary timetable.

Littéraire – published in The Hague and much occupied with publicity for English natural philosophy – carried a notice sent from London of the rewards on offer: 'we are here in a great impatience to have an opening into this method; the skill of Mr Whiston and Mr. Ditton in mathematics is recognised, and they are not in danger of being confused with that crowd of charlatans that we see appear from all sides'. The following month, the *Journal* carried a detailed summary of the scheme proposed by Whiston and Ditton.[69]

The two English projectors were ebullient, proclaiming that their 'Method has been so far approved by this present Parliament, that they have passed an Act, ordering 20000 l. Reward for such a Discovery'.[70] In principle, the Act seemed tailor-made for them. It had been shaped by their suggestions and those of associates and supporters. These included the experts who had testified to the parliamentary longitude committee and authors of the new legislation including Joseph Addison and General Stanhope.[71] In March, Stanhope had joined Addison and Robert Walpole as advocates for Richard Steele when the publisher was accused of writing seditious material.[72] He would become one of the first Commissioners of Longitude and then Secretary of State in September 1714, and had extensive experience in international trade and relevant legislation.[73] As Stewart and others have suggested, Whiston's favour with such powerful Whigs aided his and Ditton's lobbying for the establishment of a longitude reward, as well as providing opportunities for their lecturing and publishing.[74] Nevertheless, the Act had also received strong Tory and governmental support in its passage through Parliament as one of a group of acts in support of Britain's commercial and maritime ambitions.

Ultimately, however, Whiston and Ditton did not win one of the new rewards. Ditton died on 15 October 1715, while Whiston continued pursuing and advertising a variety of longitude and other projects for decades. Nor were Whiston's ensuing decades of work on longitude schemes remunerated, although he did obtain one of the earliest of the smaller rewards issued by the Commissioners (see Chapter 3). Nevertheless, the events the two projectors set in motion breathed new life into the search for longitude.

[69] Mobbs and Unwin, 'The Longitude Act', p. 164; Hall and Tilling (eds.), *The Correspondence of Isaac Newton*, VI, pp. 163–165; *Journal Littéraire* 4, part 1 (May–June 1714), pp. 235–236 and part 2 (July–August 1714), pp. 443–453. Newton received a longitude scheme via Claude Jordan, historiographer at Bar Le Duc, who proposed the Commissioners visit the proposer Romuald Le Muet, a mathematician and priest in Metz; and a scheme from a mechanical inventor André Dalesme, associated with the Académie Royale des Sciences.

[70] *Daily Courant*, 8 July 1714; Whiston and Ditton, *A New Method*, introduction dated 7 July (p. 27).

[71] Howse, *Greenwich Time*, p. 57.

[72] Cobbett, *Cobbett's Parliamentary History*, VI, p. 268.

[73] Hanham, 'Stanhope, James'.

[74] Stewart, *The Rise of Public Science*, pp. 91–97.

3 Launching the Eighteenth-Century Search for Longitude

Alexi Baker

The Act of 1714 was born of diverse influences, stretching back centuries. Whiston and Ditton's lobbying dovetailed with national and political interests to prompt the establishment of new longitude rewards shortly before the death of queen Anne. However, the Act's authors drew upon their expert knowledge and institutional memory of the longer history of the search for longitude, resulting in significant continuity with the past. In addition to the suggestions made by Whiston and Ditton and prominent men of science including Newton and Halley, two specific precedents most strongly shaped the legislation's requirements for judging and rewarding proposals. These were Charles II's appointment of commissioners to examine a longitude proposal in 1674 and the reward established by the will of the Somerset gentleman Thomas Axe in 1691.

A close rereading of the Act of 1714 within this broader historical context allows a significant reinterpretation of the legislation and the events that occurred between its passage and the first communal meeting of Commissioners of Longitude in 1737. Many modern scholars have viewed this period as one of official inactivity and missed opportunities, on the assumption that the Act established a standing Board of Longitude that failed to meet for twenty-three years. This chapter argues that the legislation did not in fact explicitly appoint a standing board but that it did immediately have the effects its authors intended. It revitalised longstanding longitude efforts, with relevant national officials and institutions in Britain aiding and judging them.

Newspapers and periodicals, books and pamphlets, institutional records and letters illuminate this rejuvenation of the search for the longitude. Reward-seekers at home and abroad began putting forward a slew of longitude-finding technologies and methods even before the passage of the Act of 1714, in print and through interpersonal and institutional channels. Key Commissioners, in particular the Astronomer Royal, and institutions including the Navy and trading companies, communicated frequently with reward-seekers. Public and private trials were also conducted for specific proposals, long before the better-known sea trials of John Harrison's timekeepers. The stage was largely set for the types of schemes that would be entertained and for the types of institutions and people who would be involved throughout the long history of the Commissioners of Longitude.

A Closer Reading of the Longitude Act

The Act of 1714 borrowed heavily from earlier precedents, although it was unusual in naming so many individuals as fit judges for the funding and somewhat unusual in offering such large rewards. To put the sums of up to £20,000 in context, it has been estimated that just 6 per cent of families had an annual income greater than £100 in this period and less than 3 per cent an income exceeding £200.[1] The rewards would, therefore, have appeared very large to most private individuals. They were, however, comparable to the construction, rebuilding or major repair of a naval vessel. They were also much smaller than the £2.5 million allocated annually to the Royal Navy in the previous decade, a time of war, or the roughly £1.25 million committed each year after 1714.[2]

The text of the Act (Appendix 1) cites the importance of finding the longitude for the 'Safety and Quickness of Voyages, the Preservation of Ships and the Lives of Men', the 'Trade of Great Britain' and 'the Honour of [the] Kingdom'.[3] These were already common refrains in the quest for longitude by the time Whiston and Ditton employed them when lobbying for the new rewards. The Act nominated Commissioners by name or office. These included key figures from politics, the Navy, astronomy and mathematics, groups already perceived as being the principal practitioners and beneficiaries of the search for longitude, bar the craftsmen and inventors who typically came from different socio-economic classes than the Commissioners.

If five or more Commissioners thought a longitude proposal promising, they could direct the Commissioners of the Navy to have their Treasurer issue up to £2000 to conduct trials, the money to come from any sums to hand that were not needed for the Navy's use (see Appendix 2). This direction would create complications and delays in the future, when there were no funds to extend.[4]

1 Hume, 'The Value of Money', pp. 376–377.

2 The first cost of the first-rate *Royal George*, launched 1715, for example, was £28,707 (including fitting); Winfield, *British Warships ... 1714–1792*, p. 2. Major repairs to the second-rate *Prince* (formerly the *Triumph*, launched in 1698) in March 1717/18 cost just over £20,266; Winfield, *British Warships ... 1603–1714*, pp. 35–36. See also Baugh, 'Parliament, Naval Spending and the Public', p. 39; Rosier, 'The Construction Costs'. In 1715, the 'extra' expenditure of the Navy – building and repairing ships, providing furniture and stores for vessels and making improvements to the Royal Dockyards – was £237,277, but had dropped to £50,200 in 1721; Winfield, *British Warships ... 1714–1792*, p. xvi.

3 Discovery of Longitude at Sea Act, 1713 (13 Anne c 14), copy at CUL RGO 14/1, pp. 10r–12r.

4 For example, the Commissioners wrote to the king on 30 July 1765 that the Secretary's salary and travel reimbursements for members outside London remained unpaid 'owing to the precariousness and uncertainty of the Fund out of which the same is directed to be paid'; TNA ADM 7/684, fols 15–16. He agreed that the sale of old naval stores be used to pay the outstanding sums.

After experiments were made, the Commissioners or 'the major part of them' were to decide whether the proposal was 'Practicable, and to what Degree of Exactness'. The Act created a three-tiered reward system: £10,000 to be given to the inventor of a method that could find longitude 'to One Degree of a great Circle, or Sixty Geographical Miles'; £15,000 if the method could find longitude to two-thirds of that distance; and £20,000 if it found longitude to half a degree. Half the reward would be paid when the Commissioners or 'the major part of them' agreed that the method could also ensure the safety of ships within 80 miles of shore. (William Whiston later claimed he had been directly involved in adding this clause, since his and Ditton's scheme seemed promising for use near dangerous coastlines.[5]) The other half would be paid when the method had been successfully tried on a voyage to the West Indies. The Commissioners, or a majority of them, could also direct that lesser rewards be given to projectors whose schemes exceeded the limits but might still be 'of considerable Use to the Publick'.

In addition to drawing upon suggestions made by Whiston and Ditton and experts including Newton and Halley, these clauses were closely patterned after two British precedents: Charles II's appointment of commissioners to examine a longitude proposal in 1674 and the reward established by the will of the Somerset gentleman Thomas Axe in 1691. Charles II had nominated eight commissioners to examine a celestial and lunar proposal made by the Frenchman Le Sieur de St Pierre, who was put forward by the king's favourite mistress, Louise de Kérouaille.[6] In wording highly reminiscent of the directions given forty years later, the king decreed that his commissioners or any four of them were to assess the French projector's skill to see 'how farre [his method] may be practicable and usefull to the Publick'. The types of individuals appointed in 1674 were notably similar to those nominated in 1714, including: Lord Brouncker, president of the Royal Society and comptroller of the Navy; Seth Ward, bishop of Salisbury and previously Savilian professor of astronomy at Oxford; Sir Christopher Wren, surveyor of the King's Works and previously Gresham professor of geometry and Savilian professor of astronomy at Oxford; mathematician Dr John Pell; Robert Hooke, the Royal Society curator and Gresham professor of geometry; Sir Samuel Morland, a mathematician, inventor and Gentleman of the Bedchamber; and Colonel Silius Titus, a Gentleman of the Bedchamber.

On 12 February 1675, Morland, Pell, Titus and Hooke met at Titus's house to consider St Pierre's method. John Flamsteed, whom they had named their

[5] Whiston, *The Longitude Discovered*, 'Historical Preface', pp. v–vi; Mobbs and Unwin, 'The Longitude Act of 1714', p. 163. The historical preface detailing Whiston's longitude endeavours, dated 1741, is bound with some copies of the 1738 treatise; see Farrell, *William Whiston*, p. 165. Whiston, *Memoirs*, I, p. 369, suggests that the preface runs to 1745.

[6] Howse, *Greenwich Time*, pp. 36–43.

assistant, noted that the proposal was inferior to existing methods. He added, however, that poor astronomical data was hindering the improvement and use of these other methods. St Pierre ultimately fell by the wayside but, as described in the previous chapter, the commissioners' opinions contributed to Charles II deciding to build the Royal Observatory, with Flamsteed as its first Astronomer Royal, in order to make more accurate astronomical observations for finding the longitude at sea.[7]

The second influential British precedent was the longitude reward established by the will of Thomas Axe, who had close connections with the Royal Society before his death in 1691. Axe's will stated that, if his wife and son died without surviving children, £1000 was to be awarded to whomever could 'make such a perfect discovery how men of mean capacity may find out the Longitude at Sea, so as thatt they can truly pronounce upon observations if within halfe a degree of the true Longitude'.[8] This was an early attempt at stipulating a specific degree of precision, when that concept was not yet well-developed in astronomy and mathematics. It was later echoed in Whiston and Ditton's suggestions in 1714.

To win Axe's reward, a projector would need the approval of the professors of astronomy and geometry at Cambridge and Oxford and the testimonials of 'at least Twenty able masters of Shippes that shall have made several experiments thereof in long voyages'. The Act of 1714 would similarly name the Oxford and Cambridge professors as Commissioners. Axe's will also allocated £40 per year to making astronomical observations to improve mapping, to be overseen by his friends Edmond Halley and Robert Hooke, recognising that longitude determination was not the only obstacle to safe and accurate navigation. His requirement for evidence of sixty or more successful sea trials was perhaps too demanding, but the authors of the later Act went to the opposite extreme by only requiring one successful sea trial.[9] By the 1760s, this would become central to the disagreements between John Harrison and some of the Commissioners.

Another way in which the adoption and adaptation of earlier ideas for the Act of 1714 would prove problematic was how the Commissioners of Longitude were described. The Commissioners of 1714 were closely patterned after Charles II's earlier Longitude Commissioners, who were appointed to examine just one proposal and would lose their commissions when four or more had done so. The Commissioners of 1714 were likewise presented as individuals identified as acceptable judges for the rewards, rather than as a standing body as most modern scholars have assumed.

The title 'commissioner' did not automatically indicate membership in a permanently constituted body such as a board or committee during this period.

[7] Howse, *Greenwich Time*, pp. 43–45. [8] Turner, 'In the Wake of the Act', pp. 120–122.
[9] De Grijs, 'Longitude Prizes. 4'.

It could also signify an individual granted a responsibility or commission. Someone chosen to represent or negotiate on behalf of the king, for example, was often labelled a King's Commissioner.[10] Neither that title nor any of the other language in the Act of 1714 pointed unequivocally to a standing body, nor did the legislation grant other attributes this might have implied, such as a communal name and meeting place or administrative resources such as a secretary. This interpretation is borne out by the way in which contemporaries talked about the officials. Only a few early projectors, such as Jane Squire and Robert Browne, suggested the Commissioners should be meeting communally.[11] Most, even when complaining about their difficulties in getting the officials to respond to proposals, voiced no such expectations.[12] Nor did publications such as newspapers describe the Commissioners as a standing body until the 1760s.

Communicating and Judging Longitude

In practice, the wording of the Act of 1714 meant there was no single official channel for longitude projectors trying to obtain advice, trials, funding or a reward. This would only change decades later when the Commissioners institutionalised themselves (see Chapter 5). Projectors approached the same sorts of individuals and institutions they had in previous decades, except when they dedicated publications to the Commissioners. This created lasting confusion over how reward-seekers *should* seek advice and judgement, and even in later decades some asked individuals and periodicals for advice.[13] Instead, projectors approached diverse individuals and institutions: prominent experts like Isaac Newton; institutions such as the Admiralty and Navy, the State Office, trading companies, the Royal Society, and livery companies such as the Clockmakers; and individuals at these institutions.

The most visible and oft-contacted of the Commissioners of Longitude for most of their 114-year history was the Astronomer Royal. The Royal Observatory and the position had been explicitly created to aid the finding of longitude, which was largely considered an astronomical problem. Many projectors, as well as the Office of Ordnance, had consulted the Astronomer Royal about longitude since the position was created in 1675 and continued to do so

[10] 'commissioner, n.', *OED Online* (Oxford: Oxford University Press, 2021) www.oed.com/view/Entry/37147 [accessed 15 April 2021].

[11] Squire, *A Proposal to Determine*, 2nd ed., pp. 26–36; Browne, *To the Honourable the Commons*, title page.

[12] See, for example, Plank, *An Introduction to a True Method*, p. 7; Gordon, *A Compleat Discovery*, p. 31.

[13] As late as 1766, a projector wrote that he 'would be obliged to any gentleman whose curiosity or good will to the public, will acquaint him how he can apply to the board of longitude'; *Gazetteer and New Daily Advertiser*, 19 February 1766.

after 1714, when he became an *ex officio* Commissioner. The degree to which the Astronomer Royal responded to any approach from a projector varied according to who held the office: John Flamsteed (Astronomer Royal 1675–1719), Edmond Halley (1720–1742), James Bradley (1742–1762), Nathaniel Bliss (1762–1764), Nevil Maskelyne (1765–1811) and John Pond (1811–1835). Perhaps the most responsive and bureaucratic was Nevil Maskelyne, who would play a pivotal role in the institutionalisation of what became the Board of Longitude.

The first Astronomer Royal, John Flamsteed, responded to and received projectors he thought deserved that much but ignored others. This was partly a result of the number of projectors wishing to impose upon his time, especially after the establishment of the new rewards. Flamsteed viewed most longitude projectors as 'pretenders' with insufficient understanding of the subject.[14] In 1704, he said that, after having seen at least twenty such projectors, 'there was but one of all this Number that really understood how it was to be found'.[15] That projector was relying on inexact astronomical tables and, once Flamsteed pointed this out, 'he modestly drop'd his Proposall, and said no more of it'. The astronomer approved of projectors who at least recognised their own ignorance and heeded his advice, but could be scathing towards others.

For example, when Flamsteed was approached in 1704 for advice by John Coster and his friend, a London merchant named Sowter, he tried to dissuade them by criticising the ignorance of other projectors. He particularly belittled naval captain Robert Rowe – an 'Antient Seaman' – who advertised but apparently never published his scheme, *Longitude made plaine and easie*. When Coster and Sowter went to consult authorities at Oxford including Edmond Halley, Flamsteed wrote to Sowter that:

> I have had another, an Antient Seaman lately with me, who wanted the Places of some fix'd stars near the Pole to make a Curious Nocturnall for the same Purpose, but he is so Ignorant, that 'tis Impossible to Convince him of his Mistake. And I fear it will be the Same with Mr Coster except he understand, how to Calculate the Configurations and Eclipses of Jupiters Satellits or Eclipses of the Sun and Moon and how to Construct the Solar. which if he does I shall readily afford him all that lies in my Power to promote his Discovery, but if he does not, it will be in Vain for him to give any further Trouble about his business either to you or Sir.[16]

It was common after the foundation of the Royal Observatory, and particularly after 1714, for institutions such as government departments to forward

[14] John Flamsteed to Abraham Sharp, 31 August 1714, in Forbes, Murdin and Wilmoth (eds), *The Correspondence of John Flamsteed*, III, letter 1360, p. 701.

[15] John Flamsteed to Mr Sowter, 20 June 1704, in Forbes, Murdin and Wilmoth (eds), *The Correspondence of John Flamsteed*, III, letter 945, pp. 83–84.

[16] Ibid. Coster appears to have been associated with Wadham College, Oxford, where there was a servitor of that name in 1657–1659.

longitude proposals to the Astronomer Royal for his comment. Irritated by being forwarded impractical proposals in 1705, Flamsteed sent the Office of Ordnance a list of questions for anyone purporting to find longitude by lunar observation, so that they could better 'manage the pretenders to the Longitude'.[17] This was prompted by prince George, queen Anne's husband and Lord High Admiral, having had the Admiralty office forward a lunar and solar scheme by a Mr Green. This may have been the natural philosopher Robert Greene, who taught at Clare College, Cambridge, after gaining an MA there in 1703. He proposed a celestial means of finding longitude in his lengthy *Principles of the Philosophy of the Expansive and Contractive Forces* (1727), which discussed and discarded other schemes including Whiston's 'bombs', timekeepers and magnetic variation.[18]

When Flamsteed met projectors such as Green(e) with silence or scorn, this might only spur them on. This could include contacting other individuals and institutions until they found some encouragement, arguing their case in print, or more rarely petitioning Parliament. The rather vague description of and directions given to the Commissioners of Longitude in the Act of 1714 simultaneously complicated reward-seekers' need for authoritative judgement and encouraged them to approach a wide range of relevant actors.

Longitude Rejuvenated

Regardless of confusion over the proper channels for submitting proposals, the rewards established in 1714 had the desired effect of increasing activity. News rapidly circulated across Britain and Europe and encouraged projectors beyond just Whiston and Ditton to begin or renew efforts towards a solution.[19] It is revealing to look at the pace and diversity of activity in the first six months after the passage of the Act. Even within this short time, there was a representative array of the types of schemes, projectors and institutions that would recur throughout the century.

Even before queen Anne gave her royal assent to the legislation on 9 July, bookseller George Strahan near the Royal Exchange in London began advertising George Keith's scheme: 'An easie Method, not to be found hitherto in any Author or History; whereby the Longitude of any Places at Sea or Land, West or East from any first Meridian may be found at any Distance, great or small, by certain fixed Stars'.[20] Keith, a Scottish-born

[17] Josiah Burchett to John Flamsteed, 30 November 1705, in Forbes, Murdin and Wilmoth (eds), *The Correspondence of John Flamsteed*, III, letter 1052, pp. 260–261.
[18] Greene, *The Principles of the Philosophy*, pp. 927–928.
[19] Wigelsworth, 'Navigation and Newsprint'; Wigelsworth, *Selling Science*, pp. 119–145.
[20] *Daily Courant*, London, 9 June 1714; Keith, *An Easie Method.*

missionary and rector, originally Quaker and later Anglican, had served as Surveyor-General of East Jersey in the American colonies in 1685–1688. His method, first published in 1709, involved the declination of fixed stars, great circles and geometry. Five years later, astronomer and mathematician Abraham Sharp mentioned the 'mistakes and fallacys' in Keith's publication to Flamsteed when discussing how Whiston and Ditton might be risking their own reputations.[21] As the search for longitude proceeded, a number of other religious figures would also participate. Some did so because of a true interest in longitude and mathematics, others mainly as a means of bringing attention to their religious beliefs. This overlapping of spheres had existed even before 1714, for example in the influential Swedish visionary theologian and natural philosopher Emanuel Swedenborg's work on longitude from 1710. Swedenborg first developed a lunar method for finding longitude at sea while visiting London in 1710–1712, about which he communicated with the Royal Society, Flamsteed and Halley. Much later in life, he would present his scheme to the Board of Longitude in 1766.[22]

William Hobbs also moved swiftly, presenting his proposal for a longitude timekeeper to Flamsteed on 9 July – the day the Act of 1714 gained royal assent.[23] Flamsteed described the timekeeper as a 'pendulum clock hung by a contrivance so as to answer all the motions of a ship so as he thought they could not affect the Pendulum'. He disapproved of Hobbs and compared him unfavourably to another projector, Case Billingsley. Regardless, Hobbs published his scheme on 12 July.[24] Hobbs's identity is somewhat of a mystery. Authors including Roy Porter have been unable to match this Philo-Mathematicus William Hobbs, who said he could be contacted via the watchmaker James Hubert near the Royal Exchange, with an eccentric natural philosopher of that name from Weymouth who published *The Earth Generated and Anatomized* in 1715.[25] However, there are striking parallels between the first William Hobbs, who as of mid-November 1714 was in London exhibiting his longitude timekeeper in coffeehouses, and the second, who in mid-December sent from Weymouth a method 'for Regulating or Equalizing Motion without a Spring', which he also wanted laid before the Commissioners of Longitude.[26]

[21] Abraham Sharp to John Flamsteed, 3 April 1714, in Forbes, Murdin and Wilmoth (eds), *The Correspondence of John Flamsteed*, III, letter 1356, p. 696.
[22] Schaffer, 'Swedenborg's Lunars'.
[23] John Flamsteed to Abraham Sharp, 31 August 1714, in Forbes, Murdin and Wilmoth (eds), *The Correspondence of John Flamsteed*, III, letter 1360, p. 700.
[24] Hobbs, *A New Discovery*.
[25] Porter and Hobbs, *The Earth Generated*, p. 33; Howse, *Greenwich Time*, p. 59.
[26] W[illiam] H[obbs] at Weymouth, 'A Method proposed for Regulating or Equalizing Motion without a Spring ... Intended to have been Laid before the Honorable Commission for the Longitude', 14 December 1714, RS CLP/3ii/11.

After the passage of the Act of 1714, William Whiston continued trying to capitalise on his and Ditton's scheme and on the role they played in the legislation. Whiston described their method as 'so far approved by this present Parliament, that they have passed an Act, ordering 20000 l. Reward for such a Discovery'.[27] He quickly incorporated the 'Discovery of the Longitude' into his public lectures on mathematical and scientific subjects.[28] It increasingly became apparent, however, that, despite their influential roles in events, Whiston and Ditton would not themselves be earning a reward. Flamsteed wrote to his friend Abraham Sharp in August:

> Mr Whistons and Mr Dittons proposealls tend onely to finding the distance of a ship at sea from any mark or coast from whence they can see the fire and hear the noyse of a gun shot of at 12 a clock at night … so their proposeall sinkes *from finding of the longitude* to finding of the distance of the ship from a seamark. I am sorry for them. … Twas their attempt that occasioned the Act of parliament that proposes a reward for any one that shall discover the Longitude.[29]

Sharp responded on 18 December that, although Whiston's longitude proposal was the only one he had recently read which had any scientific merit,

> I very much fear his project being so chargeable and impracticable at Sea will scarce turn to an advantageous account to himself: the News Paper gave notice of his proposeing to make the experiment by causing a Ball of fire to be shott up into the air every Saturday evening at eight of the Clock upon Black Heath which being so near, you cannot but see or hear it, tho you are too near and we too farr of to make any advantage of it yet would gladly know if possible <u>whether any others at a competent distance may have observ'd it</u>.[30]

Whiston continued advertising and working on this and other longitude schemes for many decades after Ditton's death the following year, sometimes asking the public to aid in observing and timing trial rocket launches. These efforts were sometimes met with scorn from the public and the intellectual establishment, in no small part because of opposition to Whiston's religious beliefs. Nevertheless, he would have regular dealings with the Commissioners of Longitude (see Chapter 4).[31]

On 24 July 1714, the inventor Case Billingsley published a scheme to find longitude at sea 'from the Sun, Moon, or Stars, and an Exact Time-Keeper;

[27] *Daily Courant*, 8 July 1714.

[28] See, for example, *Post Man and the Historical Account*, 5–7 August 1714; *Daily Courant*, 10 August 1714.

[29] John Flamsteed to Abraham Sharp, 31 August 1714, in Forbes, Murdin and Wilmoth (eds), *The Correspondence of John Flamsteed*, III, letter 1360, p. 701.

[30] Abraham Sharp to John Flamsteed, 18 December 1714, in Forbes, Murdin and Wilmoth (eds), *The Correspondence of John Flamsteed*, III, letter 1372, pp. 719–720.

[31] Farrell, *William Whiston*, pp. 134–142; Stewart, *The Rise of Public Science*, pp. 189–192; Werrett, 'Both by Sea and Land', pp. 200–201.

with such necessary Improvements as have not yet been described by any other Person, and (with respect to the Term of any ordinary Voyage) may properly be call'd a Perpetual-Motion'.[32] Billingsley pointedly mentioned that his method did not involve 'firing Guns', as did that of Whiston and Ditton, nor 'the most curious Spring-Clocks or Watches' upon which many other methods relied. Perpetual motion, like longitude, had been long sought but never perfected and would feature in a number of eighteenth-century longitude schemes. On 26 July, 'R.B.' – who said that he had served as secretary to Sir Francis Wheeler, an officer of the Royal Navy in the Nine Years War – also published a scheme for finding longitude at sea or on land with a newly invented instrument. His method was 'extracted from the 3 Years Observations made at Islington by Dr. Edmund Halley, Savilian Professor of Geometry in Oxford, for knowing the true Place of the Moon, and which are now inferred in Mr. Street's Caroline Tables … With a better Method for discovering Longitude than that lately proposed by Mr. Whiston and Mr. Ditton.'[33]

On 30 July 1714, Mr Daille of Gensac in Guyenne sought the help of Matthew Prior, a British diplomat and secret agent stationed in France, in putting his longitude proposal for land and sea before Parliament.[34] Five days later, Prior's secretary Adrian Drift sent it to the State Office in London. The British newspapers tracked Daille's efforts until in February 1715 the French Académie Royale des Sciences reportedly examined his longitude machine, 'which he thinks capable of Perpetual Motion, and other extraordinary Effects', judging it impractical at sea although not in theory.[35] Throughout the eighteenth century, foreign projectors would also vie for the British rewards, trying to make contact through a variety of individuals and institutions and appealing to the public in print.

On 4 August, the natural philosopher William Derham sent his 'Observations concerning the motion of chronometers' to the Royal Society, although it appears not to have been read there until 4 November.[36] In this and the third edition of his *Artificial Clock-Maker*, published the same year, Derham may have coined the use of the English term 'chronometer' to indicate a precision timekeeper in the modern sense.[37] The term was not, however, widely used in this sense until much later in the century. Terms including clock, watch, machine and timekeeper were more usual in the earlier decades, as well as occasionally terms such as 'engine'.

[32] *Post Boy*, London, 24–27 July 1714; Billingsley, *The Longitude at Sea*; Stewart, *The Rise of Public Science*, pp. 194–196.

[33] *Daily Courant*, 24 July 1714.

[34] 'Memorial on proposals of M. Daille for calculating longitude' (f.30 French Copy), 30 July 1714, TNA SP 78/159.

[35] *Weekly Journal with Fresh Advices Foreign and Domestick*, 5 February 1715.

[36] William Derham, 'Observations concerning the motion of chronometers', 4 August 1714, RS Cl.P/3ii/10.

[37] Derham, *The Artificial Clock-Maker*, p. 125.

On 24 August, the aforementioned Case Billingsley sent a printed letter to each Commissioner further explaining his proposal and advertised that the public could buy a copy at booksellers. The letter promised that he could have a newly designed platform for keeping either a timekeeper or an observer stable at sea ready for trial within ten days.[38] He brought his treatise in person to Flamsteed, who judged him 'a man of better sence then Mr Hobs'. While Flamsteed had summarily dismissed William Hobbs, he took the time to show Billingsley the error of his ways:

I had shewed him that what ever he imagined to the Contrary the pendulum would be affected with every shake or Motion of the ship and the Vibrations made wider or Narrower than they ought to be, and this in the company of some very intelligent persons who assented to what I urged he held his peace which is as much as I can expect from persons that have swelled themselves with the hopes of getting twenty thousand pounds.[39]

On 30 August, Conyers Purshall – apparently a physician of Bewdly, Worcestershire – left a proposal for finding the longitude at Greenwich for Flamsteed.[40] Purshall had matriculated at Pembroke College, Oxford, in 1675 at age 18 and called himself an M.D.[41] He had previously published two editions of an anti-Newtonian book that included his method 'to find out the Exact Rate that a Ship Runs', which he revived after the passage of the Act of 1714.[42] Flamsteed wrote to a friend that Purshall's method employed 'a water-wheele that should follow the ship and Measure its way which it will doe like the way-wiser over every wave and swell as well as when its smooth, I have not seen this ingenious gentleman. but allowing his contrivance to be liable to no other inconvenience by this, tis enough to cast it and render it ridiculous.' Unable to gain encouragement from the Commissioners of Longitude, Purshall elaborated on his invention six years later in a newspaper advertisement. Like Whiston and Ditton before him offering a cut to Edmond Halley for expert assistance, Purshall proposed that:

if any Captain or Master of a Ship, or other Person, will undertake to get this Engine made, as above described, and make the Experiment of it at Sea, so as to satisfy the Government of the Truth of its Performance, if it takes, the first that does it shall have

[38] Billingsley, 'Since the publication', advertised in the *Daily Courant*, 23 August 1714. A copy in the Beinecke Rare Book and Manuscript Library, Yale University (BrSides 1988 110), is signed by Billingsley and addressed to Thomas, Earl of Pembroke, named as a Commissioner in 1714.

[39] John Flamsteed to Abraham Sharp, 31 August 1714, in Forbes, Murdin and Wilmoth (eds), *The Correspondence of John Flamsteed*, III, letter 1360, p. 700.

[40] Purshall, *To the Lords Commissioners*; John Flamsteed to Abraham Sharp, 31August 1714, in Forbes, Murdin and Wilmoth (eds), *The Correspondence of John Flamsteed*, III, letter 1360, p. 700.

[41] Schofield, *Mechanism and Materialism*, pp. 115–117; indenture from 23 March 1699/1700, McFarlin Library of the University of Tulsa, 2007–006.

[42] Purshall, *An Essay at the Mechanism*; second edition published as *An Essay on the Mechanical Fabrick*.

the tenth Part of any Praemium or Reward that shall accrue to me from the Government by Act of Parliament, for finding out the Longitude; and of all other Praemiums from all foreign Nations that I shall obtain thereby.[43]

On 4 September, Benjamin Habakkuk Jackson published his thoughts on subjects including the longitude, which he dedicated to the Fellows of the Royal Society.[44] Jackson was, like many other longitude reward-seekers, a wide-ranging inventor or projector, patenting in 1704 and 1715 a method of keeping coaches and similar vehicles upright during accidents, as well as a swimming engine.[45] On 7 September 1714, William Hall published his 'new and true Method to find the Longitude, much more exacter than that of Latitude by Quadrant … by one who hath been Commander and Owner of several Vessels. Prov'd by Experience, and recommended to publick Consideration.'[46] It was, unsurprisingly, common for merchant and naval seamen to propose longitude schemes highlighting their experience at sea.

In mid-September, the Nonconformist preacher Isaac Hawkins published an account that suggested using a barometer to determine the timing of local high tides (via the rise in altitude) and thus the Moon's location at a given place and time. It also detailed the requirements for a marine timekeeper.[47] He paired his longitude lobbying with public support of the new German-born king George I, offering a sermon on his ascension to the throne – 'an Expedient to Remove the Groundless Fears and Jealousies of honest People concerning his Majesty'.[48] It appears that Hawkins and another Nonconformist preacher had visited Flamsteed the previous month.[49] Flamsteed wrote that the two were from Worksop in Derbyshire and had travelled 150 miles to see him. He was amused by their schemes, which involved keeping time by air trickling back into a vessel evacuated with an air pump. Flamsteed recorded that he 'treated them very Civilly but refused to give them my hand to testifie that I had seen their proposealls and advised them not to print them'.[50] Like other projectors faced with silence or disapproval, Hawkins ignored his advice. On 15 September, the aforementioned William Hobbs also sent Flamsteed his proposal for a 'Movement with a Decimal Horologe', requested a trial and was ignored.[51]

[43] *London Gazette*, 22–26 November 1720. John Flamsteed to Abraham Sharp, 31 August 1714, in Forbes, Murdin and Wilmoth (eds), *The Correspondence of John Flamsteed*, III, letter 1360, pp. 702–703.
[44] Jackson, *Some New Thoughts*.
[45] Woodcroft, *Subject-Matter Index*, p. 186; Rogers, *Documenting Eighteenth Century Satire*, p. 56.
[46] *Daily Courant*, 7 September 1714; Hall, *A New and True Method*.
[47] Hawkins, *An Essay for the Discovery*.
[48] *Post Man and the Historical Account*, 11–14 September 1714.
[49] John Flamsteed to Abraham Sharp, 31 August 1714, in Forbes, Murdin and Wilmoth (eds), *The Correspondence of John Flamsteed*, III, letter 1360, p. 700.
[50] Ibid., pp. 700–701.
[51] William Hobbs, broadsheet, 15 September 1714, CUL RGO 1/69/F, fol. 160r.

Digby Bull sent a proposal combining a 'shipwatch' and 'Ship Dial', first published in 1706, to the Commissioners on 29 September 1714. Admiralty Secretary Josiah Burchett forwarded it to Flamsteed, but Bull's offer to teach others to make his devices was not taken up.[52] By 5 October, Samuel Watson had renewed earlier efforts to gain recognition for a method of finding longitude at sea and on land with a 'small Machine'. Watson, a successful and innovative clockmaker in Long Acre in London, and previously of Coventry, had made timekeepers for royalty and was associated with leading figures including Newton. As early as 1707 he had sought financial reward for a navigational instrument for finding latitude as well as for a longitude timekeeper:

> The other being a Time-keeper, is made so by Art, that no Sea, tho' disturb'd n'ere so much, nor any alteration of air, will hinder its shewing the motion of the sun, in his true Longitude in Degrees and Minutes, always moving as the Sun, faster and slower, and at all times; you have the true distance of Degrees and Minutes of Terrestrial Longitude from the Place of Departure when the Sun appears. If the Author hereof does finds an Encouragement equivalent to this Projection (which will be of so great and Universal Good) he promises to perform what is here before Proposed for the benefit of the Publick.[53]

On 3 November 1712, Watson offered to sell his inventions to the Worshipful Company of Clockmakers, but its Court was not interested.[54] He repackaged his scheme in 1714, dedicating it to the new Commissioners:

> That great Secret of Longitude both by Sea [and] Land, which hath so much puzled the World for many Years past, is now perfectly discovered, and made so very easy that the meanest Capacity may be Master of it in half an hour … This Machine is not liable to be any way disorder'd, neither is there any attendance required, but is fit for Service at all Times either Night or Day. Invented and made Sam[uel] Watson … who will lay it before the Honourable Commissioners appointed by Act of Parliament, when they shall think fit.[55]

On 9 October, John Coster published his 'Practicable Method for finding the Longitude at Sea, by a Marmeter, or Instrument for measuring the exact Run of the Ship', also dedicated to the Commissioners.[56] Coster had visited Flamsteed and others in Oxford about a longitude scheme in 1704, without success. On 14 October 1714, Francis Haldanby of Yorkshire published a two-page proposal dedicated to James Stanhope, one of the secretaries of state.[57]

[52] Digby Bull to the Commissioners of Longitude, 29 September 1714, in Forbes, Murdin and Wilmoth (eds), *The Correspondence of John Flamsteed*, III, letter 1365, pp. 710–711. On 6 January 1772, Nevil Maskelyne read and annotated Bull's letter: 'the only hint in it that might have been useful is the strong and rapid motion of his watch, wherein it agrees with Mr Harrison's'. Bull, *A Letter of Advice*.

[53] *Post Boy*, 17–19 July 1707.

[54] Atkins and Overall, *Some Account of the Worshipful Company of Clockmakers*, p. 250.

[55] *Post Man and the Historical Account*, 5–7 October 1714.

[56] *Post Man and the Historical Account*, 9–12 October 1714; Coster, *A Practicable Method*.

[57] *Post Boy*, 14–16 October 1714; Haldanby, *An Attempt to Discover*.

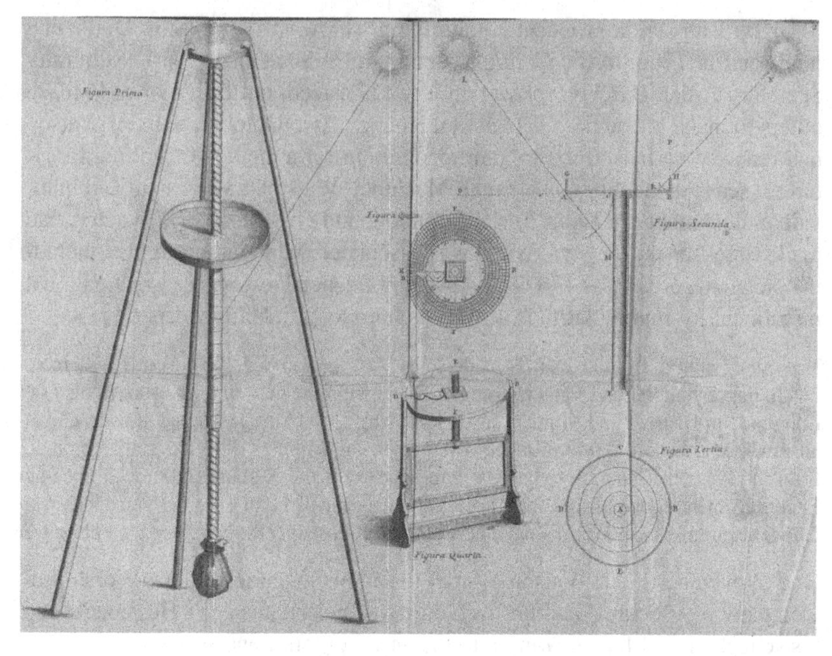

Figure 3.1 Dorotheo Alimari's proposed observing instrument for finding longitude, from Alimari, *Longitudinis aut terra aut mari investiganda methodus* (London, 1715). ETH-Bibliothek Zurich https://doi.org/10.3931/e-rara-1355

The scheme employed grids for each latitude which would show how much longitude would be gained per distance of latitude sailed (if there were no ocean current). The same day, a pamphlet in English and Latin and dedicated to the Commissioners was published in Venice describing a scheme by military and naval writer Dorotheo Alimari.[58] The Venetian painter Sebastiano Ricci had informed the Commissioners of Alimari's method soon after the passage of the Act of 1714. The complete Latin treatise was published in early 1715, with solar tables and images of instruments (Figure 3.1).[59] It required the production of an exact almanac for a given meridian and observations of the Sun's local position, or an accurate clock.[60] Flamsteed dismissed it: 'Seignr. Alimari has sent me a Description of his Instrument, with a Figure of it, which was needless, for it was easy to apprehend without the figure, that it was one of the worst contrivances for taking the height of the Sun or Stars that was ever thought of.'[61]

[58] *Post Man and the Historical Account*, 14–16 October 1714.
[59] Ricci, *The New Method*; Alimari, *Longitudinis aut terra.*
[60] Schaffer, 'Swedenborg's Lunars', p. 7.
[61] John Flamsteed to Abraham Sharp, 22/23 October 1714, in Forbes, Murdin and Wilmoth (eds), *The Correspondence of John Flamsteed*, III, letter 1366, p. 712.

On 15 October, James Clarke of Calne, Wiltshire, finished his proposal for keeping time to find longitude at sea by pouring mercury between different glass vessels in the vein of an hourglass.[62] Clarke was one of the projectors criticised in November by pamphleteer Jeremy Thacker of Beverley in Yorkshire, which led to Clarke publishing a longer illustrated pamphlet in 1715, which hit back at Thacker.[63] On 9 November, William Hobbs began supplementing his lobbying for a longitude reward by advertising that at various London coffee-houses he would exhibit:

> to the publick View of all Mathematicians, and such-like Watchmakers, the new invented Movement or Machine … so contrived (with a Decimal Horologe) as that it does and will keep exact Time with the Sun, (or Earth) whether its Motion be Swift or Slow, or whatsoever Effects the Seasons of Heat or cold may have, either to excite or retard the Motion of the said Movement.[64]

Hobbs produced a longer version of the pamphlet two years later, adding to the diagrams and mathematical working and suggesting how sea trials might be conducted.[65]

By 9 November, Thacker also began advertising *The Longitudes Examin'd*, which described the deployment of an adapted clock inside a vacuum (Figure 3.2).[66] As Rogers has suggested, generating some controversy, this appears to have been a satire by John Arbuthnot, although it reveals a strong understanding of horology and of other longitude schemes. There was a prevalent early modern tradition of producing mock pamphlets, advertisements, lectures and similar including many on longitude. There are parallels between the wording and content of Thacker's pamphlet and texts by the 'Scriblerians' – authors including Arbuthnot, Alexander Pope and Jonathan Swift who ridiculed 'false tastes in learning'. In addition, Thacker employed the new term 'chronometer' in this publication some months *after* the Scriblerian William Derham had submitted a paper to the Royal Society that employed the word in the modern sense.[67]

Further clues which strongly point to satire include Thacker's incorporation in his publication and advertisements of the phrase 'Quid non pectora mortalia cogis, Auri sacra Fames'. This Latin motto from the *Aeneid* roughly translates as 'Accursed thirst for gold, what dost thou not compel mortals to do?'

[62] Clarke, *An Essay*; will of James Clarke, 16 February 1721, TNA PROB 11/582/14.
[63] Clarke, *The Mercurial Chronometer*. Barrett, *Looking for Longitude*, pp. 62–67, discusses the printed exchanges between Clarke and Thacker.
[64] *Daily Courant*, 9 November 1714.
[65] Hobbs, *A New Discovery*. Barrett, *Looking for Longitude*, pp. 58–67, discusses Hobbs's visual presentations of his invention and response to Thacker.
[66] Thacker, *The Longitudes Examin'd; Post Man and the Historical Account*, 9–11 November 1714.
[67] Rogers, 'Longitude Forged'; Betts and King, 'Jeremy Thacker'; Lynall, 'Scriblerian Projections', pp. 4–5.

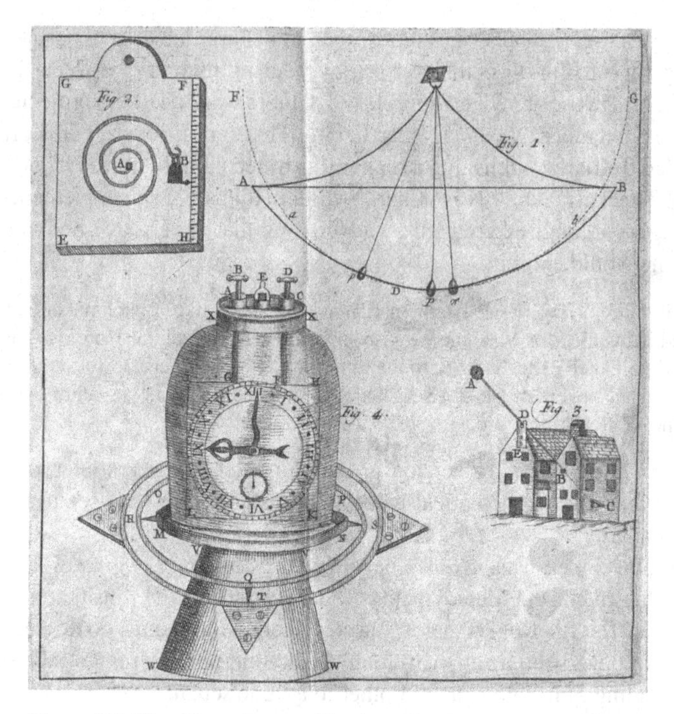

Figure 3.2 Plate from Jeremy Thacker, *The Longitudes Examin'd* (London, 1714). By permission of the Master and Fellows of St John's College, Cambridge

Other wording appears to mock the conventions of longitude proposals, such as Thacker calling his machine 'pretty' and saying that he is 'almost' certain it will win him one of the new rewards:

The Longitudes examined. Beginning with a short Epistle to the Longitudinarians, and ending with the Description of a smart pretty Machine of my Own, which I am (almost) sure will do for the Longitude and procure me the 20000 l. By Jeremy Thacker of Beverly in Yorkshire. Quid non mortalia pectora cogis Aura sacra Fames.[68]

On 30 November, famed polymath Christopher Wren delivered to the Royal Society a ciphered description of 'three Instruments proper for discovering the Longitude at Sea'. These were later deciphered as a marine timekeeper in vacuum with a magnetic balance, observing Jupiter's moons with a telescope, and a log for measuring a ship's speed.[69] Wren, by then eighty-two, had been one of Charles II's longitude commissioners in 1674. An anonymous author also

[68] *Post Man and the Historical Account*, 9–11 November 1714.
[69] Bennett, *The Mathematical Science*, pp. 44–54; Howse, *Greenwich Time*, pp. 59–61.

published a scheme based on lighthouses.[70] This was mentioned favourably in the *Weekly Packet* on 4 December with respect to a longitude book published in Hamburg and dedicated to the (Hanoverian) George I:

We are likewise advis'd from Hamborough, that a certain Person has found out a Method to discover the Longitude, and has printed a Book there on that Subject, dedicated to the King of Great Britain; but among all the Pretenecs to that Knowledge, none seems more capable of being put in Practice, than that of erecting Light-Houses, describ's in a Book publish'd by Way of Advertisement in this Day's Packet; which the Reader must allow to be extreamly beneficial.[71]

Sometime in the second half of 1714, Stephen Plank of Spitalfields, London, published his longitude scheme.[72] It employed a dial for local time and some good watches set at the port of departure and kept in cotton in a brass box over a small stove to keep them at constant temperature, helping them keep time accurately. Six years later, he would publish a tract about the lunar method and claim to have built at 'Great Trouble and Expense' a device to accurately measure the Moon's positions and to have attempted a sea trial to Jamaica.[73] As a result, in 1734 Plank challenged John Hadley's application for a patent for a mariner's quadrant, saying that he had invented the same instrument in 1717 and showed it to various learned people before describing it in his next longitude pamphlet.[74] He continued proposing longitude-related technologies until at least the 1740s.[75]

Also in 1714, Robert Browne of Wapping first published his methods for finding 'The Longitude at Sea. By Caelestial Observations only. And also by Watches, Clocks, &c. and to correct them and know their Alterations'.[76] Browne, an instrument maker and carpenter of Wapping, promoted his scheme to the Commissioners of Longitude and the South Sea Company. He was a long-time longitude projector who later published on spiritual matters and was one of the few projectors to express expectations of the Commissioners holding communal meetings. Browne petitioned Parliament about poor treatment at the hands of Astronomer Royal Edmond Halley in 1732.[77]

Interestingly given the modern debate over Thacker's existence, at the turn of the year the Chester- and previously London-based mathematician John Ward claimed in print that Thacker had plagiarised his design for a longitude

[70] Anonymous, *An Essay Towards a New Method.*
[71] *Weekly Packet*, 4–11 December 1714. [72] Plank, *An Introduction to the Only Method.*
[73] Plank, *An Introduction to a True Method.*
[74] Plank, 'Affidavits of above petitioner and others in support of Stephen Plank who entered a caveat against the passing of the patent. Extract of minutes of the Royal Society dated 2nd March 1732, and a copy of Stephen Plank's paper delivered to the Royal Society the same day', 17 May 1734, TNA SP 36/31.
[75] *Daily Advertiser*, 15 March 1745; *General Advertiser*, 20 September 1746.
[76] Browne, *Methods, Propositions and Problems.*
[77] Browne, *To the Honourable the Commons*; Stewart, *The Rise of Public Science*, p. 198.

timekeeper.[78] Ward was an inventor of mathematical instruments and chief surveyor and gauger to the Excise. His pamphlet, dedicated to the Commissioners, concerned a timekeeping automaton in a vacuum – 'an *Artificial Movement*, made either by Wheel-work, or some other contrivance equivalent to it' – which the mathematical instrument maker John Rowley of Fleet Street in London had made for him. Rowley also supplied instruments to the Office of Ordnance and the Royal Mathematical School at Christ's Hospital and was Master of Mechanics and Engine Keeper to the Board of Works. Here the sometimes farcical elements of the search for the longitude came to the fore, as a real and respected longitudinarian crossed swords with someone who may have been a satirical ghost.

Further Developments before 1737

In the first six months following the establishment of the new rewards in July 1714, we can see the diversity of longitude efforts which ensued, the 'Swarme of *hopefull Authors*' about whom Flamsteed complained in October.[79] Respected and unknown figures alike participated, as did young and old, wealthy and impecunious. Reward-seekers came from all over Britain, Europe and their colonies. They hailed from all manner of professions and pursuits largely related to seafaring, mathematics, astronomy, technology and inventing but also included other learned individuals such as physicians and religious figures. These diverse projectors pitched their schemes to the Commissioners and particularly the Astronomer Royal, as well as to government offices and officials such as at the Admiralty and State Office. Many made use of printing and sometimes of public lectures and exhibitions to target these actors and the broader public.

This first half year encompassed many of the types of longitude schemes that had long been and would continue to be entertained across the eighteenth century, reflecting the fact that the problem was not one of abstract principle but of its application in maritime practice. They included myriad celestial schemes, rockets as suggested by Whiston and Ditton, technologies including what were essentially waywisers for ships, improved charts and almanacs, platforms and other devices to steady astronomical observers or timekeepers at sea, and marine timekeepers from complex clocks to what were essentially hourglasses using everything from mercury to air running into a vacuum. In the years before the first known communal meeting of some of the Commissioners in 1737, these dynamics persisted in the keen pursuit of the large rewards. The most relevant of the Commissioners, who had not been envisioned and constituted as a formal body, were key players in the events of this period. These

[78] Ward, *A Practical Method.*

[79] John Flamsteed to Abraham Sharp, 22/23 October 1714, in Forbes, Murdin and Wilmoth (eds), *The Correspondence of John Flamsteed*, III, letter 1366, pp. 712–713.

included not only continued communication and attempts at communication between projectors and relevant institutions and individuals but also sea trials, private and officially supported, and attempts to alter both the Act of 1714 and the behaviour of its appointed judges.

During the earlier decades in particular, it was common for projectors to approach and often to circulate between all manner of institutions and individuals in an effort to gain an authoritative hearing and approbation. One of many examples is found in the tireless attempts of London clockmaker Peter Laurans during the 1720s. Laurans, probably a French immigrant or of French descent, has left little beyond a signed movement and dial in a longcase clock sold in 2001.[80] He may have been the Peter Laurent – son of Moses Laurent, late of Issunden in Berry, France – apprenticed to Solomon Bouquett of the Clockmakers' Company on 17 May 1700.[81] From at least 1721 to 1723, Laurans contacted the Royal Society and the Treasury repeatedly, seeking approval and increasingly a financial reward for his longitude method. A number of the Commissioners would have been aware of these attempts. Newton mentioned his proposal in a letter to Secretary of the Admiralty Josiah Burchett in 1721, concluding that it was unlikely that longitude could be found at sea with a method not yet proved successful on land (i.e. 'watch-work').[82] Newton received numerous longitude proposals both before and after 1714, as a result of the high esteem in which he was held as president of the Royal Society from 1703 and then as an *ex officio* Commissioner as a result of that role.

Laurans presented 'A Scheeme of the Longitude, or a peice of Watche Worke of New Construction that shal goe upon Sae as reguler as the best pandulum Clock shal goe upon Land' to the Royal Society on 14 July 1721, read to the Fellows on 19 October.[83] Failing to obtain much encouragement, he went on to petition the Treasury repeatedly until at least 1723, '[h]aving form'd an imagination there is a piece of mony allow'd by the government, or other ways, for the incouragement of any parsone that shal produce a machine tending to the discovery of the Longitude upon Sea'.[84] Whereas the proposal sent to the Royal Society was straightforward and placed no great emphasis on possible rewards, the Treasury petitions reveal ever-increasing anxiety about

[80] Peter Laurans, mahogany longcase clock, Christie's, South Kensington, Sale 9204, 3 October 2001, lot 134 www.christies.com/lot/lot-an-english-mahogany-longcase-clock-3049198 [accessed 8 April 2021].

[81] Worshipful Company of Clockmakers, *Register of Apprentices*, p. 176. With thanks to Rory McEvoy for this information.

[82] King, 'John Harrison, Clockmaker at Barrow', p. 184.

[83] Peter Laurans, 'Scheme of the Longitude or a peice of Watch worke of new Construction that shal goe upon Sea as regular as the best pendulum Clock shal goe upon Land', 14 July 1721, RS CLP/3ii/13.

[84] Peter Laurans to the Lords of the Treasury, 27 July 1723, quoted in Hart, 'Gleanings', pp. 297–298.

finances – perhaps the main motivation for his efforts, although he also cited public utility, as projectors usually did. By 14 October 1723, Laurans wrote quite plainly of his financial troubles:

> [T]he said petitioner having for a very considerable time endeavour'd to fix his talant in this Country, and having through Losse of time and expences plung'd himself in extreme bad sircumstances, in so much that he is destitute of all visible ways of subsisting, the said humble petitionner being inform'd that some Nobles gentlemen in this towne having taken notice of his miserable Condition, out of their goodnesse and Charity have gathered among themselves a sum of mony which sum they have desseigned to releave him in his necessities, the said humble petitionner being also inform'd that the said sum has been Lodged in your Lordships hands for that purpose, the said humble petitionner supposing his information wright, with humble respect and submission taketh the Liberty to begg that your Lordships may be pleas'd to grant him the said sum.[85]

Unfortunately, the Lords of the Treasury seem to have known nothing about the money purportedly collected on his behalf and repeated that they were unable to offer him funding without the recommendation of the Commissioners of Longitude.

Many other projectors sent their proposals, at least initially, to the Royal Society. The president was *ex officio* a Commissioner by virtue of the position, in addition to which the Society was a known learned authority. Some presidents were more active in the search for the longitude than others, notably Newton and later Joseph Banks (president 1778–1820). In between, it was common for the Society to forward proposals to trusted individuals such as James Hodgson for comment. Hodgson became a Fellow of the Royal Society in 1703 and was master of the Royal Mathematical School at Christ's Hospital from 1709 to his death in 1755. He read and commented on several proposals after Newton's presidency of the Society, including that sent from St Petersburg by P. I. Rocquette in 1732, a scheme perhaps inspired by a lecture on the longitude problem given at the Imperial Academy of Sciences in 1727.[86]

Rocquette said that he was clockmaker to the Russian empress Anna and hoped to gain the Fellows' approval for his longitude scheme and their aid in gaining recompense for it through the Longitude Act.[87] He had heard that John Harrison claimed to have discovered longitude but knew that this had not yet been proven at sea, and believed his own timekeeper might be better. Rocquette's *Nouveau sisteme de la sphere celeste suivi d'un moyen pour trouver les longuitudes sur*

[85] Peter Laurans, petition to the Lords of the Treasury, 14 October 1723, quoted in Hart, 'Gleanings', p. 298.

[86] Werrett, 'Perfectly Correct', pp. 113–114.

[87] Letters from P. I. Rocquette to the Royal Society about his method for calculating longitude at sea, RS LBO/20/62 (24 June 1732); RS LBO/20/63 (15 July 1732); RS LBO/20/203 (29 November 1733); RS LBO/20/204 (c. 1732–1733); RS LBO/20/226 (31 May 1733); P. I. Rocquette to Peter Collinson on 'Methode pour trouver la longitude sur mer', RS LBO/20/204 n.d. (c. 1732–1733); RS LBO/20/226 (31 May 1733).

mer, which he also sent to the Académie Royale des Sciences, described finding longitude at sea by managing two portable pendulum clocks.[88] Hodgson was scathing about the proposal and the clockmaker's claim to have been the first to discover the effects of different climates on the running of clocks. He doubted Rocquette's proposal could meet the terms of the Act of 1714:

He hopes, that the Virtuosi of the first rate, for whom this is done, will supply what is wanting: since his design is not to teach the Sciences to those, who do not understand them, but only give the Learned an Idea of his great discovery, for which he expects they will do them justice ... What answer must be given to a Man, who is so very ignorant of the first Principles of Astronomy and Philosophy, who has asserted so many falsehoods, and calls them Demonstrations; and is so vastly fond of his Performance; I leave you, Gentlemen, to determine.[89]

Hodgson was not always so harsh, as evidenced in letters exchanged in 1735 and 1737–1738 with John Philip Baratier, a prodigious teenage polymath from Brandenburg. Baratier contacted the Royal Society with a longitude scheme having just turned fifteen and corresponded with the scientific academies in France and Berlin, becoming a member of the latter. His biographer claimed that it was the British queen, Caroline of Ansbach, who directed the Society to examine his proposal at the behest of a foreign noble.[90] Hodgson gently dismissed Baratier's lunar-distance method as being somewhat outdated and unworkable at sea, concluding that:

I flatter myself, that what I have said with regard to his Method of discovering the Longitude, will be no prejudice to a Youth, who so far excels in Science the generality of his Age, and who has given such early Proofs of an extraordinary Genius, by such a Progress in astronomical Knowledge, at these Years, when others scarcely have attain'd its first Rudiments.[91]

When in late 1737 Baratier sent a magnetic variation scheme to the Society, he was told the Fellows could not pass comment without knowing more about

[88] Rocquette, *Nouveau sisteme*.

[89] James Hodgson, 'Abstract concerning Mr Roquette, watchmaker to her Imperial Russian Majesty, with his proposal to find the longitude at sea with the help of two portable clock or watches', 1732, RS CLP/22ii/59, fols 285–290 (fol. 289v), quoted in Werrett, 'Perfectly Correct', p. 114.

[90] Formey, *The Life of John Philip Baratier*, p. 257.

[91] 'Account of method proposed by John Philip Barerius for finding the longitude at Sea as communicated by him to the Royal Society with some remarks by James Hodgson', 1735, RS CLP/22ii/66; John Philip Baratier to Philip Henry Zollman, 10 December, 1737, RS EL/B3/64; translation of a letter from John Philip Baratier to Philip Henry Zollman, 1737, RS EL/B3/65; John Philip Baratier, to the Royal Society, 10 December 1737, RS EL/B3/66; translation of a letter from John Philip Baratier, to the Royal Society, 1737, RS EL/B3/67; T. Stack to Philip Zollman concerning John Philip Baratier's proposal for measuring longitude at sea, 10 February 1737, RS EL/B3/68; T. Stack concerning John Philip Baratier's proposal for determining longitude at sea, 8 March 1737, RS EL/B3/69; copy of a minute of a Royal Society meeting concerning Baratier's proposal for determining longitude at sea, 26 January 1737, RS EL/B3/70.

it but was sent a copy of the Act of 1714 and assured that 'the Act of Parliament for a Reward, for the Discovery of the Longitude, is still in Force, and that there are Commissioners appointed as the proper Judges'.[92] Tired of longitude, the prodigy turned to creating a sea chart based on the newest geographical and magnetic variation observations.

Whether a projector was abroad like Baratier and Rocquette or in Britain, interpersonal networking proved crucial in gaining a hearing or any recompense. The Georgian world was predicated on interpersonal connections and interactions and the use of trustworthy go-betweens. Reward-seekers approached all manner of influential individuals who might be able to put them in touch with Commissioners or provide influential approval and funding. The contacts beyond the Commissioners could range from officials in the London and national governments all the way to the king. On 1 October 1717, for example, William Whiston laid a proposal before the Lord Mayor and the Court of Aldermen of London, 'but the ingenious Mr. Halley, and some other Mathematicians, being call'd in, disapprov'd the same; especially Whiston's Maggot of throwing Bombs in the Air on Hampstead and Black-Heath'.[93]

In 1715, Henry de Saumarez of Guernsey petitioned George I after the Royal Society declined to provide feedback on his invention of a ship's wheel and dial which would ring a bell to indicate distance travelled, which he thought would do for the longitude rewards.[94] The petition reportedly gained de Saumarez some noncommittal commentary from Newton through the intervention of the Admiralty, and the Lords of the Admiralty directed him to Trinity House. He presented Trinity House with drawings and wood and copper models of the invention but met with objections. He continued to pursue the scheme and published an account of his interactions with the Royal Society, Trinity House and Admiralty in 1717. This was intended to 'vindicate his Reputation, and to assert the Reality and Practicableness of the said Project' and included another plea for help from the king.[95]

Three years later, de Saumarez approached the Lord Chancellor, Thomas Parker, and reportedly gained his approval while trying to secure government support for a sea trial of his 'Marine-Surveyor'.[96] This followed tests on the canal in St James's Park with the assistance of John Theophilus Desaguliers, then Demonstrator at the Royal Society.[97] In 1725, de Saumarez described his invention and trial results in the *Philosophical Transactions* with characterful engravings

[92] Philip Henry Zollman to John Philip Baratier concerning his proposal for determining longitude at sea, 13 February 1738, RS EL/B3/71.
[93] *Weekly Journal or British Gazetteer*, 5 October 1717.
[94] Duncan, *The History of Guernsey*, pp. 591–592.
[95] De Saumarez, *An Account of the Proceedings*.
[96] *London Journal*, 30 July–6 August 1720.
[97] Carpenter, *John Theophilus Desaguliers*, p. 68; *Original Weekly Journal*, 28 May 1720.

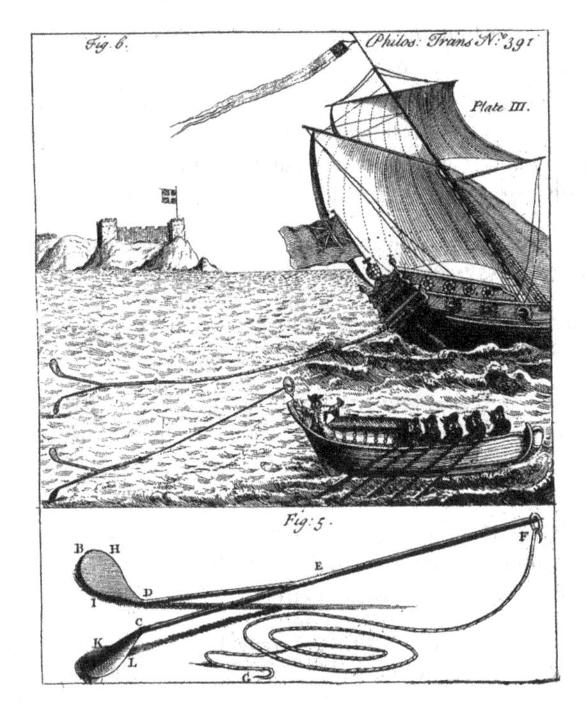

Figure 3.3 Henry de Saumarez's Marine Surveyor, from 'An Account of a New Machine, called the Marine Surveyor', *Philosophical Transactions*, 33 (1725)

of its use at sea (Figure 3.3) alongside testimonials from ships' captains and mates.[98] This was read to the Royal Society that year, and he later presented an expanded account in person in 1729.[99] De Saumarez was not the only longitude projector to appeal directly to Parliament or the reigning monarch. In 1720, a gentleman from Hampshire named George Ham reportedly had his scheme put before Parliament. One newspaper claimed his unspecified methods had 'been so well approved that an Account of it has been dispatched abroad to France and Holland, as well as to Hanover'.[100] That same year, William Whiston laid before George I his latest work on the use of variation in magnetic dip to find longitude, which resulted in some of the earliest isoclinic maps and garnered £100.[101]

[98] De Saumarez, 'An Account'.

[99] 'A further Account of a new machine call'd the marine Surveyor … by M Henry de Saumarez', read to the Royal Society, 23 January 1729, RS RBO/13/73.

[100] *Weekly Journal or Saturday's Post*, 20 August 1720; *Weekly Journal or Saturday's Post*, 5 November 1720. Knoppers, 'The Visits of Peter the Great', pp. 65–66, notes Peter the Great spent time with Ham and his invention in The Hague in 1717.

[101] *London Journal*, 12 November 1720; *Weekly Journal or British Gazetteer*, 26 November 1720; Treasury, *Calendar of Treasury Papers*, VI, p. xxii.

It was not just copious longitude proposals and hearings thereof that emerged during the years between the Act of 1714 and the first formal meeting of Commissioners. Some sea trials were also conducted, with varying degrees of official support and approval, well before the better-known trials of Harrison's timekeepers. One example is that of the young wig-maker John Bates, whose somewhat unlikely progress the periodicals followed keenly:

> A young Man, who was bred a Peruke-maker, having by a close and tedious Study, made some Steps towards discovering the Longitude, has been examined before the Lords of the Admiralty, Dr. Halley, and others; we hear he is to be furnish'd with Books, Instruments, and other Necessaries, and to be sent to Sea under the Care of a Commander in the Royal Navy.[102]

Bates left London on 25 December 1728 on the man-of-war *Blandford*, sailing for Lisbon and the Straits.[103] While he was away, schoolmaster John Wells of the Hand and Pen in Carey Street repeatedly advertised that he had taught Bates arithmetic.[104]

Bates's case also emphasises the importance of the East India Company to the search for longitude, since he approached its directors on 7 October 1730 to seek funding or a reward after his naval trials:

> I having for some time Appli'd my self to find a Method whereby the Longitude might be found, have of Late found a Method for Obtaining it by Celestial Observations and New Calculations of the Moon and fixed Stars which Method I am here Present with, to lay before your Hon:rs If it is your pleasure so as to see it.[105]

The East India Company had a natural interest in longitude, given that its fortunes were made at sea and its ships conducted more long-distance voyages than did naval vessels.[106] As a result, it often assisted with trials and transportation and offered another potential source of funding. John Harrison was able to construct his first sea clock (H1) in part because clock-maker George Graham convinced the East India Company and Charles Stanhope to grant him additional sums.[107] When Nevil Maskelyne returned from his voyage to observe the Transit of Venus from St Helena in 1762, he reported to the Royal Society and to the Court of Directors of the East India Company, since the Company had assisted him in travelling to and from the island and in making observations.[108] Perhaps he also hoped for some funding for improving the lunar-distance method. When it came to

[102] *Daily Post*, 11 December 1728. [103] *Daily Journal*, 27 December 1728.
[104] *Daily Post*, 2 January 1729.
[105] John Bates to East India Company Directors, 7 October 1730, BL IOR/E/1/21, fols 263r–263v, letter 145.
[106] See, for example, Cook, 'Establishing the Sea-Routes'.
[107] Harrison, *A Description Concerning Such Mechanism*, 1st ed., p. 21; Quill, *John Harrison*, p. 36.
[108] Bennett, 'Mathematicians on Board'; Howse, *Nevil Maskelyne*, pp. 42, 218–219.

longitude-related innovations such as Halley's magnetic variation charts, new navigational and lunar-distance publications and tables, and eventually marine chronometers, the Company also deployed these tools earlier and more widely than did the British Navy.

In addition to this early period seeing a constant flow of new proposals and schemes getting as far as sea trials, there were also periodic efforts to alter the Act of 1714. One failed attempt in 1720 would have allowed the Commissioners to dispense rewards of up to £10,000 for the general improvement of navigation. This would reportedly have been for the benefit of the inventor and marine salvager Jacob Rowe, whose 'fluid quadrant' the Admiralty had tested at sea.[109] There also exists a mysterious undated draft bill formerly from Isaac Newton's papers, labelled a 'Proposal abt Masters of ships/Draught of a Bill in Parl^{t} for a Reward for finding Longitude'. It cites the encouragement given by the Lord Mayor and the Court of Aldermen of London for an easily practicable method of finding longitude. A duty on the tonnage carried by British and visiting foreign ships, collected by the aforesaid officials, would be used to 'deliver to all Captains Comanders M.^{rs} or owners of any ships, Bookes of Instructions for the practicing of the arte of Longitude for the 1st yeare gratis, and afterwards as often as shall be desired at the price of _ s for Each Booke & no more'. Ships that did not pay the duty would be detained.[110]

Beyond attempts at further legislation, some reward-seekers sought different behaviour from the Commissioners appointed by the Act of 1714. Authorities including individual Commissioners most often directed reward-seekers to either take their proposal to the Astronomer Royal or gather expert opinions and proof of success through publication and trials. Some Commissioners specifically stated that they could not be expected to meet communally to consider every proposal, given the large number put forward each year. Indeed, they did not have financial or logistical support for doing so. Newton further commented that the only way the officials could have considered every proposal communally was if they had begun doing so directly after the passage of the Act.[111] However, a few reward-seekers including Robert Browne and Jane Squire (discussed in Chapter 4) publicly and vocally challenged this.[112]

[109] Rowe, *Navigation Improved*, pp. 9–10; *Journals of the House of Commons*, 19, pp. 210, 222, 228, 231; Barrett, *Looking for Longitude*, pp. 67–72.

[110] Draft purchased by the Karpeles Manuscript Library Museum at Bloomsbury Auctions, 14 February 1991. It might have belonged to manuscript collector Edward W. Hennell, given other documents in the sale. Photocopies of the draft and manuscripts sold with it are at BL RP 4713.

[111] Isaac Newton, 'Papers on Finding the Longitude at Sea', 1697–1725, CUL MS Add.3972, fols 35v–37r.

[112] Browne, *To the Honourable the Commons*, title page.

Conclusion

The frustrated attempts by Browne and Squire to gain an audience with a 'Board of Longitude' that did not yet exist emphasise that the Act of 1714 named individuals as acceptable judges for the new rewards but did not create a standing body. Not only was the legislation shaped by the proceedings prompted by Whiston and Ditton's lobbying, it was also informed by centuries of precedents, notably those of Charles II's one-off commissions in 1674 and Thomas Axe's reward established in 1691. However, this did not result in official inactivity or creative stagnation, as is sometimes the modern interpretation. Rather, the establishment of longitude rewards had the enduring, desired effect of expanding activity in the centuries-long search. The judging, trial and funding of proposals was almost always overseen by the most relevant individuals and institutions as well, albeit not in a clear, bureaucratic manner.

The first twenty-three years of the longitude rewards also exhibited the dynamics that would persist throughout the eighteenth century. The events of this period furthermore facilitated and provided precedents for the more familiar horological, astronomical and bureaucratic developments from the 1730s. Proposals given credence were never restricted to timekeepers and the lunar-distance method alone, nor were the clockmaker John Harrison and the astronomer Nevil Maskelyne the only interested parties. However, as discussed in the following chapters, those actors did play crucial roles in the Commissioners' transformation from independent experts to the first science and technology funding body in Britain almost fifty years after modern historians previously assumed was the case.

4 The Early Commissioners in Transition

Alexi Baker

The middle decades of the eighteenth century were a transformative period in the history of the Commissioners of Longitude, seeing them gradually transition from individual judges who acted independently to a standing and increasingly organised Board. This transformation effectively created the first British state funding body for scientific inquiry and technical innovation, which would become a respected authority across Europe. There are two distinct stages within this period: from 1737 to 1757, when small groups of Commissioners were called together sporadically for ad hoc meetings, predominantly to agree funding for specific projectors, and from the 1760s, when the Commissioners began to meet together at least annually and to adopt a more bureaucratic structure against the backdrop of the British naval and colonial expansion that followed the Seven Years War. One of the key drivers for these changes was the Commissioners' evolving relationship with the clockmaker John Harrison, which was fruitfully collaborative during the first period and increasingly confrontational during the second. This chapter looks at the first period of transition, from the 1730s to the 1750s, in order to explore how the Commissioners worked when they first began to meet together communally and issue funding.

Although the Commissioners' activities became more visible and organised as their working practice changed, public opinion concerning the search for the longitude was becoming more mixed. The Act of 1714 had prompted an initial burst of optimism about a solution to the longitude problem being imminent but, as the years passed, commentators and the broader public increasingly returned to the centuries-long tradition of invoking longitude negatively. This included associating longitude with other 'impossible' endeavours and with the accusations of fraudulence and even madness that were often made against projectors. Finding longitude was frequently listed alongside squaring the circle, producing the Philosopher's Stone (Figure 4.1), perfecting perpetual motion, finding a Northeast and later a Northwest Passage, and achieving human flight or deep-sea diving. These associations were most often expressed in text but occasionally in images as well, as when William Hogarth included

Figure 4.1 William Hunt, *The Projectors: A Comedy* (London, 1737), frontispiece. The bowing projector holds a document entitled 'Longitude', while another marked 'Philosopher's Stone' lies on the ground

a 'longitude lunatic' in the culminating Bedlam scene of his series of prints *A Rake's Progress* (1735).[1]

Regardless of this increasing public scepticism, the search for longitude in general continued to flourish, even as decades passed without the rewards of 1714 being allocated. This continued momentum was fuelled by the allure of the three large rewards and of the smaller rewards and funding that the Commissioners of Longitude began to issue for improvements to longitude and navigation. During the earlier decades after 1714, British and foreign projectors continued to propose a wide variety of schemes and to approach relevant individuals and institutions. A few of these aroused official interest, including John Harrison's marine timekeepers, William Whiston's varied ongoing endeavours, lunar distances as improved by Tobias Mayer and (more briefly in the 1760s) the keen self-promoter Christopher Irwin with his marine chair for observing Jupiter's satellites.

[1] Barrett, *Looking for Longitude*.

From the 1730s to the 1750s, the dynamics of communicating and judging longitude schemes remained similar to those outlined in the previous chapter, with only eight communal meetings of the Commissioners known to have taken place in twenty-three years. Most activities, networking and decision-making remained focused on individual officials such as the Astronomer Royal and on institutions including the Admiralty and the Royal Society. Such individuals and institutional authorities would indeed remain prominent participants throughout the history of the Commissioners, while interpersonal networking and independent expertise remained important parts of communicating, judging, and funding longitude.

First Formal Meetings

The period from the 1730s to the 1750s sowed the seeds for the increasing bureaucratisation of the decade that followed, which will be discussed in Chapter 5. These years saw the first known communal meetings of some of the Commissioners, initially in response to the keen interest surrounding the work of John Harrison, alongside key Commissioners' continued activities as individual judges. The period encompassed more than two decades of collaborative accord between the officials and the Yorkshire-born clockmaker, as well as important work done on astronomical and magnetic schemes and new technologies.

During these decades, there continued to be a steady issue and reissue of longitude proposals, which were met with mixed reactions from the Commissioners and other authorities. While some were one-off attempts, many were the work of dedicated longitude projectors who were motivated to keep trying for years by the potential rewards, self-belief or religious conviction. For example, the long-time projector Robert Wright was still trying to attract public and official attention for his lunar-distance method at the start of the 1730s. Wright earned a BA from Jesus College, Cambridge, in 1697, and was master of Winnick Grammar School in Lancashire from 1717 to 1735.[2] He had first sent manuscript copies of his 'Viaticum Nautarum or the Sailor's Vade Mecum' to Isaac Newton and other individual Commissioners in 1726.[3] Two years later, following Newton's death and a lack of official encouragement, he published a treatise addressed to the Commissioners. Wright repeated parts of his original manuscript – 'since probably the Book may be long since thrown aside, or your more necessary Thoughts may have thrust them out of your Memory' – and said that he did not think his hard work 'ought to be thrown a-side unregarded,

[2] Robson, *Some Aspects of Education*, p. 183.

[3] *Evening Post*, 16–19 April 1726; a copy of the manuscript said to have come from Newton's library is at NMM NVT/5.

and buried in Silence'.[4] In 1730, the projector sought subscriptions to publish his lunar tables with a description of an altitude instrument said to be superior to any quadrant, which were intended to complement his original publication: 'These two Books in Quarto bound up together, one for finding the Moon's Place by Observation, the other by Calculation, will be a compleat System of all that is yet wanting in Navigation'.[5] They were published two years later, but Wright never achieved any significant success in the search for longitude.[6]

Another example of long-term dedication to longitude is William Whiston, who had played such an integral part in the shaping and passage of the Longitude Act. Long after the death of his collaborator Humphry Ditton in 1714, Whiston continued to pitch different longitude schemes to the public and to individual and institutional authorities in the 1730s and 1740s. Having moved on from his and Ditton's rocket-based scheme, he published about magnetic dip in 1719, securing more than £470 from subscribers including the royal family, the Duke of Chandos and Martin Folkes, and had his dipping needles put to test on a number of voyages in 1722–1723. Although he abandoned this scheme due to the unsuccessful results, in 1724 he published a treatise on solar eclipses that included not only a discussion of the dipping needle work but also a method for determining geographical longitudes from the duration of the eclipses. Then, six years later, he put forward another longitude scheme, this time based on observations of the eclipses of Jupiter's satellites, in which he tried to interest the Royal Society and which he also incorporated into his public lectures.[7]

Another determined projector at work at the same time was Richard Locke of Pitminster in Somerset, who from 1730 published descriptions of a lunar-distance method and new instrument, which were often sold alongside his books about mathematics and squaring the circle.[8] Receiving no encouragement from the Astronomer Royal, Edmond Halley, Locke became a naval chaplain and served as a priest in the American colonies, where he conducted sea trials of his schemes. In 1750, he returned to England a middle-aged man and resumed his appeals to the public and the Commissioners, fruitlessly

[4] *Daily Post*, 28 May 1728; Wright, *An Humble Address*, p. 22.

[5] *London Evening Post*, 26–28 May 1730. [6] Wright, *New and Correct Tables*.

[7] Whiston, *The Longitude and Latitude*; Whiston, *The Calculation of Solar Eclipses*; William Whiston, 'A Description of Mr. Whiston's Reflecting Telescope, for the Discovery of the Longitude at Sea', 6 November 1734, NMM MSS/79/130 (letter received by Sir Charles Wager on 7 November, noting that Whiston presented his scheme to the Royal Society on 22 October 1730). Whiston, *Memoirs*, I, pp. 298–299, gives details of the 'large Subscription' made in November 1721 to support both his family and his work on discovering longitude by dipping needle. See also *Monthly Chronicle*, October 1730; *Daily Post*, 23 October 1730; *Daily Courant*, 2 February 1731; Farrell, *William Whiston*, pp. 147–165; Köberer, 'German Contributions', p. 273; Stewart, *The Rise of Public Science*, pp. 189–192.

[8] Locke, *The Circle Squared*; *Country Journal or The Craftsman*, 27 February 1731; *Daily Journal*, 2 April 1734; *London Evening Post*, 24–26 July 1733; *London Evening Post*, 27 February–1 March 1739.

hoping that 'he shall be entitled at least to some Advantage of the promis'd Reward; as it hath been very difficult and expensive to him in the Discovery, by Study, Travelling, and hindering him in his temporal Affairs'.[9] These men are only a few examples among many long-term longitude devotees.

Sources including letters, periodicals, projectors' publications and institutional archives reveal how relevant individuals and organisations responded to the steady flow of proposals. From 1737, these existing avenues of communication and authority were supplemented by the Commissioners' occasional communal hearings and decision-making.[10] This gradual change in practice came about initially because of the public and institutional interest surrounding John Harrison's first marine timekeeper.

The first known communal meeting of some of the Commissioners of Longitude took place on 30 June 1737, with Harrison present to make his case for official funding, but the clockmaker and his work were already well-known to the majority of those present. Eight Commissioners attended the meeting: Sir Charles Wager, First Lord of the Admiralty; Sir Arthur Onslow, Speaker of the House of Commons; Lord John Monson, First Commissioner of Trade; Sir John Norris, Admiral of the White; Sir Hans Sloane, president of the Royal Society; Edmond Halley, Astronomer Royal and Savilian professor of geometry; James Bradley, Savilian professor of astronomy; and Robert Smith, Plumian professor of astronomy. Many of these Commissioners had been aware of Harrison's work for at least two years; Halley had supported it for almost a decade.

Harrison's work followed a long line of European efforts to develop a reliable marine timekeeper, which had first begun to seem promising in the seventeenth century in the work of Christiaan Huygens, Robert Hooke and others.[11] They continued into the early eighteenth century through the work of clockmakers including Henry Sully and possibly Daniel Quare.[12] Sully, a Briton working in France with well-known craftsmen such as Julien Le Roy, first presented a sea clock to the Académie Royale des Sciences in 1716. He then produced a new design in the early 1720s, which had been influenced by Huygens's earlier work trying to combine the advantages of both pendulum and balance. Sully sent his scheme to the renowned London clockmaker and Fellow of the

[9] Taylor, *The Mathematical Practitioners*, p. 34; *London Evening Post*, 22–24 January 1751; *General Evening Post*, 7 May 1751; Locke, *A New Problem*.

[10] BoL, confirmed minutes, 30 June 1737, CUL RGO 14/5, pp. 3–5. It is possible that some Commissioners met before 1737, as some later documents hint. However, evidence of meetings has been found neither in surviving Board minutes, which were compiled decades later, nor in other archives or publications.

[11] Betts, *Marine Chronometers at Greenwich*, pp. 9–14; Dunn and Higgitt, *Finding Longitude*, pp. 57–63; Gould, *The Marine Chronometer*, pp. 19–30; Jardine, *Going Dutch*, pp. 263–290; Leopold, 'The Longitude Timekeepers', pp. 101–114.

[12] Betts, *Harrison*, p. 26; Betts, *Marine Chronometers at Greenwich*, pp. 16, 37, 120–123; Gould, *The Marine Chronometer*, pp. 27–39.

Royal Society George Graham and published an account of it in 1726, trying to gain the interest of the Royal Society and thus the Commissioners of Longitude.[13] Graham responded that the design would likely still suffer from the motion of the ship affecting the inertia of the oscillator, and indeed the sea trials the inventor conducted before his death two years later did not produce satisfactory results.[14] Sully was not alone in trying to attract the attention of the Commissioners of Longitude with a scheme involving a mechanical timekeeper. In addition to a number of projects announced in British publications, Lotharius Zumbach de Koesfelt – a Dutch physician, mathematician and musician – described a sea clock in a publication addressed to the Commissioners in 1714. It was not, however, until the 1740s that his son Conrad had one made.[15]

In 1726, John Harrison first heard of the rewards available under the Longitude Act and turned his attention to designing a timekeeper that would perform accurately at sea.[16] He came to London to seek support in about 1727–1728 and met with Edmond Halley, by then an acknowledged longitude expert. Though the Astronomer Royal received Harrison warmly, he felt unable to judge his work and sent him to George Graham.[17] This proved to be a key moment, since until his death in 1751 Graham would become Harrison's most powerful supporter. He made it possible for the younger man to finish his first sea clock by not only giving him a loan but also convincing the East India Company and Charles Stanhope to grant him funding.[18] Stanhope was the son of the first Earl Stanhope, who had been instrumental in the passage of the Act of 1714. Graham also served as an influential mediator between Harrison and first the Royal Society and then the Commissioners of Longitude.[19] At this time, probably as a result of the discussions with Halley and Graham, Harrison

[13] Sully, *Description abregée*; Betts, *Marine Chronometers at Greenwich*, pp. 121–122; Bertucci, *Artisanal Enlightenment*, pp. 84–88, 93–98. The same year Sully published a French translation of the Longitude Act of 1714; see Turner, 'L'Angleterre, la France', p. 152.

[14] Andrewes, 'Even Newton', pp. 192–194; Betts, *Marine Chronometers at Greenwich*, pp. 120–123.

[15] Hooijmaiers, 'A Claim for Finding the Longitude'. See Chapter 3 for examples of British schemes published at this time.

[16] Anonymous, *The Case of Mr. John Harrison*, fol. 1r; John Harrison, petition to Parliament, 2 April 1773, *Journals of the House of Commons*, 34 (1773), p. 244.

[17] On Graham's status, see Bennett, 'Instrument Makers'; Sorrenson, 'George Graham'.

[18] Harrison, *A Description Concerning Such Mechanism*, 1st ed., p. 21; Howse, *Greenwich Time*, p. 72; Quill, *John Harrison*, p. 36.

[19] Harrison's other high-profile supporters included James Short, Sir John Cust, John Bevis, Walter Williams, an attorney, and Taylor White, treasurer of the Foundling Hospital, who later helped him make a case to the public for a reward. See Bennett, 'James Short and John Harrison'; Forbes, *Tobias Mayer*, p. 199; Quill, *John Harrison*, pp. 111–112, 136, 144, 208; Turner, 'L'Angleterre, la France et la navigation', p. 157. For Cust's aid to Harrison, see 'Harrison Journal', 1817, State Library of Victoria H17809, pp. 193–195. On Walter Williams and the putative authorship of the 'Harrison Journal', see Horrins, *Memoirs of a Trait*, pp. xxx–viii, although Williams in fact appears primarily as a witness to the final incident described.

produced a handwritten description of his ideas that reads as if it had been intended for publication.[20]

Funded by the loans and grants from his supporters, Harrison spent the ensuing years in Barrow-upon-Humber building a sea clock (now known as H1), which he completed and brought to London in 1735.[21] It was probably first shown at Graham's workshop, which added credibility, and soon gained the approval of the Royal Society. A group of Fellows issued a certificate declaring that the 'Machine for Measuring Time at Sea … highly deserves Publick Encouragement', including a sea trial.[22] Significantly, the signatories included not only Graham but also three Commissioners of Longitude: Halley, Bradley and Robert Smith. There would be overlaps, and later frequent collaborations, between the Royal Society and the Commissioners throughout their 114-year history.

In 1736, H1 underwent a semi-official trial to Lisbon on the *Centurion*, returning on the *Orford*. Although it appears to have gone poorly in rough seas on the outward journey, its performance on the *Orford* was far more encouraging.[23] Again, the Admiralty officials involved in the organisation of this trial included two of the Commissioners who would attend the ensuing meeting in 1737, Sir Charles Wager and Sir John Norris.[24] In other words, five of the eight Commissioners who responded to Wager's call for a communal meeting on 30 June had by then already shown support for Harrison and his timekeeper.[25] It is highly likely that others were at least aware of them, in particular Hans Sloane as president of the Royal Society. Thus the meeting represented more of a technicality than a significant change in the Commissioners' behaviour, carried out so that Harrison could receive funding to improve his designs.

[20] John Harrison, untitled description of ideas for making clocks for use on land and at sea, 10 June 1730, Worshipful Company of Clockmakers, 6026/1.

[21] John Harrison, one-day marine timekeeper (H1), c. 1735, NMM ZAA0034; Betts, *Marine Chronometers at Greenwich*, pp. 134–147. According to Betts, *Time Restored*, pp. 86 and 318, although Gould first referred to the fourth timekeeper as 'H4' in 1939, the appellations H1, H2, H3, H4 and H5 for Harrison's five marine timekeepers were established after his death, from the 1950s.

[22] The text of the Royal Society certificate appears in anonymous, *An Account of the Proceedings*, p. 17.

[23] Lieutenants Montague Bertie, John Draper, Rowland Cotton and Captain George Proctor, logs of the *Centurion*, 1735–1739, NMM ADM/L/C/82; Lieutenant Isaac Barnard, log of the *Orford*, 1734–1736, NMM ADM/L/O/27; Captain Robert Man, log of the *Orford*, NMM ADM/L/O/28; Lieutenants Ormond Thomson and Charles Knowler, logs of the Orford, 1734–1736, NMM ADM/L/O/29A. Proctor's log records longitudes determined from H1; see Ereira, 'The Voyages of H1'.

[24] Sir Charles Wager to Captain Proctor, 14 May 1736, in anonymous, *An Account of the Proceedings*, pp. 17–18.

[25] Charles Wager to Hans Sloane, 27 June 1737, BL Sloane MS 4055, f.126, mentions the Lisbon voyage and Harrison's 'several Experiments in his Passage thither' as the reason for the Commissioners to meet 'as the said Act directs'.

Harrison exhibited H1 at the 1737 meeting and sought £500 'by reason of his necessitous Circumstances' to make a smaller improved design within two years for a trial voyage to the West Indies. The Commissioners readily agreed, commenting that the sea clock might 'tend very much to the Advantage of Navigation'. The implication was that this award would be drawn from the £2000 specified in the Act of 1714 'for making the Experiments'.[26] The Commissioners ordered that half of the requested sum be paid to the clock-maker immediately and that the other half be paid on production of a certificate proving that the new timekeeper had undergone a sea trial to the West Indies. They instructed the Navy to draw up an agreement with Harrison to that effect, which also required that he turn over both machines to the Commissioners after the trial, for the use of the public.[27]

Although official documents recorded these proceedings in 1737, it is significant that many longitude projectors at the time, and even years later, were unaware that a communal meeting of the Commissioners had taken place. It was similarly not unusual as the years passed for projectors to be uncertain whether or not the longitude rewards were still on offer. For example, Anthony Thompson – a British chargé d'affaires at the French court – forwarded that very question to London on behalf of an unnamed French projector in 1741.[28] The projector said that he had worked for a number of European diplomats and hoped to present his scheme involving a clepsydra or water clock to the British Parliament, but appears to have been ignored by the relevant authorities.[29] Projectors could be forgiven for their uncertainty about the status of the longitude rewards and for having missed the first communal meeting of the Commissioners, given the limited newspaper coverage of these events. Reports of the initial meeting in 1737 did not even directly refer either to the Commissioners or the Act of 1714:

On Thursday the Right Hon. Arthur Onslow, Esq; Speaker of the House of Commons, the Right Hon. the Lords Monson and Lovel, Sir John Norris, Sir Charles Wager, and several Persons of Distinction, view'd a curious Instrument for finding out the Longitude, made by Mr. Harrison of Leather-lane, which he has been six Years in

[26] BoL, confirmed minutes, 30 June 1737, CUL RGO 14/5, pp. 3–5.

[27] 'Certificate authorising the Commissioners of the Navy to advance £500 to John Harrison to enable him to make a machine for discovering the longitude at sea', 30 June 1737, TNA TS 21/19; 'Indenture of Covenants that John Harrison shall make a machine for better keeping of time at sea, mount it in one of His Majesty's warships and test it at sea off the West Indies. The Commissioners in return covenant to supply him with a sum of money, he being destitute and unable to make the machine otherwise', 27 July 1737, TNA TS 21/19.

[28] 'Memorial from, or on behalf of, an inventor of a ship's chronometer for determining longitude, enquiring if a reward is still being offered. (f.75 enclosed in f.74 French)', 1741, TNA SP 78/226.

[29] Authors including Guillaume Amontons had previously proposed repurposing the clepsydra for finding longitude at sea; see Amontons, *Remarques et expériences phisiques*.

finishing. They all express'd the greatest Satisfaction at it, order'd him 250 l. and gave him Directions to make another, to be kept at the Admiralty, for which he is to have 250 l. more.[30]

At least another seven times between 1737 and 1757, five or more of the Commissioners of Longitude would similarly assemble. These early meetings were mainly called to extend further funds to Harrison, who attracted almost all the officials' attention and expenditure before the 1760s (see Appendices 2 and 3). In 1740, Parliament also passed a new 'Act for surveying the Chief Ports and Head Lands on the Coasts of Great Britain and Ireland, and the Islands and Plantations thereto belonging, in order to the more exact Determination of the Longitude and Latitude thereof'.[31] This allowed the Commissioners to apply up to £2000 to better ensure ships' safety by surveying the British, Irish and colonial coasts and more accurately determining their latitude and longitude. Coastal positions were another key concern in navigation, as had been noted in the original Act.

Accordingly, the next known communal meeting of the Commissioners on 16 January 1742 represented the first time the officials voted to extend their powers beyond finding longitude at sea.[32] The seven men who assembled that day agreed that William Whiston – taking advantage of the Act published two years earlier – be awarded £500 for his ongoing mapping of the British coasts by magnetic dip.[33] They cautioned, however, that he should not 'without particular orders for so doing, put the Public to any Further expence on any Account whatever in relation to his said Instruments in the making any Trial or Trials of the same'.[34] Given Whiston's prominent role in bringing about the Act of 1714, it is intriguing to speculate whether the skilled lobbyist might also have contributed to the passage of an Act twenty-six years later that benefited him monetarily within two years. There is, for instance, evidence in draft petitions from the 1730s intended for the Commissioners of Longitude that Whiston was aiming to enlist the support of Halley, Bradley and others in his proposal for a coastal survey. At least one of these petitions seems to have reached the Commissioners in 1739.[35]

This second communal meeting of the Commissioners was also used to check on Harrison's progress, since he had agreed to finish his second sea

[30] *London Evening Post*, 30 June–2 July 1737; see also *Old Whig or The Consistent Protestant*, 7 July 1737.

[31] Longitude and Latitude Act, 1740, 14 Geo 2 c 39; copy at CUL RGO 14/1, pp. 13r–14v.

[32] BoL, confirmed minutes, 16 January 1742, CUL RGO 14/5, pp. 6–9.

[33] The results of the survey, carried out by John Renshaw under Whiston's guidance, were published in a chart, 'An exact Trigonometrical Survey of the British Channel from the North Foreland to Scilly Islands and Cape Clear on the South-West part of Ireland. (The River Thames from the Buoy of the Noure to London Bridge.) Performed in the year 1741 and 1742 by J. Renshaw under the directions of W. Whiston', c. 1745, BL Maps * 1068.(11.).

[34] BoL, confirmed minutes, 16 January 1742, CUL RGO 14/5, p. 7.

[35] Farrell, *William Whiston*, pp. 166–173.

clock by 1739. 'H2' had been built but abandoned, and Harrison had since embarked upon a third timekeeper, now known as H3.[36] George Graham represented to the Commissioners on Harrison's behalf that he needed another £500 to make it ready for sea trial by August 1743. Harrison's request was supported by another testimonial signed by prominent Fellows of the Royal Society.[37] Six of these signatories were *ex officio* Commissioners of Longitude, and three attended the meeting – Smith, Bradley and Martin Folkes, then president of the Royal Society. Three additional attendees had been present at the meeting with Harrison in 1737 – Wager, Norris and Lord Monson. Sir Thomas Hanmer, the only Commissioner remaining from those personally named in the Act of 1714, attended the meeting as well, although he had not attended the first. The 1742 meeting was also essentially a formality in order to issue funds. The Commissioners granted Harrison's request under the same conditions as in 1737, including that H3 and the earlier machines would ultimately be placed in their possession for the public good.[38] They added that, should the clockmaker win one of the rewards established in 1714, any funds already extended to him would be subtracted from the total.

With the business of Harrison's funding concluded, Norris introduced a letter from Richard Graham FRS, who served until his death in 1749 as comptroller of the project to build Westminster Bridge. Graham wrote that he had invented an altitude instrument which could find longitude at sea without the need for either calculation or exact time. He had previously presented an equinoctial dial and an instrument for finding latitude to the Royal Society in 1731.[39] It was relatively common for men with connections and standing like Graham to be able to get their proposals laid before a key Commissioner, or more rarely before one of the communal meetings. However, as appears to have been the case here, most proposals were not deemed worthy of further consideration.

[36] John Harrison, one-day marine timekeeper (H2), c. 1739, NMM ZAA0035; John Harrison, one-day marine timekeeper (H3), c. 1759, NMM ZAA0036; Betts, *Marine Chronometers at Greenwich*, pp. 147–174.

[37] The certificate was signed by Martin Folkes, Robert Smith, James Bradley, John Colson, George Graham, Edmond Halley, William Jones, the Earl of Macclesfield, James Jurin, Charles Cavendish, Abraham de Moivre and John Hadley; see anonymous, *An Account of the Proceedings*, pp. 19–21.

[38] 'Indenture of Covenants to make another machine', 1 March 1742, TNA TS 21/19.

[39] Richard Graham, 'Description and Use of an improv'd equinoctial Dial for shewing the hour by the Sun or any of the Stars, with great Exactness Contrived and finish'd Anno 1727 by Richard Graham', read on 16 March 1731, Royal Society RBO/17/7, CLP/3ii/34; Richard Graham, 'A description of an Instrument for finding the Latitude from two observations at any time of the Day', read on 9 December 1731, Royal Society RBO/16/32, CLP/8ii/32; Graham, 'The Description and Use'.

Jane Squire: Gender, Religion and the Early Commissioners

Given the sporadic nature of the Commissioners' early communal meetings, there were far more vital routes for communicating and judging longitude from the 1730s to the 1750s, including interpersonal networking and publication. One of the most illuminating sources for understanding these dynamics, before the advent of a standing board, is the life story of Jane Squire.[40] Squire published two editions of a book that, in addition to explaining her complex longitude scheme, outlined in unusual detail her efforts to secure the encouragement of the Commissioners during the 1730s and 1740s. She was also the only woman to openly pursue the British rewards and funds. Her experiences shed light on the roles played by gender, as well as other factors including religion, in the Georgian search for the longitude.

Modern historians have all too often treated activities focused on knowledge and technology in early modern Europe as having been largely or entirely male-dominated. However, within the past few decades scholars have been able to reveal that in fact many women (or people who presented as women) made vital contributions to these and similar fields, by applying new insights from gender studies to their apparent presences and absences in the surviving historical record.[41] For example, research in instrument studies is increasingly indicating that women were deeply involved in making and selling instruments in London in a host of ways, despite long-held assumptions to the contrary.[42] Much research remains to be done on the participation of women in other longitude-adjacent activities, but there are a number of evocative examples.

The case of Mary and Eliza Edwards of Ludlow, initially described by Mary Croarken, points to the vital work which women might have done behind the scenes.[43] Both were long-serving 'computers' or calculators for the annual *Nautical Almanac*. Although it was Mary's husband John who was first officially hired as a computer in 1773, she handled most of the calculations even then and officially took over the position after he passed away eleven years later.[44] Edwards continued working as a *Nautical Almanac* computer as her sole livelihood until she passed away in 1815, whereupon her daughter Eliza took up the work until the formation of the Nautical Almanac Office in 1831. The surviving Board of Longitude and Royal Observatory archives include

40 Baker, 'Squire, Jane'.

41 Erickson, 'Married Women's Occupations'; Phillips, *Women in Business*; Reinke-Williams, *Women, Work and Sociability*; Erickson, *Women and Property*.

42 Baker, 'The Business of Life', pp. 182–184; Baker, 'This Ingenious Business', pp. 226–240; Baker, 'Interpreting Silences'; Desborough and Clifton, 'Science and the City'.

43 Croarken, 'Mary Edwards'.

44 The Board of Longitude also awarded John Edwards £20 and £200 for his work on metals for reflecting telescope mirrors in 1778 and 1780, respectively.

many references to the women's work, as well as to the family's relationship with Nevil Maskelyne. For example, in 1810 the archives show the Astronomer Royal sending Mary and three other valued computers a pricey twelve-inch celestial globe with a quadrant of altitude and accompanying book. The globe was to assist their work with lunar eclipses by allowing them to mark the Moon's location on the instrument with red ink, which could then be rubbed off with a cloth.[45]

Elizabeth Johnson of Torrington shows how women might have also been involved anonymously in the public dialogue over longitude and in the competition for rewards and patronage. Johnson, one of the sisters of the painter Sir Joshua Reynolds, began anonymously publishing religious pamphlets in Oxford in 1781 after her husband abandoned her and their seven children. One of the Bodleian Library's copies of her first pamphlet, *The Explication of the Vision to Ezekiel*, has a manuscript note naming her as the author. Her fourth pamphlet, *The Astronomy and Geography of the Created World*, concluded that only she should be given 'the palm for finding the longitude'. Johnson anonymously sent this tract to the Board of Longitude a year later but received no encouragement from the Commissioners or from reviewers.[46] Some subjects related to longitude such as astronomy were being increasingly viewed as appropriate female pastimes in nations such as Britain and Italy. However, in addition to the diverse other reasons that early modern women might have sought to avoid public scrutiny, longitude's unique confluence of navigation, mathematics and financial interest was far more male-coded. 'Projecting' in general was often associated with negative qualities including delusion or dishonesty as well, even for men.

Finally, Jane Squire provides the ultimate example of a woman who was *publicly* known in her time for her involvement with the search for longitude, although she was later ignored or underestimated by historians. She was born into an affluent and influential Yorkshire family in about 1686 and was left a well-off heiress after her father's death in 1707. However, Squire ultimately fell into debt after moving to London and investing heavily in different projects. She spent the early 1720s pursuing litigation against a large group of men over her failed investments in a marine salvage expedition, including Edward Harley, Earl of Oxford and Earl Mortimer. It was likely at the behest of Harley or his representatives that an accusation of being a Roman Catholic was

45 Nevil Maskelyne to Henry Andrews, 19 March 1810, CUL RGO 4/149, p. 75r; W. & S. Jones, receipt for items including four globes, 17 March 1810, CUL RGO 14/16, p. 438r.

46 Anonymous (Elizabeth Johnson) to BoL, n.d. (received 15 November 1786), CUL RGO 14/53, p. 4 (tract at pp. 6–44); Johnson (attr.), *The Astronomy and Geography*, p. 68. Wendorf, *Sir Joshua Reynolds*, pp. 67–68; Cotton, *Sir Joshua Reynolds*, p. 251; Johnson (attr.), *Explication of the Vision*; *The Critical Review, or Annals of Literature*, 56 (1783), p. 394 (listing for *The Explication*).

brought against her in 1725. Although the accusation was waived, it resulted in the dismissal of Squire's lawsuit, and she was imprisoned for debt from 1726 to 1729.[47] Within a few years of leaving prison, she began applying the same determination, outspokenness and refusal to accede to gendered limitations that she had displayed during her lawsuits to pursuing the longitude rewards. Squire turned much of her time and attention to developing a new universal way of comprehending the celestial and terrestrial spheres. Although this scheme was also intended to facilitate the finding of longitude at sea and to thus hopefully obtain one of the rewards, its main purpose was to move humanity closer to the state it had occupied before the fall of the Tower of Babel. In seeking this redemption for humanity, Squire achieved a remarkable degree of socio-economic redemption for herself.

Squire was not unusual in the early modern period in intertwining religion and longitude. A number of longitude projectors and other correspondents of the Commissioners of Longitude were clearly inspired by religion, whether in the general sense of God having granted them the necessary skills and insights or in their developing entire systems and world views rooted in the Bible. It would have been more unusual for a philosopher or mathematician, or for that matter a projector, of this period to have been uninfluenced by religious thought and feeling in their efforts and publications. For example, William Whiston's visionary schemes for longitude and millenarian prophecy were often linked – including in the public slights frequently made against his reputation – as were the those of Emanuel Swedenborg with his theology.[48] There were also strong institutional connections between church affairs and the administration of matters such as longitude. The Savilian professors, who were *ex officio* members of the Board of Longitude, had to be ordained ministers of the Church of England. Several Astronomers Royal, including Bradley, Bliss and Maskelyne, were in holy orders as well. Similarly, many Fellows of the Royal Society and other respected researchers and theorists incorporated biblical events such as the deluge into their accounts of the natural world or based their philosophy on religious foundations.[49] The belief that God's words provided direct insights into the nature of the world also echoes the suggestion made by a number of protagonists in the search for the longitude that an answer to the problem would only be found through astronomy – because God had created the skies as a natural timekeeper or

[47] *Squire v. Burnaby*, 1722–1725, TNA C 11/2689/17, C 11/303/19; Middlesex session rolls, Jane Squire's accusation, 20 September 1722, LMA MJ/SR/2394; 'The joint &c. several plea of the Right Hon'ble Edward Harley Esqu'r', c. 1725, King's Inns, Dublin N3/7/4/2; Jane Squire's prison commitment records, TNA PRIS 1/3 127.

[48] Schaffer, 'Swedenborg's Lunars'.

[49] Cantor, *Quakers, Jews, and Science*, pp. 301–302; Wilde, 'Hutchinsonianism, Natural Philosophy'.

orrery for the enlightenment and use of humans. The longitude furthermore lent itself to being twinned with visionary religious schemes and with apocalyptic and anti-Catholic warnings because it was viewed as a mysterious and perhaps insoluble problem rather than as a dry, manmade construction of lines criss-crossing maps and globes. In addition, longitude was such a notorious problem – its fame heightened by the rewards established in 1714 – that hitching one's star to it could provide a larger pulpit for expressing any number of beliefs.

The number of lectures and publications that purported to address both longitude and religion, in some cases satirically, reflect how common these interconnections were. For example, in November 1734, the eccentric preacher and showman John Henley advertised that the speaker at his 'Oratory' at the corner of Lincoln's Inn Fields in London would include 'Mr. [Benjamin] Parker of Derby, Author of a Scheme of the Longitude, best approved by Dr. Halley, – on his new Discovery of the Place of Hell; on removing the Waters of the Deluge; and whether Hell be Punishment of Body or Mind'.[50] Religion and longitude had first intersected like this well before the establishment of the rewards of 1714, as when Edward Harrison went on at length in his *Idea longitudinis* of 1696 about having received his abilities and ideas from God and suggested that 'if the Heads of our Church and State, think it good, let there be made a Figure, representing a first Meridian, and Erected over St. Pauls Church in London, with this Inscription, Glory be to God, good Will towards Men'.[51] Later, in 1715, the clockmaker and long-time longitude projector Samuel Watson advertised that his method had been 'Invented with God's Assistance'.[52]

Such thoughts and approaches to longitude were not uncommon. However, other longitude actors were perceived as rather more disturbed in their theories, aligning more closely with the common negative stereotypes of projectors. An example is Digby Bull (born c. 1648), a former rector of Sheldon in Warwickshire with an MA from Sidney Sussex College, Cambridge, who had lost his place through loyalty to the Stuarts.[53] He spent his final years writing letters and publications that warned of 'popery' and the coming Day of Judgement, as well as a longitude proposal first published in 1706.[54] Bull wrote on 29 September 1714 to the East India Company, an important institutional agent in the quest for longitude, in the wake of the first Longitude Act.

[50] *Daily Journal*, 8 and 13 November 1734. A few years earlier, Parker produced 'A Projection of the Longitude at Sea; whereby the Mariners may be enabled to correct their Accounts as often as they shall have the Benefit of a clear Sky and calm Sea at the Time of the Moon's visible Southing, communicated to and approv'd of by Dr. Halley, Astronomer Royal'; *Daily Post*, 11 May 1731.

[51] Harrison, *Idea longitudinis*. [52] *Post Man and the Historical Account*, 23 June 1715.

[53] Faris, *Plantagenet Ancestry*, p. 56; anonymous, *The Shropshire Gazetteer*, p. 315.

[54] Brady, *The Contribution of British Writers*, p. 216; Bull, *An Exhortation to Trust*; Bull, *A Farther Warning*; Bull, *A Letter of Advice*.

He offered its merchants horological innovations including a revolutionary 'shipwatch' and 'Ship Dial' combination for finding longitude at sea before attempting to save the souls of the East India merchants:

But above all I earnestly desire that you & all men would take great Care to save your Souls & Lives in this great time of Tryal & Danger that is at hand: For I know that Popery & the Dreadfull judgment of God are both certainly at hand, as I have shewed in my Book written upon this account.[55]

The poet Christopher Smart, who was confined to St Luke's Hospital for Lunatics between 1757 and 1763, incorporated the longitude problem into his poem *Jubilate Agno*. This argued that natural philosophy needed to incorporate religious beliefs into a correct interpretation of nature. Smart suggested finding the longitude by finding due east, 'For due East is the way to Paradise, which man knoweth not by reason of his fall', by means of the glass substitute Gladwick, 'a substance growing on hills in the East, candied by the sun, and of diverse colours', and by measuring the Sun as it crossed the meridian with an astrolabe while kneeling as if in prayer.[56] Examples such as these show how it was common for religious beliefs and the search for longitude to intermingle in a variety of ways, and with varying degrees of credibility, even before Squire began developing a new world view rooted in her enduring and unusually public Catholicism.

In 1731, Squire had the Attorney General, Sir Philip Yorke, read the Act of 1714 aloud to her prior to beginning work on her longitude proposal. She discussed her ideas with Hans Sloane, president of the Royal Society and *ex officio* Commissioner of Longitude, and with the mathematician Abraham de Moivre. Both directed her to Edmond Halley, by then Astronomer Royal and *ex officio* Commissioner, but he did not acknowledge her approaches. In response, Squire sent a printed copy of her proposal to each Commissioner and to de Moivre in August 1731.[57] This described her plan for a new universal language and system of measurement as what she would later call a 'simple easy Method' of navigation suitable for sailors with little mathematical knowledge.[58] It involved dividing the heavens into more than a million equal segments, each centred on a constellation and its associated zenith position. This would allow sailors to take a precise reading as they navigated, with an astral clock to help correct the difference between apparent and mean solar time. Once the meridian was reset to

55 Digby Bull to the East India Company, BL IOR/E/1/5 fols 242–243v, letter 135.

56 Christopher Smart, 'Jubilate Agno', 1759–1763, Houghton Library, Harvard University, MS Eng 719, Fragment B, seqs 5, 13; Barrett, *Looking for Longitude*, pp. 145–146; Browne, *A System of Theology*; Kuhn, 'Dr. Johnson', p. 51.

57 Squire, *A Proposal to Determine* (1st ed.), of which a copy in BL (1397.d.43) includes edits in Squire's hand, which were incorporated into her published books more than a decade later.

58 Squire, *A Proposal for Discovering*, p. 8.

Bethlehem, ships that stayed true to their course could be piloted by matching the skies to specialist cards based on John Flamsteed's star catalogues. Squire also proposed the use of a 'Speaking-Trumpet' to announce local time from 'the Top of a Church or some other appointed Place', and the deployment of artificial sea creatures as regularly spaced buoys to aid mapping.[59] Though such details may now sound fantastical, they reflect Squire's familiarity with contemporary intellectual and utilitarian pursuits. Attempts to develop universal languages and systems of measurement were relatively common, as was the proposed use of regularly spaced seamarks to address the longitude and similar problems, notably in the proposal by Whiston and Ditton that sparked the Act of 1714.

Responding to Squire's proposal, de Moivre forwarded an opinion from an unnamed friend that her proposal was not entirely feasible. Sloane reportedly found that Halley agreed.[60] Squire, however, strongly objected to the suggestion that Halley had put himself in sole charge of judging longitude proposals and yet was pursuing the lunar-distance method himself.[61] In the meantime, she had written to the well-known maker of mathematical instruments Thomas Heath of the Strand about making her 'astral clock' and 'two little Water-Engines; to give our Longitude and Latitude separately'.[62]

In early 1733, Squire exchanged letters with her acquaintance Sir Thomas Hanmer (1677–1746), who was one of the few Commissioners personally named in the 1714 Act still living. Squire expressed anger that a sitting Board of Longitude had not met since 1714 to consider proposed solutions, as she and an apparently small number of other projectors expected. In response to her ongoing complaints, Hanmer encouraged Squire to publish her ideas because he could not ever see the Commissioners judging all proposals as she expected. Instead, any proposal 'must undergo the Scrutiny of all the great Professors of the Sciences of Astronomy and Navigation, and not only that but it must stand the Test of Practice'.[63] However, he added, 'if they carry that Demonstration which you think they do, Mankind will greedily receive them'.[64] This exchange reinforces how many of the early Commissioners saw publication as a necessary step to gaining credibility and support for a longitude proposal and thus

[59] Ibid., p. 4.

[60] Jane Squire to Sir Philip York, 15 September 1732; Abraham de Moivre to Jane Squire, n.d. [1732]; Jane Squire to Abraham de Moivre, n.d. [1732], in Squire, *A Proposal to Determine*, 2nd ed., pp. 37–43.

[61] Jane Squire to Sir Philip York, 15 September 1732, in Squire, *A Proposal to Determine*, 2nd ed., pp. 38–39.

[62] Jane Squire to Lord Torrington, 24 December 1731, in Squire, *A Proposal to Determine*, 2nd ed., p. 20.

[63] Jane Squire to Sir Thomas Hanmer, 10 and 29 January 1733; Sir Thomas Hanmer to Jane Squire, n.d. [1733]; Jane Squire to Sir Thomas Hanmer, n.d. [1733], in Squire, *A Proposal to Determine*, 2nd ed., pp. 26–36 (quote at p. 33).

[64] Sir Thomas Hanmer to Jane Squire, n.d. [1733], in Squire, *A Proposal to Determine*, 2nd ed., p. 33.

encouraged projectors to put their ideas into the public domain. Squire told Hanmer that she believed the obstacles and opposition she faced were in large part due to her gender. In January 1733, she wrote to him of a lifelong interest in more than the traditional female pastimes:

> I do not remember any Play-thing, that does not appear to me a mathematical Instrument; nor any mathematical Instrument, that does not appear to me a Play-thing: I see not, therefore, why I should confine myself to Needles, Cards, and Dice.[65]

Hanmer later agreed with her assertion about gender, telling Squire 'that you are to expect to lye under some Prejudice upon account of your Sex'.[66]

Like so many other longitude projectors, Squire was not aware of the first communal meeting of the Commissioners that took place in 1737. In the wake of the second Longitude Act in 1740, she resumed efforts to interest the Commissioners in her scheme, but with little success. On 16 January 1742, she wrote angrily concerning rumours that George Graham was to present them with an astral clock patterned after her own design. Graham was indeed with the Commissioners that day, but only to support John Harrison. Later that year, Squire finally turned to publishing her method. *A Proposal for Discovering our Longitude* elaborated on her explanation of 1731, and included selected correspondence with individual Commissioners and other relevant authorities. It was published in English and French, followed by a revised version under the original title and in English only in 1743.[67] Both editions were printed and bound to an unusually high standard for a longitude proposal and included a large fold-out chart summarising the scheme (Figure 4.2). Their leather bindings, featuring star symbols devised by the author, may have been the first in England to be decorated with symbols specific to the text.[68] The *Proposal* attracted much attention, given its production quality and Squire's gender. It garnered praise for its author's learning and dedication but also generated some confusion. For example, the poet Elizabeth Carter complained to Catherine Talbot of the book's incomprehensibility:

> I am told the project is thought ingenious, and if you should happen to be of that opinion, 'tis ten to one but I may take up the book again, which I have at present thrown by in a great rage (at my own stupidity) and study myself half mad to find out the meaning of it.[69]

Squire died in London on 4 April 1743, shortly after the publication of her second edition. An obituary notice in the *Daily Post* commended her as 'a Lady

[65] Jane Squire to Sir Thomas Hanmer, 10 January 1733 and 29 January 1733, in Squire, *A Proposal to Determine*, 2nd ed., p. 31.

[66] Sir Thomas Hanmer to Jane Squire, 27 July 1741, in Squire, *A Proposal to Determine*, 2nd ed., p. 54.

[67] Squire, *A Proposal to Determine*, 2nd ed.

[68] Miller, *Books Will Speak Plain*, pp. 10–11.

[69] Elizabeth Carter to Catherine Talbot, 13 July 1743, in Carter, *A Series of Letters*, I, p. 35.

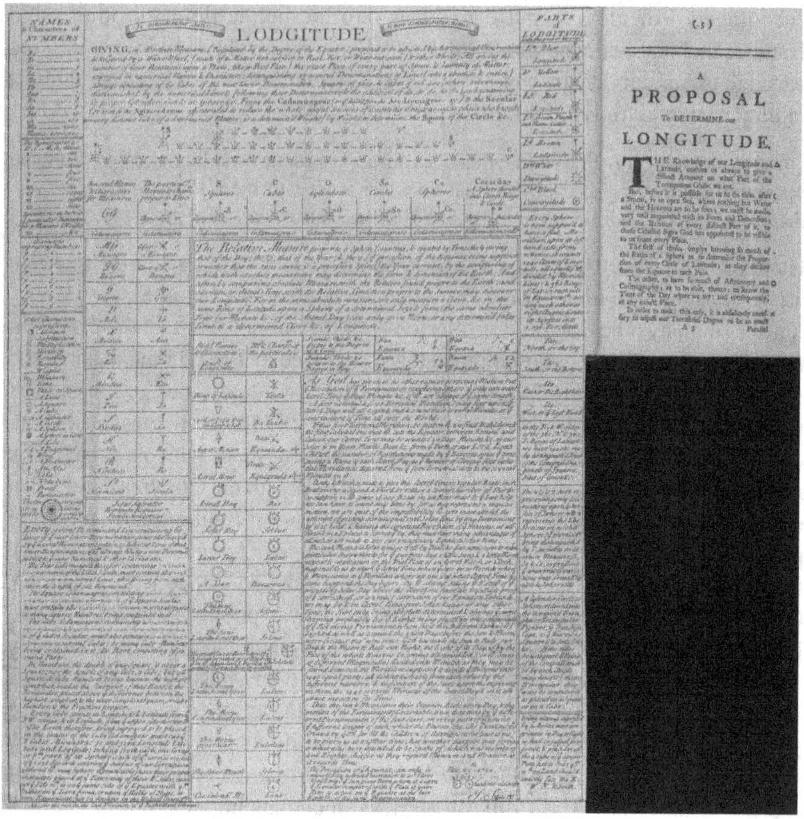

LODGITUDE

(5)

A

PROPOSAL

To DETERMINE our

LONGITUDE.

THE Knowledge of our Longitude and Latitude, enables us always to give a distinct Account on what Part of the Terraqueous Globe we are.

But, before it is possible for us to do this, after a Storm, in an open Sea, where nothing but Water and the Heavens are to be seen, we must be made very well acquainted with its Form and Dimensions; and the Relation of every distinct Part of it, to those Celestial Signs God has appointed to be visible to us from every Place.

The first of these, implys knowing so much of the Ratio of a Sphere as to determine the Proportion of every Circle of Latitude, as they decline from the Equator to each Pole.

The other, to have so much of Astronomy and Cosmography, as to be able, thence, to know the Time of the Day where we are: and consequently, at any nam'd Place.

In order to make this easy, it is absolutely necessary to adjust our Terrestrial Degree to be an exact

A 3 Parallel

Figure 4.2 Summary chart, from Jane Squire, *A Proposal to Determine our Longitude*, 2nd ed. (London, 1743). Reproduced by kind permission of the Syndics of Cambridge University Library

excellently well vers'd in Astronomy, Philosophy, and most Parts of polite Literature [who] led a most moral Life, and was often more charitable than her Circumstances would permit'.[70] Meanwhile, a copy of the late projector's *Proposal* had been sent to the court of Pope Benedict XIV, either by Squire herself before her death or by someone else shortly thereafter. The Pope sought the opinion of the Bologna Academy of Sciences, whose members – unaware of her death – advised Benedict not to fund the scheme but to inform its author of his support for women's participation in the mathematical sciences.[71] Squire had remained committed to her proposal to the very end. Her will directed that, after payment of her debts, sums of £1000 each were to be left to a maternal

[70] *Daily Post*, 13 April 1743. [71] Findlen, 'Calculations of Faith', p. 263.

relative and to three friends 'out of the premium or reward I look on my Self Intitled to for the Method for discovering the Longitude'.[72]

Early Encouragement for Longitude Projects

While Jane Squire and other longitude projectors went round in circles fruitlessly hoping to interest the Commissioners in their ideas, by the time of Squire's death in 1743 an increasing number of schemes had gained official support. The most prominent was John Harrison's proposal for marine timekeepers, which had been garnering official funds and logistical support since 1737. Indeed, the Commissioners' third known communal meeting, which took place at the Admiralty on 4 June 1746 with seven Commissioners attending, was called to consider allotting another £500 to Harrison's work on H3.[73] The clockmaker testified that his extended longitude efforts had rendered him 'quite incapable of following any gainfull employment for the support of himself & family'.[74] The Commissioners revisited the terms laid down, as well as the Royal Society testimonial received, at their previous meeting in 1742. They then examined the current state of H3 and were assured of its progress by Martin Folkes, Robert Smith and James Bradley before agreeing to the clockmaker's request.

The fourth meeting, on 5 March 1748, is the first known to have been called for the communal presentation and consideration of proposals from projectors other than Harrison or the equally well-known Whiston.[75] Seven Commissioners met with Richard Heaton at the Admiralty that day but dismissed his plan, which is not discussed in further detail in the surviving records. They also reviewed and dismissed a written description of an improved refracting telescope for finding longitude from eclipses of Jupiter's satellites, which had been invented by the Reverend Samuel Hardy, a curate and lecturer from Suffolk and later Norfolk. Hardy had begun publicising his longitude scheme the previous year, and would continue to publish and propose astronomical schemes and new instruments until at least the 1770s.[76]

It is important also to note that John Harrison's early champion, George Graham, died in 1751 before the next meeting of the Commissioners. He fulfilled a crucial role as a trusted mediator between the younger clockmaker, the longitude officials and the Royal Society, whose Fellows produced influential

[72] Copy of the will of Jane Squire, 14 March 1743, Doncaster Archives CWM/2/10/50.

[73] BoL, confirmed minutes, 4 June 1746, CUL RGO 14/5, pp. 10–14.

[74] BoL, confirmed minutes, 4 June 1746, CUL RGO 14/5, p. 14.

[75] BoL, confirmed minutes, 5 March 1748, CUL RGO 14/5, p. 15.

[76] *The Dublin Journal*, 8–11 August 1747; *London Evening Post*, 24–26 May 1748; *Whitehall Evening Post or London Intelligencer*, 26 December 1751; Hardy, *The Theory of the Moon*; Hardy, *A Translation of Scherffer's Treatise*.

certificates in support of the younger man's work. Harrison later wrote that his mentor had 'proved a very great Friend to me, viz. not only by his Assistance at the Board of Longitude, &c., but also in his so willingly lending me Money, as without any Security or Interest'.[77]

In January 1753, a new Longitude Act created several additional *ex officio* Commissioners, since all those personally named in the Act of 1714 were by then dead.[78] The new Commissioners included the governor of the Royal Hospital for Seamen, the judge of the High Court of the Admiralty, the Secretaries of the Treasury, the Secretary of the Admiralty and the Comptroller of the Navy. The Act of 1753 also allowed the Commissioners to draw up to an additional £2000 from the Treasurer of the Navy for 'Experiments' (i.e. sea trials), since the pot set aside for that purpose by the initial Act of 1714 had been diminished by the funding already allocated to Harrison and Whiston.

On 17 July 1753, ten of the Commissioners met again at the Admiralty to hear a petition from Harrison requesting further funds for perfecting H3. By this time, Harrison had also been awarded the Royal Society's Copley Medal (in 1749) for his work on the sea clock.[79] As at previous meetings, the Commissioners examined the timekeeper and talked with the clockmaker before agreeing to extend another £500. Meeting again two years later, they awarded a further £500 for Harrison to finish H3, but also so that he could strike out in an entirely new direction. This was 'to make two watches one of such a size as may be worn in the Pocket & the other bigger'. He added that he had already built a promising watch to this design and believed that these watches could 'be purchased at a much cheaper Rate than his superior Machine'. This was a significant moment in hindsight, being the first meeting at which lower cost was flagged as a desideratum for marine timekeepers, and it was Harrison rather than the Commissioners who raised the issue. Harrison seems to have been hoping that the new watches might merit a reward in their own right, over and above any reward resulting from the performance of H3. During the following decade, it would by contrast be the Commissioners who would bring questions of cost into the debates over whether or not Harrison's inventions had sufficiently met the Act of 1714's requirements for a longitude 'solution'.[80]

When the Commissioners next met, on 6 March 1756, they began a second crucial project in the institutional history of longitude. Lord Anson, the First Lord of the Admiralty, introduced 'new invented Tables of the Moon's

[77] Harrison, *A Description Concerning Such Mechanism*, p. 21.
[78] Discovery of Longitude at Sea Act 1753, 26 Geo 2 c 25.
[79] BoL, confirmed minutes, 17 July 1753, CUL RGO 14/5, pp. 16–17.
[80] BoL, confirmed minutes, 19 June 1755, CUL RGO 14/5, p. 18–19. The watch was made by London watchmaker John Jefferys to Harrison's specification in about 1751–1752 and belongs to Trinity House Hull; see Betts, *Harrison*, pp. 63–65.

Motions and some other papers relating to the method of finding the Longitude at sea' from Tobias Mayer of the University of Göttingen. These tables had first been sent to Anson in 1754 through Johann David Michaelis, secretary of Hanoverian affairs in Göttingen, and his cousin William Philip Best, a private secretary of George II. A memorial from Mayer requesting consideration for a reward under the Act of 1714 was also sent through the same channels in the autumn of 1754 and was passed to the Earl of Macclesfield, president of the Royal Society.[81] By the time the Commissioners met in 1756, Astronomer Royal James Bradley had examined Mayer's tables and method and 'very strongly recommended that Trials & Experiments thereof should be made at sea'. This would require the making of 'proper instruments ... Hadleys' Quadrants not being, in his opinion, altogether fit for that purpose'.[82]

Mayer's potential claim to a longitude reward was dependant on achieving three interlinked developments: a lunar theory that could account for the irregular motions of the Moon and allow the production of tables of its future positions; a method for performing the necessary observations and calculations while at sea; and an instrument for making accurate shipboard observations of the lunar distance, the measurement of which was sensitive and critical.[83]

The instrument Mayer proposed took as its basis the double-reflection principle of the Hadley quadrant, which was coming into common usage at sea. Instruments based on this principle, and in particular on the use of mirrors, had already been developed by Robert Hooke and Isaac Newton and tested by Edmond Halley. However, more successful designs were announced in both England and America in the 1730s. Thomas Godfrey of Philadelphia developed an observing instrument that harnessed the principle of double reflection and was tested on the sloop *Trueman* in 1730. He had an intermediary write to Edmond Halley the following year with a view to a possible longitude reward.[84] John Hadley, by then vice-president of the Royal Society, suggested a similar invention at a meeting in May 1731.[85] Successful practical trials took place the following year.[86] While the Royal Society later judged this to be a case of simultaneous invention, it was Hadley's version that became commercially successful, in no small part because he chose to patent and exploit his

[81] Forbes, *Tobias Mayer*, pp. 164–168, 199–201; Forbes, 'Tobias Mayer's Claim'.

[82] BoL, confirmed minutes, 6 March 1756, CUL RGO 14/5, pp. 20–21.

[83] Forbes, *Tobias Mayer*, pp. 151–172, 191–205. For a detailed analysis of Mayer's lunar theory, see Wepster, *Between Theory and Observations*.

[84] John Logan to Edmond Halley, 25 May 1732, Royal Society EL/L6/59; Bedini, *Thinkers and Tinkers*, pp. 118–123; Bennett, 'Catadioptrics and Commerce', pp. 266–274.

[85] Hadley, 'The Description of a New Instrument'; Bennett, 'Catadioptrics and Commerce'.

[86] Hadley, 'An Account of Observations', p. 351.

Figure 4.3 Tobias Mayer's reflecting circle, drawn from the original instrument made by John Bird for testing at sea by John Campbell in 1757, from Abraham Rees, *The Cyclopaedia; or, Universal Dictionary of Arts, Sciences, and Literature, Plates Vol. I* (London, 1820)

idea rather than offer it up for a possible longitude reward. His invention came to be called the Hadley quadrant, and later the octant.[87]

In the 1750s, Tobias Mayer proposed an instrument that was based on his reading of the principles behind the Hadley quadrant but which was a complete circle for reasons of increased observational accuracy.[88] At their meeting in 1756, the Commissioners instructed James Bradley to have three such instruments made for testing at sea and to oversee the execution of trials. Bradley commissioned the London instrument maker John Bird to make a brass copy of Mayer's repeating circle (Figure 4.3). Trials of the instrument and of Mayer's lunar tables took place in early 1757 under Captain John Campbell, but they were limited by Britain being at war and had to be carried out during naval blockade duty off the French coast. Campbell found the circular instrument rather cumbersome and proposed an alternative design, comprising just one sixth of a circle and able to measure up to 120 degrees. Bird made this, the first marine sextant, and Campbell tested it in 1758 and 1759. Bradley later reported to the Commissioners that a good instrument of this sort and Mayer's tables

[87] Although it was introduced as a possible aid for determining longitude, the Hadley quadrant mainly came to be used for finding latitude and local time; Mörzer Bruyns, *Sextants at Greenwich*, pp. 23–35; Clifton, 'The Adoption of the Octant'.

[88] Forbes, *Tobias Mayer*, pp. 163–164.

could determine a ship's longitude to within one degree.[89] The Commissioners were now pursuing the lunar-distance approach to finding longitude at sea with as much dedication as they had marine timekeepers.

At the same time, Harrison continued his work on H3 and the proposed watches. On 28 November 1757, thirteen Commissioners met and agreed to extend another £500 so that he could finish H3 and send it to sea within a year.[90] If possible, he was to also finish and send the two watches, so that the performance of all three timekeepers could be compared.

By this time, Harrison had been collaborating amicably with the Commissioners for twenty years and had often been the focus of their attention. The officials had happily awarded the clockmaker a total of £3000 in funding to continue perfecting his designs and to bring them to trial as directed by the Act of 1714 (Appendix 3). Their relationship had always been cordial and productive, as the Harrisons themselves noted even after the disputes of the 1760s had taken place. John's son, William, wrote in 1772 that:

> Till about the year 1761, we had no difficulties to encounter besides those which nature threw in our way; for until the Invention was brought to maturity, we were happy in the uniform countenance and protection of the Board of Longitude.[91]

In his last published work, John Harrison similarly acknowledged the support of individual Commissioners including Anthony Shepherd, Plumian professor of astronomy, as well as the Astronomers Royal Edmond Halley and James Bradley before the 1760s.[92] The clockmaker had furthermore received financial support from the *ex officio* Commissioners Martin Folkes and Lord Barrington, the Treasurer of the Navy.[93] Folkes, as president of the Royal Society, had also delivered the formal address when Harrison was awarded the Copley Medal in 1749 for his work on H3 and seems to have shown considerable favour towards the clockmaker. Indeed, he was criticised during his presidency for refusing to countenance papers on astronomical schemes pertaining to longitude, a situation that initially continued at the Society after a stroke forced Folkes to relinquish the role.[94]

[89] James Bradley to Mr Clevland, Secretary of the Admiralty, 14 April 1760, in Mayer, *Tabulæ motuum solis*, pp. cxi–cxv; see also Dunn, 'A Bird in the Hand', pp. 83–85; Mörzer Bruyns, *Sextants at Greenwich*, pp. 37–38.

[90] BoL, confirmed minutes, 28 November 1757, RGO 14/5, pp. 22–23.

[91] William Harrison to Stephen Demainbray, 31 January 1772, in Horrins, *Memoirs of a Trait*, pp. 191–200 (p. 192).

[92] Harrison, *A Description Concerning Such Mechanism*, pp. 42 (Shepherd), 49 (Bradley's support until the 1760s), 59 (Halley's support and belief that the lunar-distance method would never come to fruition, according to Graham).

[93] Ibid., p. 21.

[94] Haycock, *William Stukeley*, pp. 226–227; Roos, *Martin Folkes*, p. 246; Rousseau and Haycock, 'Voices Calling for Reform', pp. 395–399.

By the 1760s, the makeup of the Commissioners was significantly different than when the clockmaker first started work, due to retirements, deaths and new appointments. James Bradley was the only attendee from the first communal meeting in 1737 still present in 1760. The next chapter explores how such changes as well as dealings with the Harrisons and their supporters contributed to the Commissioners' transformation into a standing body for the first time, at the same time dramatically expanding their activities and authority.

5 The Birth of the Board of Longitude

Alexi Baker and Richard Dunn

Projectors such as Jane Squire largely dealt with individual Commissioners of Longitude as independent authorities. This was arguably the intent of the original Longitude Act, even if some contemporaries suggested otherwise. From the pivotal decade of the 1760s, this state of events changed rapidly. The Commissioners transformed into a board that met regularly, although much activity still took place outside meetings. The Board of Longitude, as it came to be known, thus became the first standing funding body for science and technology in Britain and an international authority in those areas. It increasingly collaborated with other bodies including the Admiralty and Navy, the Royal Observatory, the Royal Society and institutions overseas. In addition to furthering John Harrison's work, albeit in ways that differed from those he expected, the Board encouraged the improvement of the lunar-distance method and considered other proposals, while also expanding into the improvement of navigation more generally.

This change in the modus operandi of the Commissioners, which included the appointment of a secretary, was in part a self-transformation wrought by proactive officials including Nevil Maskelyne, an *ex officio* Commissioner as Astronomer Royal from 1765. An immediate cause was the change in the relationship between the Commissioners and Harrison and his son, William, from cooperation to confrontation, a change provoked by disagreements over trials of Harrison's fourth marine timekeeper (H4) and of other new methods for finding longitude. These trials and the arguments over them brought about a series of existential crises for the nascent Board, since it was neither inevitable that its work should continue beyond successful trials of a longitude method nor clear what its remit would be should it continue. Thus, the period to 1774 can be seen as one in which the nature of a new body, the Board of Longitude, was established. At the same time, one should recognise transformation into a standing board as symptomatic of broader changes in British government during and after the Seven Years War (1756–1763), notably the growth of bureaucracy and what has become known as the military-fiscal state.[1]

[1] On the military-fiscal state, see Brewer, *The Sinews of Power*. Subsequent analyses, which have coined the term 'contractor state', include Ertman, *Birth of the Leviathan*; Graham and

A range of sources, none entirely objective, reveal aspects of this period in which the Board's dealings were subject to increasing public scrutiny and open debate. The Board's minutes, which the Commissioners ensured were put into proper order with the appointment of a secretary in 1763, are just that: authorised minutes of meetings during which discussions ranged more widely, and equally silent about dealings and correspondence outside those meetings.[2] Regarding the long dispute with Harrison, significant information comes from published works in support of his case and from the manuscript known as the Harrison Journal.[3] Covering the period 1761–1766, the journal appears to be a draft for another publication intended to further his case, and includes not only transcripts of Board minutes, correspondence and memorials by Harrison but also descriptions of other incidents and discussions not mentioned elsewhere, including the conduct of Board meetings.

Imagining the Board

The transformation from individuals to collective was never a given. In addition to the different early modern significances of the term 'commissioner', the appellations 'board' and 'board of longitude' were employed quite differently before the 1760s. They crop up occasionally in the minutes of the Commissioners' meetings from 1737 to 1757. Read in context, however, it is clear that each meeting and the specific Commissioners attending were considered a separate board. This usage endured into the 1760s; for example:

> That a Board of Longitude having seen & approved that Machine, were pleased to encourage him to make another of the same kind but with several Improvements …[4]
>
> … praying that his Lordship would please to assemble a Board of Longitude to consider the state of the said Machine …[5]
>
> … sometime since laid before a Board of Longitude …[6]

The earliest known use of 'the Board of Longitude' in the sense of a standing body appears in an outlying reference in a letter from Astronomer Royal James Bradley to the Secretary of the Admiralty in 1756. Bradley suggested that, since improvements had recently been made to the tables to be used in the lunar-distance method, 'it may well deserve the attention of my Lords

Walsh, *The British Fiscal-Military States*; Harding and Ferri (eds), *The Contractor State and Its Implications*; Knight and Martin, *Sustaining the Fleet.*

[2] Barrett, '"Explaining" Themselves', discusses surviving papers (NMM BGN/1-10) revealing some of the background discussions in 1765.

[3] 'Harrison Journal', 1817, State Library of Victoria H17809, a copy of a manuscript written for the Harrisons. Two earlier copies, one with corrections in William Harrison's hand, survive in private hands.

[4] BoL, confirmed minutes, 4 June 1746, CUL RGO 14/5, p. 10.

[5] BoL, confirmed minutes, 17 July 1753, CUL RGO 14/5 p. 16.

[6] BoL, confirmed minutes, 4 August 1763, CUL RGO 14/5, p. 53.

Commissioners of the Admiralty (as also the Board of Longitude)'.[7] It appears that Bradley might have meant to refer to a standing body (*the* Board), which met when the need arose, such as to consider improved lunar-distance tables, rather than to an individual meeting (*a* Board), as characterised earlier usage of the term.[8] A review of over 1500 mentions of longitude in the newspapers of eighteenth-century London suggests that those publications did not use the term 'Board' in reference to the Commissioners until at least 1762, when some began to talk of 'a board of longitude at the Admiralty'.[9]

Actions as well as terminology show the independent Commissioners transforming into a standing Board at this time. In contrast to the eight recorded meetings of the Commissioners from 1737 through 1757, more than fifty meetings of what became the Board took place from 1760 to 1774 alone, occurring at least annually and often more frequently. Again, these were largely a response to evolving discussions and arguments over Harrison's work.

Harrison's Fourth Marine Timekeeper

At a meeting on 18 July 1760, Harrison told the Commissioners that 'he had perfected his Third Large Machine and was ready to make Trial of the same'. At the same time, however, he 'produced a Large Watch' (Figure 5.1), which 'answered beyond his expectation' but needed further adjustment. He asked, therefore, to be given until the following April to prepare for sea trials. This signalled a radical change in Harrison's approach, although the Commissioners were content to grant another £500 in the expectation that the two timekeepers would be trialled together.[10]

By the following February, Harrison had informed Lord Anson that the timekeepers were ready. Meeting in March, the Commissioners approved a trial, with John's son, William, to accompany the timekeepers.[11] But this took time to organise, exacerbated by the exigencies of war. As a result, William Harrison waited fruitlessly in Portsmouth for some time before returning to

[7] James Bradley to Mr. Clevland, Secretary of the Admiralty, 10 February 1756, in Mayer, *Tabulæ motuum solis et lunæ*, pp. cix–cx (p. cx).

[8] Before hiring a secretary, it was usually the First Lord of the Admiralty who wrote to his fellow Commissioners to call a meeting at the Admiralty; for example, 'Papers related to the Board of Longitude', 1713–1775, TNA ADM 7/684, pp. 1v, 8–10.

[9] *Lloyd's Evening Post and British Chronicle*, 16–18 August 1762; see also *St. James's Chronicle or the British Evening Post*, 3–5 June 1762; *London Chronicle*, 17–19 August 1762; *London Evening Post*, 17–19 August 1762; *Gazetteer and London Daily Advertiser*, 19 August 1762; *London Evening Post*, 19–21 August 1762; *St. James's Chronicle or the British Evening Post*, 19–21 August 1762; *London Chronicle*, 24–26 August 1762.

[10] John and William Harrison, one-day marine timekeeper (H4), c. 1759, NMM ZAA0037; BoL, confirmed minutes, 18 July 1760, CUL RGO 14/5, pp. 24–25; Betts, *Marine Chronometers at Greenwich*, pp. 175–185.

[11] BoL, confirmed minutes, 12 March 1761, CUL RGO 14/5, pp. 26–27.

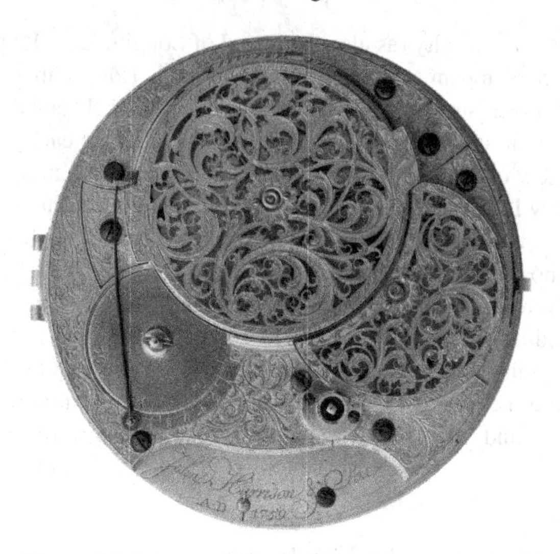

Figure 5.1 John and William Harrison, one-day marine timekeeper (H4), c. 1759, movement, NMM ZAA0037 © National Maritime Museum, Greenwich, London

London. By October, however, there was progress and the Commissioners had the advice of the Council of the Royal Society about the trial's conduct. Their recommendations included: that local time be determined at Portsmouth and used to set the timekeepers before the voyage; that the ship's officers act as witnesses when winding and recording the time from the timekeepers during the voyage; that a competent observer be sent to determine local time and time by the timekeepers upon arrival at Jamaica; that the same or another observer determine Jamaica's longitude from observations of Jupiter's satellites, with similar observations organised in Portsmouth; and that the telescopes used in both places be of the same focal length and magnifying power.[12]

By the meeting, Harrison had sent a memorial moderating the Royal Society proposal and implying that only the smaller, fourth timekeeper be put to trial. Noting that it was too late in the year to observe Jupiter's satellites, he suggested that the trial consist only of determinations of local time and watch time at Portsmouth (by John Robertson of the Royal Naval Academy) and in Jamaica (by William Harrison). A supporting paper from renowned telescope-maker and astronomer James Short added that the existing determination of the longitude difference between Greenwich and Kingston, Jamaica, was 'as exactly determined by several Observations mentioned in the said Paper, as it ever can be by the Eclipses

[12] BoL, confirmed minutes, 13 October 1761, CUL RGO 14/5, p. 30; 'Harrison Journal', pp. 5–6; Bennett, 'The Travels and Trials', pp. 76–77.

of Jupiter's Satellites'.[13] The Commissioners considered and further altered the general plan, adding that the watch should have four locks, the key for each held separately, thus requiring witnesses for daily winding. At Admiral Knowles's recommendation, they nominated John Robison as official observer on the voyage. Robison, a graduate of the University of Glasgow, had previously been tutor to Knowles's son and had surveying experience. But the Commissioners offered no substitute for Jupiter's satellites for determining Jamaica's longitude.[14]

In retrospect, this would mark a turning point in the relationship between the Harrisons and the Commissioners. According to the Harrisons' account, signs of change came just after the October meeting, when they went to the London workshop of instrument maker John Bird to inspect the instruments for the trial. They met Astronomer Royal James Bradley, with whom they had previously had good relations, but later commented that:

> The D^{r}. seemed very much out of temper and in the greatest passion told M^{r}. Harrison that if it had not been for him and his plaguey Watch M^{r}. Mayer and he should have shared Ten thousand Pounds before now. This gave M^{r}. Harrison an Opportunity of seeing what sort of a Friend D^{r}. Bradley had been at the Board, who formerly had been one of the best M^{r}. Harrison had, but now self Interest seemed to be the Principle.[15]

Accusations of self-interest on the part of the Commissioners would become regular complaints.

The Jamaica Trial and Aftermath

William Harrison and H4 sailed on the *Deptford* in November 1761, returning on the *Merlin* the following March at the end of what he considered a successful voyage.[16] The Commissioners met the following June to begin the trial's assessment and in August to discuss the results. This latter meeting threw up significant concerns, since the Commissioners judged that 'the experiments already made of the Watch have not been sufficient to determine the Longitude at sea', in large part because the longitude of Port Royal was not sufficiently well known. There was an additional problem in that the watch's rate, the amount of time it gained or lost each day, had not been declared at the start of the trial. However, the Commissioners conceded that H4 was 'an Invention of considerable Utility to the Public' and awarded Harrison £2500, of which £1000 would be paid after another successful trial.[17]

[13] BoL, confirmed minutes, 13 October 1761, CUL RGO 14/5, pp. 30–31.
[14] BoL, confirmed minutes, 13 October 1761, CUL RGO 14/5, pp. 31–33; Bennett, 'The Travels and Trials', p. 78.
[15] 'Harrison Journal', p. 17. [16] Quill, *John Harrison*, pp. 95–105.
[17] BoL, confirmed minutes, 17 August 1762, CUL RGO 14/5, p. 38; 'Harrison Journal', pp. 40–45; Bennett, 'The Travels and Trials', pp. 80–81; Quill, *John Harrison*, pp. 111–114.

This judgement marked a divergence between the Harrisons and the Commissioners. According to William's later account, the Harrisons felt that:

> no sooner was it evident that we had succeeded, than by a strange fatality the Board turned against us, and has ever since acted, as if it had been instituted for the express purpose of preventing our Invention from being made useful to mankind.[18]

The letter of the law was crucial, they argued. According to their reading of the Act of 1714, H4 had successfully proved itself at sea, performing within the limits required for the maximum reward. Yet the Commissioners took it as their right to interpret the Act and emphasised that any longitude method had to be 'tried and found practicable and useful at Sea'.[19] Contrasting interpretations of the difference between a method and a specific instrument, and the implications of a method being 'practicable and useful at sea', framed the resulting dispute.

It is significant that the Commissioners saw the need for new institutional trappings at this moment of confrontation. The meeting of 17 August 1762 resolved to request funds to employ a secretary to oversee their records:

> The Board taking notice also that all the Minutes & papers, since the first appointment of Commissioners of the Longitude to the present time, are in great disorder & Confusion there never having been any Person appointed to take Charge of the same; And it being of great Consequence that the said Minutes & Papers should be forthwith digested & put into the best method their present disorder'd state will admit of; And that for the future, some fit Person should be appointed to attend them at their Meetings, to take minutes, write Letters & transact such other Business as may occur.[20]

Threatened with a significant challenge, the Commissioners recognised that having a solid bureaucratic trail of evidence was essential. At the same time, they asked that the professors from Oxford and Cambridge be reimbursed for attending meetings. Previously, communal business was so infrequent that neither institutional record keeping nor the cost of attending meetings had seemed pressing.

On 17 August 1763, the king officially agreed to the Commissioners' requests for £40 per year for a secretary and £15 per meeting for those coming from outside London. These were to be paid by the Navy from the sale of spare supplies to merchant ships. The Navy decreed that this should be enacted on 10 September 1763, with junior Admiralty clerk John Ibbetson becoming

[18] William Harrison to Stephen Demainbray, 31 January 1772, in Horrins, *Memoirs of a Trait*, pp. 191–200 (quote at p. 192).

[19] Discovery of Longitude at Sea Act (13 Anne c 14); Bennett, 'The Travels and Trials', pp. 80–81. Siegel, 'Law and Longitude', offers a recent perspective from an American Professor of Law.

[20] BoL, confirmed minutes, 17 August 1762, RGO 14/5, pp. 42–43.

the first of five secretaries to the Board, serving until 1782.[21] By 1765, the Commissioners even felt compelled to request a raise for the Secretary since the existing salary was 'very inadequate to the trouble and nature of his employment the Business thereof having been considerably increased of late'.[22]

By this time, the Harrisons and their supporters were employing new strategies in the escalating dispute. They notably opened the discussion to public scrutiny through petitioning and publication, signalled by the printing of several broadsheets in December 1762.[23] It appears these were written not by John or William Harrison but by one of their friends, probably James Short. Short had become Harrison's most vociferous supporter after the death of George Graham in 1751 and, as noted above, had already written a supportive paper for the Commissioners' meeting in October 1761.[24] One of the broadsheets, the *Memorial Concerning Mr. Harrison's Invention for Measuring the Time at Sea*, significantly suggested that Parliament be petitioned to offer a reward if Harrison were to disclose the principles of the timekeeper's construction, on the proviso that no other claimant be allowed until a final adjudication had been made concerning a full reward under the original Longitude Act.[25] The same request was made to the Commissioners and approved on 26 February.[26] Presumably responding to Harrison's lobbyists, it was resolved that the watch could be considered a practicable and useful method only if the principles of its construction were understood and repeatable by other craftsmen. Around the same time, the first of several longer publications supporting Harrison appeared, maintaining the public visibility of the dispute. Penned by 'a member of the Royal Society' (perhaps Short or Taylor White), it detailed events to date and reproduced key documents as evidence.[27]

[21] 'Papers related to the Board of Longitude', 1713–1775, TNA ADM 7/684, pp. 11–15.

[22] Copy of order agreed at the King's Council, 1 August 1765, TNA ADM 7/684, pp. 15r–16v.

[23] Anonymous, *Memorial Concerning Mr. Harrison's Invention*; anonymous, *Proposal for Examining Mr. Harrison's Time-Keeper*; anonymous, *A Calculation Shewing the Result of an Experiment*. These appear to have been printed at the same time and are bound together in the British Library (shelfmark 717.5.15) with a copy (seemingly printed at the same time) of the testimonial of 16 January 1742 signed by Folkes, Smith, Bradley and others regarding Harrison's work.

[24] Bennett, 'The Travels and Trials', p. 82; Bennett, 'James Short and John Harrison'.

[25] Anonymous, *Memorial Concerning Mr. Harrison's Invention*, p. 1.

[26] BoL, confirmed minutes, 26 February 1763, RGO 14/5, pp. 45–46.

[27] Anonymous, *An Account of the Proceedings*, 1st ed., p. 4. The first edition, preface dated February 1763, ran to forty-six pages; a second edition of ninety-eight pages (preface also dated February 1763) appeared later in the year. Jérôme Lalande suggested that it was by Taylor White, treasurer of the Foundling Hospital (of which William Harrison was a governor); see Turner, 'L'Angleterre, la France', p. 157. Besides pieces in periodicals, subsequent publications in support of Harrison included: Anonymous, *A Narrative of the Proceedings*; Anonymous, *Some Particulars Relative to the Discovery*; Anonymous, *The Case of Mr. John Harrison*, published in three versions, one undated (c. 1766), two dated (1770, 1773); Harrison, *Remarks on a Pamphlet*, published in two editions the same year. Late in life, Harrison, *A Description Concerning Such Mechanism*, included his account of the longitude affair.

Harrison's parliamentary petition resulted in 'An Act for the Encouragement of John Harrison, to publish and make known his Invention of a Machine or Watch, for the Discovery of the Longitude at Sea'. This nominated eleven 'Commissioners for the Discovery of Mr Harrison's Watch', who would witness Harrison's disclosure and ensure that the details were published to allow other clockmakers to reproduce the designs. Once they or a majority of them certified that Harrison had done so, the Treasurer of the Navy was to release £5000 from any unapplied funds. If Harrison later won one of the larger longitude rewards, the £6500 he would by then have received was to be deducted from it.[28]

Agreeing the format of the 'discovery' proved an intractable problem, however, with the new Commissioners expecting more than Harrison was willing to concede, even though Short was among those chosen. Harrison particularly objected to the stipulation that further watches be made under his supervision and successfully tested before any certificate could be issued. Unable to persuade the Commissioners, in particular the increasingly influential Earl of Morton, that these demands were unreasonable, Harrison withdrew from the debate and requested a second sea trial under the terms of the Act of 1714.[29]

Meeting in August 1763, the Board discussed Harrison's proposals for a trial. They were content that only H4 be sent but stipulated that he declare its rate beforehand. They were not, however, willing to accede to his concerns about the use of Jupiter's satellites to determine terrestrial longitudes. Nor, when the observing instruments to be used were examined (on 6 August), were they willing to replace the reflecting telescopes made by John Bird with others made by James Short. Although Short's might indeed have been better, those by Bird had already been purchased at public expense for the trial.[30] Following a second meeting, on 9 August, it was agreed that the trial would take place to Barbados and that the official observers would be astronomers Nevil Maskelyne and John Robison, although by 5 September Charles Green had replaced Robison.[31] In addition, the trial was to include two additional schemes: Tobias Mayer's lunar tables and Christopher Irwin's marine chair for observing Jupiter's satellites from a ship.

As noted in Chapter 4, Mayer's work had been known to the Commissioners since the mid-1750s and had undergone promising trials. It had also gained

[28] Discovery of Longitude at Sea Act, 1762, 2 Geo 3 c 14; copy at CUL RGO 14/1, pp. 22–25. James Short appeared with William Harrison and John Bevis as an expert witness at the examination of Harrison's petition by a parliamentary committee; *Journals of the House of Commons*, 29, pp. 515, 546–553.

[29] Bennett, 'The Travels and Trials', pp. 82–84; 'Harrison Journal', pp. 62–72; Quill, *John Harrison*, pp. 115–128.

[30] BoL, confirmed minutes, 4 August 1763, RGO 14/5, pp. 49–53; Bennett, 'The Travels and Trials', p. 85; 'Harrison Journal', pp. 97–101.

[31] BoL, confirmed minutes, 9 August and 5 September 1763, RGO 14/5, pp. 54–61.

further vindication on two voyages in 1761: by Carsten Niebuhr, astronomer and cartographer on the Royal Danish Expedition to Arabia in 1761–1767; and by Nevil Maskelyne and Robert Waddington on a voyage to St Helena to observe the Transit of Venus.[32] The latter voyage would come to hold particular significance for the later work of the Commissioners as Maskelyne became a vociferous proponent for the lunar-distance method.

Mayer had died the year before the Commissioners met on 4 August 1763, although they had received a memorial on behalf of his widow, Maria, who had been advised by Lalande to press the astronomer's claim.[33] They therefore decided that the voyage's official observer would 'make Observations of the Moon's Motions, during the Voyage, to try the Accuracy of the said Tables'.[34] The next order of business concerned Irwin's marine chair.

Christopher Irwin: Beyond Clocks and Lunars

The marine chair invented by Christopher Irwin from Roscommon, Ireland, offers a telling example of the other types of proposal that had gained traction. Irwin is interesting as one of the earliest reward-seekers to whom the Commissioners gave financial and logistical support, as well as an example of the ability of savvy self-promotion to take a person quite far with the Commissioners. His invention, a marine observatory or chair, had deep roots in the history of longitude. The concept went back to at least the sixteenth century in Europe, when seafaring nations such as Portugal and France saw a number of navigational innovations proposed. These included inventions to provide shipboard observers with a stable viewing platform from which to make astronomical measurements.[35] The designs suggested over the centuries varied but often involved gimbals or suspended weights, placed either on the ship's deck or floated on a platform nearby. At least in later centuries, these 'chairs' were often for longitude observations, particularly lunar distances or of Jupiter's satellites. Such schemes remained of interest right up to the Board of Longitude's dissolution.[36]

Christopher Irwin first appeared in relation to longitude in 1758, when he was living in Suffolk Street, London. His patent for a 'Marine Observatory'

[32] Baack, 'A Practical Skill'; Bennett, 'The Rev. Mr. Nevil Maskelyne, F.R.S. and Myself'; Bennett, 'Mathematicians on Board'; Higgitt, 'The Projects of Eighteenth-Century Astronomy'; Howse, *Nevil Maskelyne*, pp. 27–39.

[33] Forbes, *Tobias Mayer*, pp. 197–199, notes that it followed negotiations between Maria Mayer and the Göttingen Scientific Society (but states it was read at the meeting of 9 August).

[34] BoL, confirmed minutes, 4 August 1763, RGO 14/5, p. 53.

[35] See, for example, woodcut of gimballed marine chair in Besson, *Le Cosmolabe*, p. 29.

[36] Van Helden, 'Longitude and the Satellites'; Dunn, 'Scoping Longitude'; Köberer, 'German Contributions', pp. 279–280.

and related telescope and almanac was granted on 2 December, with a further description added on 14 March the following year. The chair comprised:

a stage or seat for an observor, his telescope or other instruments, fixed over deck at right angles, with a stiff pendulum, which passes under deck with a heavy weight fixed to the lower end thereof near the seat. The pendulum is suspended very glibly by cross axes or balls and sockets, and otherwise, so as by the centripetal force of the weight below to procure for the seat, &c., a center of motion distinct from that of the ship, and thereby to counteract the tossings of the ship, so as to keep the observatory horizontal, or tranquil enough for the purposes of observation.[37]

Irwin had London instrument maker Jeremiah Sisson construct a working example, which those interested could visit at Sisson's shop near the Royal Society. Several visitors described it in diaries, letters and university lectures, which helped spread news of it across Europe. Swedish astronomer Bengt Ferrner, for example, visited Sisson in October 1759 and wrote that he had placed the chair 'on top of his house' with a hole through the roof 'for the balance'. The chair seemed 'polite and comfortable' and Ferrner asked for a drawing and estimated costs, despite being unconvinced that it would work at sea.[38]

Ferrner's visit took place only a few weeks after Irwin had conducted sea trials in the English Channel, described in glowing reports in newspapers and periodicals.[39] Irwin likely planted some of these laudatory accounts, as readers suspected at the time. A satire in the *Busy Body*, for instance, mocked a particularly complimentary report from earlier that year in *The London Magazine*, that Irwin had found longitude and everyone from Lord Howe to the young prince Edward had approved of his chair.[40] News of these successful sea trials alarmed one of Tobias Mayer's correspondents sufficiently for him to write to Göttingen, concerned that 'Erwin's Easy Chair' was now a serious challenger

[37] Christopher Irwin, 'Marine Observatory and Telescope, and Almanack for Ascertaining Longitude at Sea', patent no. 731; patent granted 2 December 1758 [33 George 2], specification enrolled 14 March 1759, BL.

[38] Orrje, 'Patriotic and Cosmopolitan Patchworks', pp. 97–98. Lalande, *Journal d'un voyage*, pp. 32, 61–62, 80, describes visiting Sisson's workshop on 29 March, 9 May (when he saw three chairs) and 7 June 1763. Albert Krayer (pers. comm.) notes that Georg Christoph Lichtenberg discussed Irwin's chair in astronomy lectures at the University of Göttingen, as shown in notes and sketches in the papers of Lichtenberg and his students; see Georg Christoph Lichtenberg to Johann Friedrich Blumenbach, 5 May 1789, in Lichtenberg, *Georg Christoph Lichtenberg: Briefwechsel*, III, pp. 699–701; Akademie der Wissenschaften zu Göttingen (ed.), *Georg Christoph Lichtenberg*, p. 696; Grosser, *Johann Friedrich Benzenberg*, p. 82.

[39] *London Chronicle*, 29 September–2 October 1759; *London Chronicle*, 1 July–31 December 1760; *Lloyd's Evening Post and British Chronicle*, 8–10 October 1760; *Lloyd's Evening Post and British Chronicle*, 29 June–1 July 1761; *Lloyd's Evening Post and British Chronicle*, 10–12 August 1761; Taylor, *The Mathematical Practitioners*, pp. 33–34.

[40] *The London Magazine, or, Gentleman's Monthly Intelligencer*, 28 (1759), p. 505; *Busy Body*, 27 October 1759.

for a longitude reward.[41] The following year, Irwin published a full account of thc invention.[42]

Irwin's successful self-promotion and testimonials from high-profile observers such as Lord Howe gained him a hearing with the Board of Longitude on 3 June 1762 (when the officials also discussed H4's Jamaica trial).[43] After questioning Irwin, they decided to consult the Attorney General and Solicitor General as to whether they could reward him under the Act of 1714. Irwin further submitted a letter to be read at the next Board meeting, explaining errors previously noted in his astronomical tables and providing testimonials from the 1759 sea trials.[44] Lord Howe, present at some of the trials, certified that the marine chair, 'notwithstanding it has not yet answered the expectations of the Inventor, is nevertheless an ingenious contrivance & will greatly facilitate the making Observations of the Celestial Phaenomena at sea, which will be of great public Utility'. He believed it warranted further trial.[45] Howe, renowned for commanding several naval actions against the French during the Seven Years War, would soon be appointed a Lord Commissioner of the Admiralty. On the basis of his certificate, the Commissioners granted Irwin £500 (a sum akin to those given periodically to Harrison) to carry out further tests and asked the Admiralty to have their officers test it at sea. Reports in periodicals again included at least one suspiciously long and admiring letter.[46]

Irwin reappeared at the meeting on 26 February 1763, when an unprecedented eighteen Commissioners attended.[47] He had sent details of improvements and a request for funds for a sea trial. Calling him in, the Commissioners informed him that they would ask the Admiralty to assign him a ship for a trial as soon as possible but were unwilling to offer further money.[48] Irwin nevertheless continued his efforts through public notices declaring that he might soon win one of the large rewards.[49]

As a sign of how much busier the standing Board had become by this time, the same meeting discussed and approved Harrison's request to seek a more immediate parliamentary reward owing to his advanced age and deteriorating health. They also considered a memorial from the family of the late Tobias Mayer, met with Henrick Schultz, a mechanic and model maker to the king of Denmark, passing his proposal to the Royal Society, met with and dismissed a

[41] William Philip Best to Johann David Michaelis, 9 October 1759, Universitätsarchiv Göttingen, Cod. MS. Mich. 320, fols 618–619 (with thanks to Albert Krayer and Wolfgang Köberer).

[42] Irwin, *A Summary of the Principles*.

[43] BoL, confirmed minutes, 3 June 1762, CUL RGO 14/5, pp. 34–37.

[44] BoL, confirmed minutes, 17 August 1762, CUL RGO 14/5, pp. 37–44.

[45] BoL, confirmed minutes, 17 August 1762, CUL RGO 14/5, pp. 40–41.

[46] *London Evening Post*, 19–21 August 1762; *London Chronicle*, 24–26 August 1762.

[47] BoL, confirmed minutes, 26 February 1763, CUL RGO 14/5, p. 44–48.

[48] BoL, confirmed minutes, 26 February 1763, CUL RGO 14/5, p. 47.

[49] *Lloyd's Evening Post*, 27–29 June 1763.

Mr Robinson, and were supposed to see a Mr Monck.[50] They considered written proposals from: Samuel Hardy of Ipswich, with a celestial scheme; John Hoskins of Bristol, with an astronomical instrument; William Ross, with new instruments and methods; John Grey of Cupar in Fife, with an instrument for observing meridian altitudes; Henry Pope of Chepstow, with a magnetic variation scheme; and John Bell, about squaring the circle. Each was advised that they should 'obtain Certificates from persons of known Skill & Judgement that their schemes have been tried & have been found to answer what they propose by them; or that there is a very great probability that they will do so upon Experiment, before this Board can take any notice of them'.[51] Other letters and projectors' accounts show that this had long been the response from individual Commissioners to proposals.

At the meeting on 4 August, which discussed the forthcoming trial to the West Indies, Irwin sent a letter requesting instructions for the trial and further funding. The latter was refused, but on 5 September the Board relented and awarded Irwin £100 to help his preparations, including bringing his son and a mathematical instrument maker in case repairs were needed.[52]

Judging the Barbados Trials

Maskelyne, Green and Irwin departed on the *Princess Louisa* in September 1763, arriving in Bridgetown, Barbados, in early November.[53] William Harrison and H4 sailed the following March on the *Tartar*, arriving in May and departing again for England the following month. The Commissioners then met in September to decide how and by whom the results of the trial were to be assessed.

Irwin's invention had fared badly. Maskelyne found that it 'affords no convenience or advantage to an observer in using a telescope for observing the celestial phenomena at sea, but rather the contrary', and wrote to his brother from Barbados that 'My friend Irwin's machine proves a mere bauble, not in the least useful for the purpose intended'.[54] Aware of Maskelyne's opinion, Irwin submitted a letter complaining of 'being ill used in the process of the last Experiment … Mr. Maskelyne had made no Observation during the Voyage & prevented Mr. Green from making more than two'. The Commissioners

[50] For Schultz, see Society for the Encouragement of Arts, Manufactures, and Commerce, *A List of the Society*, p. 82.

[51] BoL, confirmed minutes, 26 February 1763, CUL RGO 14/5, p. 48.

[52] BoL, confirmed minutes, 4 August and 5 September 1763, CUL RGO 14/5, pp. 53, 60–61.

[53] Before Irwin returned to England, another suspiciously laudatory item claimed he had revolutionised the design of sugar cane mills in Barbados; *London Chronicle*, 28–31 July 1764.

[54] Nevil Maskelyne to BoL, n.d. (draft), RGO 4/320, p. 7:2v; Nevil Maskelyne to Edmund Maskelyne, 29 December 1763, NMM MSK/3/6.

responded that, based on the testimonials, Irwin should 'be acquainted the Board do not think there is the least probability of his Chair answering the purpose for which it was designed therefore recommend it to him to desist from pursuing the said Invention any further'.[55]

Unsatisfied, Irwin was almost certainly behind public calls for greater transparency that followed the meeting. These questioned whether H4 had proved its merit, hinted that the Harrisons should not have received an additional £1000 (a sum agreed upon beforehand) and suggested that 'Longitude-feasts' were held before the meeting to school the witnesses from the sea trials. The anonymous author concluded that the public should hear the cases for H4, the lunar-distance method and Jovian moons 'observed by the means of Irwin's chair (the last one recommended by a noble Peer of this realm for its utility)'.[56] The piece had little effect and when Irwin again requested money from the Board he was refused and 'asked to withdraw'.[57] He was never readmitted to meetings but was mentioned by contemporaries and later marine chair inventors, presumably due to the reference to his scheme in successive editions of the *Nautical Almanac*.[58] For example, when William Lester contacted the Lords Commissioners of the Admiralty, the king and Secretary of State in 1820–1821 about his own chair, he compared it to Irwin's and claimed to have been involved in the 1763 trial.[59]

Much of the remainder of the meeting in September 1764 dealt with the payment of expenses from the trials and organising the next steps for the assessment of Harrison's timekeeper. The Board noted that Harrison had, as requested, declared its rate before the voyage and nominated 'three Gentlemen of Skill' to carry analyse the observations: Captain John Campbell, John Bevis and George Witchell.[60] Called into the meeting, William Harrison was told that his father could nominate an additional expert.[61] Unsurprisingly, they chose James Short.

At the next meeting on 19 January 1765 (the last to which Irwin was admitted), the Commissioners received the calculations, as well as formal notification of Nevil Maskelyne's appointment as Astronomer Royal.[62] Both these matters seem to have turned their attention to the future, notably the question of the continuation (or not) of the Board. Their response was to seek legislative

55 BoL, confirmed minutes, 18 September 1764, CUL RGO 14/5, pp. 67–68.
56 *London Evening Post*, 23–25 October 1764.
57 BoL, confirmed minutes, 19 January 1765, CUL RGO 14/5, p. 73.
58 For example, *Nautical Almanac ... for the Year 1771*, pp. 153–154; Dunn, 'Scoping Longitude', pp. 144–145.
59 William Lester to the Lords Commissioners of the Admiralty, 25 December 1820, CUL RGO 14/40, pp. 340r–341v; William Lester, petition to the king, 18 January 1821, CUL RGO 14/12, p. 297r.
60 Clerke, 'Bevis, John'; Howse, 'Campbell, John'; McConnell, 'Witchell, George'.
61 BoL, confirmed minutes, 18 September 1764, CUL RGO 14/5, pp. 67–68.
62 Howse, *Nevil Maskelyne*, pp. 53–58.

changes that would grant additional funds to cover existing commitments in the short term, allow 'further Experiments if necessary' and 'give rewards to any other person or persons who may hereafter discover the Longitude at sea, by any other method than that invented by Mr. Harrison'. It might originally have been assumed that the Commissioners' role would cease once a successful solution had been successfully tested, but now they saw a longer-term role in the development of longitude-finding. In addition to addressing these prospective considerations, they resolved to send the lunar-distance observations made by Maskelyne and Green to Witchell and Bevis so that they could perform the calculations to assess Mayer's lunar tables.[63]

The following month, with Maskelyne in post as Astronomer Royal and attending as *ex officio* Commissioner, the Board met to pass formal judgement. They agreed that Harrison's timekeeper had performed within the limits of the Act of 1714. At the instigation of the Earl of Morton, however, they also wished to consider the issues of disclosure and replication that had surfaced after the Jamaica trial, ruling that,

> in regard the said Mr. John Harrison hath not yet made a discovery of the Principles upon which the said Timekeeper is constructed, nor of the method of carrying those principles into Execution, by means whereof other such Timekeepers might be framed of sufficient correctness to find the Longitude at sea within the Limits by the said Act required whereby the said Invention might be adjudged practicable and usefull in terms of the said Act & agreeable to the true Intent & meaning thereof, They do not therefore think themselves authorized to grant any Certificate to the said Mr. John Harrison until he shall have made a full & clear discovery of the said principles & method, and the same shall have been found practicable & usefull to their Satisfaction.[64]

The resulting resolution, enshrined in a new Longitude Act three months later, stipulated that Harrison would receive £10,000 (less rewards already received) once he explained the construction of his timekeepers and handed them over to the Commissioners, the remaining £10,000 to be paid when he had had copies made and successfully trialled.[65]

The meeting also considered Mayer's lunar tables, with testimony from Maskelyne and four East India Company officers as to the efficacy of the tables when tested on the 1761 voyage to St Helena. For some months, letters from William Philip Best on Maria Mayer's behalf had been putting pressure on the Commissioners, as had communications from Gerlach Adolph von Münchhausen, prime minister in Hanover, whose personal contacts allowed him to make private representations.[66] The Commissioners recommended a

[63] BoL confirmed minutes, 19 January 1765, CUL RGO 14/5, pp. 69–74.
[64] BoL confirmed minutes, 9 February 1765, CUL RGO 14/5, p. 77.
[65] Discovery of Longitude at Sea Act, 1765, 5 Geo 3 c 11; copy at CUL RGO 14/1, pp. 29–33.
[66] Forbes, *Tobias Mayer*, p. 200.

reward of up to £5000 to Tobias Mayer's widow for his work on the lunar-distance method and that powers be granted to have annual nautical tables calculated and printed.

By the time the legislation passed through Parliament, the reward to Mayer's widow had been reduced to £3000, with an additional reward of £300 for the Swiss-born mathematician Leonhard Euler. The change seems in part to have been a response to a letter from Alexis Clairaut to John Bevis, which was published in *The Gentleman's Magazine*. Clairaut claimed that his and Euler's analyses were more rigorous than Mayer's, and that Mayer's theories depended on Euler's earlier work. Clairaut's own lunar tables, he added, were equally worthy of the Commissioners' consideration and had been published earlier.[67] These were not the sorts of claim the Commissioners' wished to encourage, as became evident at the next meeting on 23 February 1765, when they heard a request from Richard Dunthorne for a reward for lunar tables published before Mayer's. The Board decided that it 'did not think fit to propose that any Reward should be given to him as it may tend to the opening a door to numberless Applications of the like kind from other persons'. The meeting also turned down James Short's request for copies of the computations from the first sea trial of H4.[68] As was the case with Harrison's marine timekeepers, Maria Mayer was required to assign to the Board of Longitude the ownership and rights to her husband's lunar tables as a condition of the reward.[69]

The meetings in early 1765 and the resulting Longitude Act marked an important turning point with respect to the timekeeper and lunar-distance methods. They also created an impasse with respect to the Harrisons. Seeking to bring an end to the debate, the continuation of which was signalled by Short's request at the meeting of 23 February, the Act enshrined in law the concerns the Commissioners had about the extent to which Harrison's marine timekeeper might be considered 'practicable and useful' and set out new conditions under which a full reward would be payable. It gave them the legal basis, which their new bureaucratic structure could support, to avoid extensive negotiation.

If ending debate was an aim, however, it failed. Harrison had already sought to head off the legislation by direct appeal to Parliament in February 1765, and he and his supporters escalated the dispute in the wake of the Act. At

[67] Clairaut, 'Copy of a Letter'; Boistel, 'From Lacaille to Lalande', p. 57; Forbes, *Tobias Mayer*, p. 201; Wepster, *Between Theory and Observations*, pp. 40–41.

[68] BoL confirmed minutes, 23 February 1765, CUL RGO 14/5, pp. 83–84.

[69] Testimonial signed by Maria Victoria Mayer, granting power of attorney to William Philip Best, 3 August 1765, CUL RGO 14/1, pp. 147r–150v; warrant signed by Maria Victoria Mayer, assigning ownership of Tobias Mayer's tables to the Commissioners of Longitude, 5 November 1765, CUL RGO 14/1, pp. 153r–154r; Forbes, *Tobias Mayer*, pp. 201–204, translates the warrant.

subsequent Board meetings, in correspondence and in print, they continued forcefully to assert Harrison's rights to the full reward without further conditions such as those imposed through the new Act.[70]

The Act had other significant consequences that served both to ensure the continuation of the Board of Longitude and expand its operations. Most importantly, it authorised the Commissioners to produce an annual nautical ephemeris (the *Nautical Almanac*), an activity that would come to dominate outgoing expenses (see Chapter 7). It also established rewards of up to £5000 for future improvements made to astronomical tables or for 'any Discovery or Discoveries, Improvement or Improvements, useful to Navigation'. In other words, it expanded the Board's remit from longitude to navigation more generally. Finally, the Act nominated an additional Commissioner, the Lowndes professor of astronomy at Cambridge.[71]

Discovery and Publication

In seeking to ensure that Harrison's 'method' would be of 'common & general Utility', the Commissioners reverted to the process of disclosure first suggested after the Jamaica trial, while also performing further tests of H4 and investigating its replicability.[72] By the summer of 1765, negotiations returned to the process by which Harrison might demonstrate the watch's workings before a new Commission for the Discovery of Mr Harrison's Watch. These soon became fraught. The Board wished Harrison to take the watch apart and explain 'by word of Mouth & experimental exhibitions' its workings and manufacture in front of six witnesses: mathematicians John Michell and William Ludlam; watchmakers Thomas Mudge, William Matthews and Larcum Kendall; and instrument maker John Bird.[73] Harrison, however, was troubled by the phrase 'experimental exhibitions', which he feared might give licence for new trials.[74] This became a flash point, Harrison declaring before leaving one confrontational meeting that 'he never would consent to it, so long as he had a drop of English Blood in his Body', leading the Commissioners to publish the minutes of the meetings in May and June 1765 as a defence of their position.[75]

[70] John Harrison, petition to the House of Commons, 25 February 1765, in 'Harrison Journal', pp. 129–131; anonymous, *A Narrative of the Proceedings*.

[71] Discovery of Longitude at Sea Act, 1765, 5 Geo 3 c 20; copy at CUL RGO 14/1, pp. 29–33. Barrett, '"Explaining" Themselves', discusses the Act's passage.

[72] BoL confirmed minutes, 9 February 1765, CUL RGO 14/5, p. 78.

[73] BoL confirmed minutes, 28 May 1765, CUL RGO 14/5, pp. 86–88; BoL confirmed minutes, 13 June 1765, CUL RGO 14/5, pp. 95–100.

[74] Bennett, 'The Travels and Trials', pp. 87–88.

[75] BoL confirmed minutes, 13 June 1765, CUL RGO 14/5, p. 99; Commissioners of Longitude, *Minutes of the Proceedings*. Versions of the minutes appeared in the *Gentleman's Magazine*, *Monthly Review* and other periodicals.

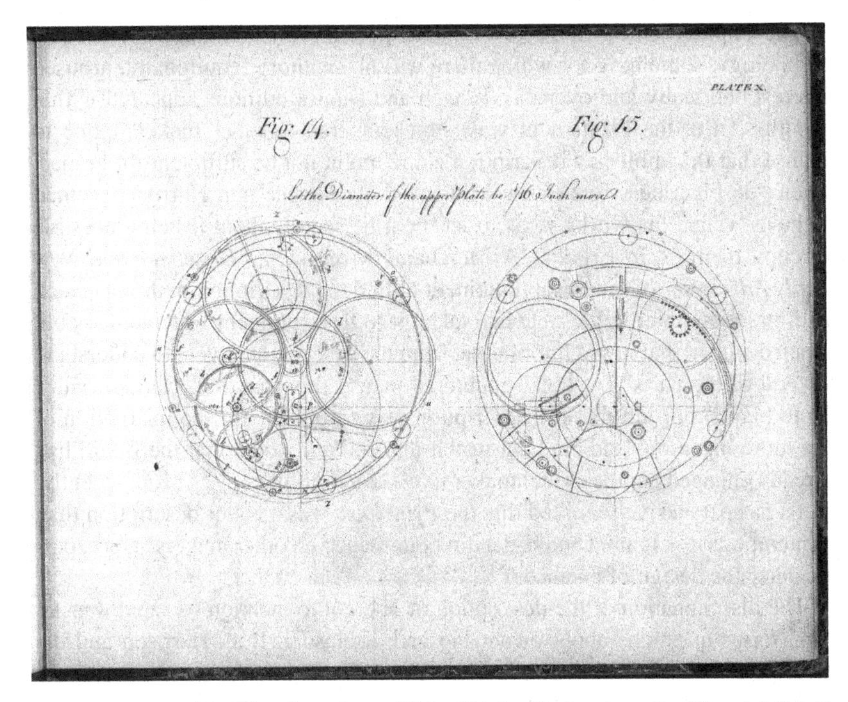

Figure 5.2 Printed versions of John Harrison's drawings for H4 as published by the Commissioners in *The Principles of Mr Harrison's Time-Keeper, with Plates of the Same* (London, 1767), plate X. Library of Congress, Rare Book and Special Collections Division

Only after further discussion and the mediation of his son did Harrison finally agree to the proposed process. Accordingly, the 'discovery' of H4 took place in August 1765 at Harrison's house and workshop in Red Lion Square, London. Maskelyne oversaw proceedings for the Commissioners and the nominated witnesses reported their satisfaction to the Board on 12 September.[76] Privately, however, three expressed concerns about the ease with which Harrison's work might be reproduced by others. A fourth, William Ludlam, made his concerns public.[77] Nevertheless, one result of the discovery was to leave the Commissioners with Harrison's drawings and written description of H4, which they published in 1767 (Figure 5.2).[78]

[76] BoL confirmed minutes, 12 September 1765, CUL RGO 14/5, pp. 108–109.
[77] Bennett, 'Science Lost'; Bennett, 'The Travels and Trials', p. 88; Quill, *John Harrison*, pp. 151–152, 154–155, 181.
[78] Harrison, *The Principles of Mr. Harrison's Time-Keeper.*

The unprecedented move to circulate publicly information about a new technology, something for which there was no statutory requirement, aroused interest nationally and overseas. French and Danish editions appeared within months.[79] But the excitement was soon tempered as other makers came to realise that the published description alone might not be sufficient. In France, Eveux de Fleurieu wrote to the Ministre de la Marine that Harrison seemed to have 'veiled his works so as to let them be seen without it being possible to copy them'.[80] In Britain, 'A Merchant' wrote in the *Gazetteer and New Daily Advertiser* of his 'disappointment to find the description both imperfect, and, to me, unintelligible'. He had taken it to three eminent watchmakers but 'each declared that it was impossible, from the description given, to understand several of the parts'.[81] 'W.C.' countered that, if this was true, Harrison must be to blame, but felt that the description was sufficient 'for the instruction of the more ingenious and eminent watch-makers' and noted that the Board had already engaged another watchmaker to make a faithful copy.[82] More recently, Betts has persuasively argued that the *Principles* was a better description than contemporaries claimed and had a direct influence on other makers' work, particularly the design of balances.[83]

The dissemination of the description of H4, not to mention its rapid translation, raises questions about ownership and disclosure. Both Harrison and the Board were clearly aware of overseas interest in Harrison's work, particularly French attempts to obtain his secrets.[84] This came to a head when it was learned that Thomas Mudge, a witness at H4's discovery, had passed details to French clockmaker Ferdinand Berthoud.[85] Mudge was called before the Board after this was revealed in the first version of *The Case of Mr. John Harrison*.[86] He admitted

[79] Harrison, *Principes de la montre de Mr. Harrison*; see Turner, 'Berthoud in England', pp. 228–234; Lous (ed.), *Historien af Mr. Harrisons Forsøg*. Lous's edition, preface dated 10 December 1767, includes a translation of *The Principles*, reproduction of the plates, and a narrative of Harrison's dealings drawn from English and French sources, notably Lalande's accounts in the *Connoissance des temps, The London Magazine or, Gentleman's Intelligencer,* 33 (1764), pp. 9–11, 316–317, and *The Gentleman's Magazine*, 35 (January 1765), pp. 34–35. Thanks to Ivan Tafteberg Jakobsen for translating and tracing Lous's sources. For a Spanish response, see Juan, 'Informe sobre el reloj de Harrison'; Barrado Navascués, *Cosmography in the Age*, pp. 284–287, translates the text.

[80] Charles-Pierre Eveux de Fleurieu to the Ministre de la Marine, 24 October 1767: 'il s'est enveloppé d'un voile qui laisse entrevoir les objects, sans qu'il soit permis de les copier', quoted in Mascart, *La vie et les travaux*, p. 321. See also Boistel, 'Esprit Pezenas', pp. 150–153; Turner, 'Berthoud in England', p. 233. Turner (p. 231) discusses errors in *The Principles* and the French edition of it.

[81] *Gazetteer and New Daily Advertiser*, 22 February 1768.

[82] *Gazetteer and New Daily Advertiser*, 26 February 1768.

[83] Betts, 'The Quest for Precision', pp. 329–331.

[84] Fauque, 'Testing Longitude Methods', p. 161; Quill, *John Harrison*, pp. 157–159; Turner, 'Berthoud in England', pp. 225–228; Turner, 'L'Angleterre, la France', pp. 155–159.

[85] Fauque, 'Testing Longitude Methods', p. 162.

[86] Anonymous, *The Case of Mr. John Harrison*, 1st ed., p. 4.

that he had indeed spoken to Berthoud and given details to ten or twelve workmen 'and to several Gentlemen curious in Mechanics besides', but considered that it was his 'duty to do it, and that it was the Intention of the Board I should do so'.[87] As with the publication, however, Berthoud considered that what he learned was insufficient for producing copies in France, while the publication of the description rendered any issues of secrecy or protection redundant.[88]

Testing Times

Despite Harrison's objections that he needed H4 to make the copies required to claim the remaining reward, the Board exercised its right to take possession of the timekeeper, a condition of the terms set out in 1765. They formally removed it from Harrison on 28 October, sealed it and delivered it to the Admiralty until such time as it would be required.[89] Coming to a decision on this took until April 1766, when the Commissioners made some important, far-reaching decisions. First, they responded to a letter from Harrison, reiterating that he would receive the rest of the reward only on completion and testing of at least two further watches. Although the minutes do not record it, Harrison's letter also requested that he either be able to draw down £800 to enable him to make two more watches or claim the remaining £10,000 in order to rent or build a house as a workshop, where he would train workmen and apprentices to make marine timekeepers.[90] The Board refused.

Second, the Board resolved that H4 be sent to the Royal Observatory for further trials. Third, the three clocks still in Harrison's possession (H1, H2 and H3) would now be moved from his workshop to the Royal Observatory. Fourth, they would print and sell 500 copies of the plates and description of H4. Finally, they made an agreement with Larcum Kendall, a witness at H4's discovery, to make a copy for a fee of £450.[91] Ten days later H4 was moved by river to the Royal Observatory under Kendall's care.

The new trial was organised in part to resolve the Commissioners' concern that H4 had succeeded in the Barbados trial by chance not design. Testing was

[87] BoL confirmed minutes, 14 March 1767, CUL RGO 14/5, p. 146; *Journals of the House of Commons*, 30, pp. 270–271, reports Mudge's evidence on 2 April 1767 about the replicability of H4 and willingness to share information with makers.

[88] Fauque, 'Testing Longitude Methods', p. 162.

[89] BoL confirmed minutes, 28 October 1765, CUL RGO 14/5, pp. 116–118. The Board issued Harrison with a certificate so that he might receive the first part of the reward.

[90] BoL confirmed minutes, 26 April 1766, CUL RGO 14/5, pp. 121–123; 'Harrison Journal', pp. 181–186.

[91] BoL confirmed minutes, 26 April 1766, CUL RGO 14/5, pp. 121–124. Further discussion took place the following month; BoL confirmed minutes, 24 May 1766, CUL RGO 14/5, pp. 134–138; 'Harrison Journal', pp. 186–199. 'Harrison Journal', pp. 201–204, describes the confrontation that arose when Maskelyne arrived on 23 May 1766 to collect the clocks, when one was damaged.

to involve daily checks of its going compared to the Observatory's main regulator, each check undertaken by Maskelyne or one of his assistants before a witness from Greenwich Hospital.[92] Beginning 6 May 1766, the tests lasted ten months, with the Commissioners resolving to publish 500 copies of the results when they met the following March.[93]

The publication proved to be one of the most inflammatory moments in the relationship with the Harrisons. As presented by Maskelyne, the results indicated that H4 had performed poorly, displaying an apparently erratic rate. Maskelyne concluded that the watch could not be depended upon to keep the longitude within a degree over the course of a voyage to the West Indies, nor to keep it within half a degree for more than a fortnight. It was, however, 'a useful and valuable invention', which, in conjunction with lunars, 'may be of considerable advantage to navigation'. He concluded that, while the watch may have met with success on a single voyage, it could not be relied upon to do so repeatedly.[94]

Harrison's response was swift and systematic. He attacked the trial's protocols and preparation, Maskelyne's integrity in its conduct, his care of the timekeepers and claims about the ease and accuracy of the lunar-distance method, in which the Astronomer Royal was 'deeply interested'. Harrison further suggested that Maskelyne's results in fact showed that, for any six-week period, save those in extreme cold or where the watch had not been kept horizontal, the results were within the limits of the Act of 1714.[95] Harrison refused to concede ground to the Board's concerns, although Bennett has noted that around the same time a short piece about the principles of H4's construction appeared in the *Gentleman's Magazine*. It concluded that 'such a time keeper goes entirely from principle, and not from chance', perhaps a response to some of the Commissioners' comments.[96]

Replication and Resolution

For the next six years of debate with the Harrisons, the Board effectively pursued parallel tracks with respect to timekeepers. They dealt with Harrison, generally through William, by sticking to the terms of the Act of 1765 and the discussions surrounding it, insisting that Harrison make two further watches

[92] The regulator, by George Graham, defined Greenwich Mean Time from 1750 and survives at the Royal Observatory, NMM ZBA0709.

[93] BoL confirmed minutes, 21 March 1767, CUL RGO 14/5, p. 152; 'Harrison Journal', p. 53.

[94] Maskelyne, *An Account of the Going*, p. 24; Bennett, 'The Travels and Trials', pp. 89–90; Quill, *John Harrison*, pp. 170–176.

[95] Harrison, *Remarks on a Pamphlet*. Quill, *John Harrison*, p. 173, suggests that Harrison had help writing it. Mudge junior, *A Reply to the Answer*, p. 99, suggests Short wrote it.

[96] Anonymous, 'The Principles of Mr. Harrison's Time-Keeper'; Bennett, 'The Travels and Trials', p. 91.

to be successfully tested before he received further rewards. At the same time, they looked to the replication of H4 by other makers, initially Larcum Kendall (see Chapter 6).[97]

Shortly after the publication of Maskelyne's trial of H4 at the Royal Observatory, Harrison requested again that he be given H4 to help him make the required copies, at the same time putting forward suggestions for trials of the new timekeepers. The Board was no longer minded to accede to Harrison's requests, replying that Kendall needed the watch and that testing Harrison's copies must comprise a ten-month trial at the Royal Observatory and two months at sea.[98] By 1770, the Board began to doubt that Harrison was going to produce the additional watches, and moved in March to have a time limit imposed by adding a clause to a proposed parliamentary Act intended to authorise further rewards for improving lunar tables and other general improvements to navigation.[99] Although the Act was passed, it did not in the end contain the clause concerning Harrison.[100]

Two months later, however, the Board became aware that Harrison had completed a copy of H4 and so suggested that he be given notice of the next meeting in order to show it.[101] William Harrison attended on 17 November and informed them that, while his father had completed the new timekeeper, it needed adjusting.[102] It was another year before he returned, when the Board suggested that the new watch ('H5'), and Kendall's copy of H4 ('K1'), be taken on the Pacific voyage being planned under the command of James Cook.[103] William declined, however, 'saying that the Trial would take up too much time'.[104] A month later, the Board heard a letter from John Harrison, complaining of the hardship he had suffered at their hands and offering various suggestions: that H4 go on the forthcoming voyage instead of K1 and that releasing the remaining reward be based on the performances of H5 and K1, for which he also outlined a new mode of trial. Again, the Board declined, telling William that neither did it 'think fit that the Watch Machine or Timekeeper made by his Father & now in their possession shall be sent out of the Kingdom', nor was it willing to alter their agreement regarding the remaining reward.[105]

Towards the end of 1772 the Commissioners faced a further challenge from the Harrisons, who informed them that H5 had undergone successful testing

[97] Jonathan Betts, 'Kendall, Larcum (1719–1790)'.
[98] BoL, confirmed minutes, 11 April 1767, CUL RGO 14/5, p. 155.
[99] BoL, confirmed minutes, 3 March 1770, CUL RGO 14/5, p. 190.
[100] Discovery of Longitude at Sea Act, 1770, 10 Geo 3 c 34; copy at CUL RGO 14/1, pp. 34–36.
[101] BoL, confirmed minutes, 26 May 1770, CUL RGO 14/5, p. 195.
[102] BoL, confirmed minutes, 17 November 1770, CUL RGO 14/5, p. 199.
[103] John Harrison, one-day marine timekeeper (H5), Worshipful Company of Clockmakers, inv. 598; Betts, *Marine Chronometers at Greenwich*, p. 35.
[104] BoL, confirmed minutes, 28 November 1771, CUL RGO 14/5, pp. 208–209.
[105] BoL, confirmed minutes, 14 December 1771, CUL RGO 14/5, p. 213.

(the results being sent with the letter) at the king's observatory, witnessed by the king and his private astronomer, Stephen Demainbray. Harrison now requested a certificate for the remainder of the reward. The Board was unwavering, telling William Harrison that it 'did not think fit to make any Alteration in the mode which they had already fix't upon for the Trial of his Father's Timekeepers And that no regard will be shewn to the results of any Trial made of them in any other way'.[106]

By this time, the Harrisons had begun moves to circumvent a Board no longer willing to negotiate by petitioning Parliament. William wrote in August 1772 to the prime minister, Lord North, and in December to John Robinson, Secretary of the Treasury. Gaining little response, the Harrisons had a new version of *The Case of Mr. John Harrison* printed in March 1773 and circulated to all Members of Parliament alongside a formal petition, which was considered at the beginning of April. Meeting three weeks later, the Board was informed of the petition by Fletcher Norton, Speaker of the House of Commons, and given the opportunity to 'revise their proceedings in relation to the s[ai]d Mr. Harrison'. Two MPs interested in the case, Mr Gray (presumably Charles Gray, MP for Colchester) and Barlow Trecothick (an MP for the City of London), also attended, as did the Harrisons and William Godschall, Fellow of the Royal Society and lawyer. The meeting went over Harrison's case, with the Board resolving to publish their proceedings to that date. William Harrison was questioned in front of the two MPs and Godschall, but his responses indicated that the Harrisons were equally unwilling to compromise.[107]

Parliamentary consideration began three days later, with copies of the Board's minutes and proceedings requested and read on 4 May. Presumably on the advice of his supporters, this led Harrison to submit a revised petition, which dwelt less on the details of having met specific conditions imposed by previous Acts and Board stipulations, emphasising instead that he had 'devoted with the closest Attention his whole Life' to the project and had succeeded. It also acknowledged that he had received 'from the Public Bounty, several Sums of Money for his Support, under the very great Expence incurred in these repeated Researches'.[108] Lord North presented the new petition to the House of Commons on 6 May, making it clear that it had the king's full support. In consequence, a bill passed through Parliament and received royal assent on 1 July 1773.[109] The Act granted

[106] BoL, confirmed minutes, 28 November 1772, CUL RGO 14/5, p. 231. Quill, *John Harrison*, pp. 187–196, discusses H5 and its testing at the king's observatory; trial results at King's College London, K/MUS 1/6.

[107] BoL, confirmed minutes, 24 April 1773, CUL RGO 14/5, pp. 239–240; Quill, *John Harrison*, pp. 197–200.

[108] *Journals of the House of Commons*, 34 (1773), pp. 244, 302.

[109] *Journals of the House of Commons*, 34 (1773), pp. 244, 285–286, 293, 298, 302, 367, 369, 383, 385, 387–389.

Harrison a further £8750 for having 'applied himself, with unremitting Industry for the Spacc of Forty-eight Years, to the making an Instrument for ascertaining the Longitude at Sea'.[110] This ended the affair as far as the Board was concerned (though not to the Harrisons' satisfaction), and at its next meeting it agreed not to publish its proceedings as the matter was now resolved.[111]

The end of formal dealings with Harrison marked a turning point in another way, since it might have been considered a natural point at which to dissolve the Board on the grounds that its founding purpose had been fulfilled. Yet the Commissioners were now engaged in activities including publishing the *Nautical Almanac* and other titles, maintaining dealings with watchmakers including Kendall, and working with instrument makers such as John Bird, not to mention receiving myriad proposals, all of which entailed ongoing financial commitments. It seemed appropriate for the Board to continue, therefore, and in 1774 a new Longitude Act ensured this through a revised set of rewards and conditions (see Chapters 6 and 8).[112]

Nevil Maskelyne: Longitude as an Astronomical Concern

Longitude at sea had long been considered first and foremost an astronomical problem. It remained so long after John Harrison's advances in horology, since it was still essential to combine astronomical observations, instrumentation and publications. The timekeeper method needed astronomical observations and tables to find local time. Lunar-distance determinations required accurate timekeepers. All methods continued to rely to some extent on dead reckoning. There were no binary choices in navigation. Even with the rise of new methods, such as line-of-position navigation in the nineteenth century, this essentially remained the case until the spread of satellite navigation in the late twentieth century.

Longitude being at heart an astronomical problem made the Astronomer Royal one of the premier longitude experts in England from the foundation of the position in 1675. Successive Astronomers Royal were regularly consulted by individuals and governmental, commercial and intellectual institutions about issues and ideas related to longitude and navigation generally.[113] Sometimes they examined related inventions and proposals or attended

[110] Supply, etc. Act 1772, 13 Geo 3 c 77. The sum granted was £1250 less than the Harrisons considered they were due.
[111] BoL, confirmed minutes, 3 July 1773, CUL RGO 14/5, pp. 241–242.
[112] Discovery of Longitude at Sea Act 1774, 14 Geo 3 c 66; copy at CUL RGO 14/1, pp. 37–42.
[113] On Flamsteed: Plumley, 'The Royal Mathematical School'; Iliffe, 'Mathematical Characters'. On Halley: Cook, *Edmond Halley*, pp. 173, 217–219, 256–291, 395–398; Perkins, 'Edmond Halley'.

shipboard trials.[114] They also performed astronomical observations, with varying degrees of regularity and openness, which might improve techniques and predictive tables.[115]

When appointed in 1765, Maskelyne followed his predecessors in assuming the role of national and international authority on longitude. He differed, however, in bringing a more bureaucratic approach, which included a more regular system for responding to unsolicited longitude communications. He also appears to have been the first Astronomer Royal interested in reshaping and strengthening the work of the Commissioners of Longitude, doing more than any other Commissioner to forge the standing board that would serve as a national and international authority in navigation, exploration, science and technology.[116]

Maskelyne initiated, organised and oversaw the Board of Longitude's activities to a remarkable degree during his 46-year tenure. This extended from the bureaucratic minutiae of its operations to the choice, planning and publication of its best-known international endeavours. Proposals and related correspondence received by the Board were most often passed to him for his opinion as to whether they merited further attention. In dealing with these, Maskelyne was often more neutral and polite in his comments than other eighteenth-century longitude experts, such as James Hodgson at the Royal Society and earlier Astronomers Royal. In 1783, William Fuller of London sent a handbill that described how he had published in the newspapers his method for finding longitude by the fixed stars and had taken copies to the Houses of Parliament, the Admiralty and coffeehouses. Although he was rather presumptuous in expecting this to induce the Commissioners to contact him, Maskelyne did meet with him. As was common, he noted succinctly on the back of the relevant documents that he was 'of opinion that they do not merit the attention of the Board of Longitude'.[117]

The surviving archive reveals in addition the many notes and drafts Maskelyne wrote for Board minutes and other documents, including background for its legal and legislative debates, correspondence and published volumes. Maskelyne was central to the Board's publishing activities. Just a year

[114] Halley was an acknowledged longitude expert before becoming Astronomer Royal. As noted in Chapter 3, in 1717 he was an expert witness about a scheme being considered by the Lord Mayor and Court of Aldermen of the City of London; see also Bennett, 'Catadioptrics and Commerce', p. 252. After becoming Astronomer Royal, Halley and Bradley took part in shipboard trials of John Hadley's new observing instrument; Hadley, 'An Account of Observations'.

[115] Cook, *Edmond Halley*, pp. 354–376 (Halley and the Moon); Forbes, *Greenwich Observatory*, pp. 80–118.

[116] The institutionalisation of the Board and expansion of its powers were fuelled and shaped by Maskelyne's personal and institutional ties; see Baker, '"Humble Servants'.

[117] William Fuller to Commissioners of Longitude, 6 March 1783, CUL RGO 14/36, p. 130v.

after his appointment, he succeeded in initiating the production of the *Nautical Almanac*, published annually from 1767 (see Chapter 7). Ultimately, the Board spent as much money producing the *Nautical Almanac* and related works as it did investigating timekeepers. Maskelyne also oversaw the production of other publications for the Board, including those related to H4 and the instrument-making methods of John Bird and Jesse Ramsden.

His bureaucratic reforms reached beyond the Board of Longitude to the reorganisation of the Royal Observatory. Previous astronomers had often failed to make the results of their observations public and sometimes let the buildings and instruments fall into disrepair. Maskelyne insisted on publication as an important part of his role, something admired by Jean Baptiste Joseph Delambre, director of the Observatoire de Paris, who wrote that 'for this he deserved to be the leader and regulator of astronomers for forty years'.[118] It was Maskelyne who encouraged the Board of Longitude to pursue ownership of Flamsteed's and Halley's observations, as well as the publication of Bradley's.[119] Increasing numbers of assistants from the Royal Observatory also went on to participate in geodetic and astronomical expeditions in their own right due to his teaching and influence.

Maskelyne's activities and interests with the Commissioners, at the Royal Observatory and with institutions such as the Royal Society constantly intermingled. This shaped and strengthened his contributions to the Board of Longitude, which increasingly collaborated with the Admiralty and Navy, Royal Observatory, Royal Society and corresponding foreign bodies. In addition to supporting work relating to timekeepers, fostering the improvement of the lunar-distance method and considering and encouraging other proposals, the Board expanded into the improvement of general navigation, science and technology. Many of these activities were in partnership with the Royal Society, of which Maskelyne was an active Fellow. He attended meetings when he could, published more than fifty articles in its *Philosophical Transactions* and was awarded the Society's prestigious Copley Medal in 1775. He also developed detailed instructions and standardised sets of equipment for the use of the observers on expeditions it supported.[120]

Maskelyne and the Board's efforts to improve observing instruments and astronomical tables, in support of both lunar-distance and timekeeper methods, had important consequences. These included the improvement and dissemination of astronomical data and the further development of observing instruments such as the sextant, with which Maskelyne was involved both before

[118] Delambre, 'Notice sur la vie et les travaux de M. Maskelyne', p. lxx: 'c'est par là qu'il a mérité d'être pendant quarante ans le chef et comme le régulateur des astronomes'.

[119] Higgitt, 'Greenwich near London', p. 303; Howse, *Nevil Maskelyne*, pp. 106–108, 121–122, 180–182.

[120] Higgitt, 'Equipping Expeditionary Astronomers'.

Figure 5.3 Portrait of John Bird, engraved and published by Valentine Green, 1776. A beam compass lies on top of Bird's *Method of Constructing Mural Quadrants*, published in 1768 by the Commissioners, and a drawing of a mural quadrant from the same work. Wellcome Collection

and after becoming Astronomer Royal.[121] The Board's activities also underpinned the improvement and propagation of the precision dividing of instrument scales, as pioneered by hand by John Bird (Figure 5.3) and mechanically by Jesse Ramsden.[122]

As well as being a trusted supplier to the Royal Observatory and observatories elsewhere, John Bird made, repaired and assessed instruments for the Commissioners during the 1750s and 1760s and was central to the creation of the marine sextant.[123] He was also one of the expert witnesses at the 'discovery' of H4.[124] Keen to capture and pass on his expertise, the Board agreed to publish two treatises detailing his methods.[125] This arose from Bird's

[121] Mörzer Bruyns, *Sextants at Greenwich*, p. 38. [122] Ibid., pp. 41–43.

[123] McConnell, 'Bird, John'; Chapman, *Dividing the Circle*, pp. 71–76; Dunn, 'A Bird in the Hand'.

[124] BoL, confirmed minutes, 4 August 1763 CUL RGO 14/5, p. 50.

[125] Bird, *The Method of Dividing Astronomical Instruments*; Bird, *The Method of Constructing Mural Quadrants*.

suggestion that they pay him £500 to build an improved workshop, in return for which he would explain his methods and teach them to an apprentice and one other.[126] The Commissioners agreed as long as he also taught mathematician William Ludlam and natural philosopher John Michell and produced a detailed explanation with illustrations and, if necessary, models. Their attitude towards Bird was notably different from that towards other artisans, an indication of the personal trust they had in him. Ironically, at the same meeting in 1766, the Board declined Harrison's suggestion of releasing funds so that he could build a workshop and train apprentices. Likewise, at a meeting on 21 March 1767, the Commissioners refused to reward the instrument maker Jeremiah Sisson, who argued that he possessed the same skills as Bird, which both had learned from Sisson's father, Jonathan.[127]

Bird, it seems, never built a new workshop and it remains uncertain whether he took on the apprentice. On 24 May 1766, the Board approved an unnamed boy whom Bird said he intended to bind to him, but nothing further has been recorded.[128] Bird also produced a model of the mural quadrant at Greenwich with an accompanying text and engraved illustration plates.[129] These became the basis of two books, *The Method of Dividing Astronomical Instruments* (1767) and *The Method of Constructing Mural Quadrants* (1768), which attempted to translate Bird's manual skills as an instrument maker into words.[130] Both suffered delays due to business demands and Bird's ill health.

A few years later, the Board became interested in mechanical methods of instrument production through the work of Jesse Ramsden, who had devised his first circular dividing engine in 1768.[131] Ramsden approached the Board in 1774, having completed a second, improved engine. He described it as an 'Instrument for dividing Sextants &c, the construction of which is as leaves no dependance upon the Workman, as a Boy can use it with the same exactness as the most experienced hand; and in the most speedy manner'.[132] He answered the Commissioners' questions and invited the professors and John Bird to visit his workshop. Recognising the potential, the Board rewarded Ramsden and encouraged other makers to build their own engines or use Ramsden's

[126] BoL, confirmed minutes, 26 April 1766, CUL RGO 14/5, p. 128–130.

[127] BoL, confirmed minutes, 21 March 1767, CUL RGO 14/5, p. 150; Bird, *The Method of Dividing Astronomical Instruments*, p. v, notes he learned from Jonathan Sisson.

[128] BoL, confirmed minutes, 24 May 1766, CUL RGO 14/5, p. 135.

[129] The mural quadrant by Bird, made in 1750, is at the Royal Observatory, Greenwich (NMM AST0971). The fate of the model remains unknown.

[130] Dunn, 'A Bird in the Hand', pp. 80–81.

[131] Jesse Ramsden, dividing engine, 1767, Musée des arts et métiers, Paris, 00100; McConnell, *Jesse Ramsden*, pp. 41–42.

[132] BoL, confirmed minutes, 25 June 1774, CUL RGO 14/5, p. 262; Jesse Ramsden, dividing engine, 1775, National Museum of American History, Washington DC, MA*215518; McConnell, *Jesse Ramsden*, pp. 43–48.

(for a fee) to divide their scales. He was later also rewarded for a straight-line dividing engine and for publishing descriptions of both engines.[133] Like Bird, Ramsden would remain a trusted supplier and authority for the Board and other institutions, including the Royal Observatory, the Royal Society and later the Ordnance Survey.[134]

The Question of Interest

Even before the disputes of the 1760s, when the Harrisons and their supporters made repeated accusations against Maskelyne and other Commissioners of Longitude, the central importance of the Astronomer Royal in the search for the longitude had raised concerns about conflicts of interest. This was particularly true when the Astronomer Royal appeared to be working on a different method. This was perhaps inevitable, since the royal warrant creating the role had directed the appointee to the improvement of methods for determining longitude at sea through astronomical observation.[135] Jane Squire and other reward-seekers criticised Edmond Halley, second Astronomer Royal, for pursuing a lunar method while acting as the ultimate judge of their proposals. Squire complained in 1731 to Viscount Torrington (George Byng), *ex officio* Commissioner as First Lord of the Admiralty, of a rumour that the astronomer would be given first chance at a reward due to his years of effort:

> it would be the highest Injustice to all … to determine them to be sacrificed, to the Honour and Interest of one; who (by having, himself undertook to make this Discovery) is certainly disqualified to be the sole Judge of the Methods by which others propose it …
>
> I know we boast the utmost Perfection in the first [astronomy], and have even idoliz'd a Name we think we owe it to; but it still remains a strong Objection against our Astronomy, that it yet enables us not to find our Longitude.[136]

The Harrisons were similarly critical of the fourth Astronomer Royal, Nathaniel Bliss. After a Board meeting in 1764, William wrote to his father-in-law that they were once again optimistic about obtaining the longitude reward, since Bliss, one of those with a particular interest in the lunar-distance method, had recently died. 'They were all as agreeable as could be, Parsons [i.e. professors] and all, as they have now lost their ringleader.'[137]

[133] Ramsden, *Description of an Engine for Dividing Mathematical Instruments*; Ramsden, *Description of an Engine for Dividing Strait Lines*; McConnell, *Jesse Ramsden*, pp. 43–51; Howse, 'Britain's Board of Longitude', p. 411; Dunn and Higgitt, *Finding Longitude*, pp. 172–175.

[134] McConnell, *Jesse Ramsden*, pp. 58–59, 105–107, 146–149, 191–218.

[135] Howse, *Greenwich Time*, p. 42.

[136] Jane Squire to Lord Torrington, 24 December 1731, in Squire, *A Proposal for Discovering*, 2nd ed., pp. 21–22.

[137] William Harrison to Robert Atkinson, 20 September 1764, quoted in Quill, *John Harrison*, p. 136.

They made similar accusations against Maskelyne. Initially, they seem not to have objected to his selection as an official observer for the second sea trial of H4, despite his known advocacy of the lunar-distance method.[138] However, William recorded in a manuscript produced some years later that, as soon as he reached Barbados,

he was told that M^r^. Maskelyne was a Candidate for the Premium for discovering the Longitude and therefore they thought it was very odd, that he should be sent to make the Observations to Judge another Scheme M^r^. Maskelyne having declared in a very Public manner that he had found the Longitude himself.[139]

The resulting altercation led to Harrison insisting that Maskelyne alternate observations with Charles Green. Under the scrutiny of several witnesses, Harrison added, Maskelyne 'was so confused … that his Observations were scarce to be depended upon for everyone that were present could see that he set some of his Observations down dubious'.[140] Maskelyne's account of these events is not known but, as Quill has written, William Harrison's objections 'reveal the spirit of suspicion and antagonism against Maskelyne that seems to have been continually in his mind, an attitude which was shared by his father, and which was to persist to the end'.[141]

Maskelyne seems, however, to have been aware of his 'delicate' position during the Barbados trials, writing of the 'weight as well the honor' he felt in passing judgement on Irwin's marine chair and his wariness at 'deceiving & injuring the public' if the correct opinion was not given. In particular, he suspected that Irwin or his supporters might turn blame onto the observer (Maskelyne) rather than the invention if the judgement was not in their favour. He proved to be right: Irwin did indeed complain about Maskelyne's conduct.[142]

Such episodes may explain why the following resolution appears to have been considered for inclusion in the Longitude Act of 1765:

That no person being a Commissioner of Longitude shall be entitled to any reward, paid by order or Certificate of the said Commissioners, on account of any discovery such person shall make tending to facilitate the method of finding the Longitude at Sea, or on account of any discovery tending in other respects to the improvement of Navigation.[143]

That it was not included in the Act suggests that it stated something long understood, and there is no evidence that Maskelyne (or any other Commissioner) intended to pursue a longitude reward on his own behalf either

[138] Bennett, 'Mathematicians on Board'; Howse, *Nevil Maskelyne*, pp. 27–39, 42–43; Maskelyne, *The British Mariner's Guide*.

[139] 'Harrison Journal', p. 112. [140] Ibid., pp. 113–114. [141] Quill, *John Harrison*, p. 133.

[142] Nevil Maskelyne, draft letter to BoL, n.d., RGO 4/320, fol. 7:2r; BoL, confirmed minutes, 18 September 1764, CUL RGO 14/5, pp. 67–68.

[143] BoL, draft resolution, n.d. [1765], NMM BGN/10; Barrett, '"Explaining" Themselves', pp. 152, 158.

before or after his appointment as Astronomer Royal.[144] His lunar-distance trials in 1761 and 1763–1764 were, after all, of Mayer's tables, not of his own work. The Commissioners understood this, as Robert Waddington, who assisted Maskelyne on the St Helena voyage in 1761, discovered when he applied for some reward following their successes with lunars and was told that, 'I Compleat only what M. Mayer has done towards it'.[145]

The Harrisons did not share this opinion and targeted Maskelyne in their published complaints about their mistreatment at the Commissioners' hands. Maskelyne was clearly stung at Harrison's criticisms in *Remarks on a Pamphlet Lately Published by the Rev. Mr. Maskelyne*. In December 1767, he requested 'attested Copies or Extracts from the Minutes of this Board for his justification from Reflections cast upon him by Mr Harrison; and that he have leave to publish the same'. The Board gave its backing and later agreed to pay for the publication of what would be a defence against Harrison's 'injurious reflections'.[146] In the end it never appeared, but was still on his mind in 1773, when he provided Lord Sandwich, then First Lord of the Admiralty, with an abstract and explanation of the Act of 1714. He had intended it, he wrote,

> as an introduction to an answer to Mr Harrison's scandalous pamphlet on my trial of his watch at the Royal Observatory; tho' I afterwards dropt the design of completing & publishing the same, thinking such abuse thrown out without probability or proof required no refutation.[147]

While Harrison and his supporters supplied the bulk of the printed discussions of the timekeeper trials and ensuing disputes, some coverage in periodicals and other publications did support Maskelyne and the Commissioners. Although the prolific publisher of natural philosophical and mathematical pamphlets Robert Heath eventually became dismissive of the whole longitude enterprise, he was for some time a critic of Harrison and defended Maskelyne's conduct in the Greenwich trial of H4. Heath wrote in 1768 that 'our Astronomer Royal, has infallibly proved, by a *Series* of correct Observations … that it [H4] can, *by no Means*, answer the *End proposed*'.[148] By 1770, he was less well disposed towards the Astronomer Royal and the 'Connoisseurs of Longitude', comparing Maskelyne to a medical projector whose 'Variety of intricate Rules

[144] Maskelyne, *An Answer to a Pamphlet*, p. 125: 'I have never shewn any inclination to become a candidate for any of these rewards.'
[145] Robert Waddington to Nathaniel Pigott, 7 January 1762, quoted in Bennett, 'The Rev. Mr. Nevil Maskelyne', p. 69.
[146] BoL, confirmed minutes, 12 December 1767, CUL RGO 14/5, pp. 165–166; BoL, confirmed minutes, 18 June 1768, CUL RGO 14/5, p. 169.
[147] Nevil Maskelyne to Lord Sandwich, 22 April 1773, NMM SAN/F/4/22.
[148] Heath, *A Supplement to the Royal Astronomer*, p. 1.

and Methods used in teaching the Use of the *Nautical Ephemeris* ... may be compared to the *Nostrums* of the *Quack-Doctors*'.[149]

Maskelyne might certainly have given Harrison more leeway in his interpretation of the results of the observatory trail of H4, but this does not appear to have been due to ill intent or prejudice, rather to his strong sense of duty.[150] Nor have historians found evidence that he had a deeply seated antagonism, citing among other things the fact that Maskelyne proposed William Harrison for election as a Fellow of the Royal Society in 1765, despite growing tensions between them.[151]

Maskelyne and the majority of his fellow Commissioners seem to have truly believed that Harrison had not yet fulfilled the requirements of the Longitude Act(s) because he had not proved that his invention could be reproduced for widespread use rather than being a brilliant one-off. The Harrisons just as surely believed that they did not need further proof and should have been given the largest longitude reward years earlier under the terms of the Act of 1714. John's petition to Parliament for recompense in 1773 effectively bypassed the Longitude Acts and allowed a final (though slightly diminished, as far as the Harrisons were concerned) sum to be awarded in light of his brilliant innovations, years of dedicated work and advanced age.

Many, Maskelyne included, thought that it might be most effective to employ lunars and clocks in tandem if reliable timekeepers could be produced; the methods were to be seen as complementary rather than in competition. Indeed, Maskelyne was instrumental in ensuring that the Board's timekeepers were put to trial on long-distance voyages in the 1770s. Looking back many years later, Maskelyne insisted that:

> He always allowed Mr. Harrison's great merit, as a genius of the first rate, who had discovered, of himself, the causes of the irregularities of watches, and pointed out the means of correcting these errors in a great degree, in the execution of a portable timekeeper, of a moderate size, to be put on board of ship, not liable to disturbance from the motions of the ship, and exact enough to keep time within two minutes in six weeks. He made no opposit[ion] to Parliament granting him the remainder of the reward of £20,000; but only to the Board of Longitude doing it; as he had not submit[ted] to trials [i.e. as dictated by the related legislation], and those sufficient to enable the Board to give it to him according to the terms of the Act.[152]

Institutional changes in the Commissioners' routine and management of decisions exacerbated the conflict between them and the Harrisons. From the 1730s to 1750s, the clockmaker was repeatedly able to secure awards of £500 with relative

[149] Heath, *The Seaman's Guide*, p. 42; see Barrett, *Looking for Longitude*, pp. 163–166.

[150] Howse, *Nevil Maskelyne*, p. 84; Quill, *John Harrison*, pp. 166–176. Forbes, 'Tobias Mayer (1723–62)', p. 19, reaches a similar conclusion, arguing the Board approached Mayer's work just as critically as it did Harrison's.

[151] Howse, *Nevil Maskelyne*, p. 77; Quill, *John Harrison*, pp. 142–143.

[152] Nevil Maskelyne, résumé of his career, n.d. [c. 1800], CUL RGO 4/320, p. 10:2r–10:2v.

ease. The Commissioners were often called together essentially to approve these grants, at a time when they were still independent agents who rarely held communal meetings. From the 1760s, however, the Commissioners – most of them new to working with the Harrisons due to the passage of time – rapidly bureaucratised. Their actions and decision-making were seldom as immediate as Harrison wanted, despite the greater frequency of meetings. The Board increasingly sought to define and apply a legal and bureaucratic framework to its operations, as well as to examine and interpret the sometimes vaguely worded directives of past legislation as the clockmaker's success seemed more imminent. The Commissioners increasingly heard other proposals and pursued an array of improvements to navigation, astronomy and technology. This bureaucracy and expanded vision no doubt seemed vexing to the Harrisons, both in comparison to the earlier decades of collaboration and due to John's advancing age.

In addition, the Commissioners and the Harrisons held different views on intellectual property in a period when such matters were the subject of ongoing debate.[153] In general, confusion existed in Georgian Britain about 'state' and 'private' ownership and rights to work that received state sponsorship. For example, the Commissioners and the Royal Society engaged in a long social and legal battle with James Bradley's heirs, and later with the Oxford University Press, over the ownership and printing of Bradley's astronomical observations at Greenwich.[154] The Commissioners believed the state owned the papers, since the astronomer was a state employee who had made observations at the Royal Observatory, which was funded (albeit poorly) by the Board of Ordnance. Bradley's relatives believed he and thus they owned the papers and at least deserved a sizeable reward if they turned them over to the Royal Society. Similarly, the Commissioners of Longitude believed the public owned Harrison's timekeepers in return for the funding and rewards they had given him over the years, a point enshrined in various Longitude Acts. The clockmaker believed them his property – as most early modern craftsmen and inventors probably would have – until he received the full reward he assumed he was due.

Conclusion

The Board of Longitude's increasing bureaucratisation and expansion of powers, reflected in new Acts of the 1760s and early 1770s, took place as a result of a combination of factors. In part, it reflected the emergence of what has been

[153] Johns, *Piracy*, pp. 109–143.

[154] BoL, confirmed minutes, 13 June 1765, 26 April 1766, 10 January 1767, 12 December 1767, 18 June 1768, 12 November 1768, 13 January 1770, 3 March 1770, 26 May 1770, 25 January 1772, 13 June 1772 and 28 November 1772, CUL RGO 14/5, pp. 102, 119–120, 141, 166–167, 173, 185–186, 189, 193, 216, 229–231; Homes, 'Friend and Foe', pp. 254–255; Reeves, 'Maskelyne the Manager', pp. 106–108.

called a proto-civil service and concurrent demands for the Admiralty to be more efficient, cost-effective and accountable. The Board's efforts were also fuelled and shaped by a small number of proactive Commissioners, particularly Nevil Maskelyne. Finally, these changes were fostered by ongoing interest in a broad range of longitude schemes. These schemes included Harrison's marine timekeepers, which motivated the first sporadic communal meetings from the 1730s and required greater communal organisation and decision-making from the early 1760s. They also included the lunar-distance method, which was pursued in practice from the 1750s with the development of the astronomical tables and technologies needed. In addition, there remained a constant flow of other proposals for the improvement of navigation, astronomy and technology, some of which attracted official attention and investment.

6 Time Trials

The Board of Longitude and the Watchmaking Trade, 1770–1821

Richard Dunn

The sea trials of the 1760s showed that mechanical timekeeping and astronomy could in principle provide methods for determining longitude at sea, yet it would be decades before such methods became the norm for seafarers. Rather than ending the activities of the Commissioners of Longitude, the trials left them with the responsibility of assisting, and in part underwriting, moves to adopt the new methods at sea. Chapter 7 shows how this played out with respect to the *Nautical Almanac*. This chapter investigates the Board of Longitude's relationships with the clock- and watchmaking trade through its role in assessing new ideas relating to timekeepers, trialling them on land and at sea, and encouraging makers' attempts to ensure that they became affordable while remaining reliable and accurate. The narrative covers the period from just before the resolution of the Board's fractious negotiations with John Harrison to the instigation of new timekeeper trials in 1821, when responsibility for the management of all naval chronometers passed to the Astronomer Royal at Greenwich.

The period was significant for both the Board and the development of the marine chronometer. It was a period for much of which Nevil Maskelyne and Joseph Banks exercised considerable control over the Board's affairs, including its work relating to timekeepers. Maskelyne was frequently responsible for judging proposals and, as Astronomer Royal, for trials at the Royal Observatory. Yet the conduct of those trials continued to attract criticism, as they had with Harrison. Banks exercised control as a presiding figurehead, spearheading the defence of the Board in the public disputes that engulfed it from the 1790s. This was also the period during which the marine chronometer gained its name and characteristic form. As Betts notes, it had 'effectively come of age' by the early 1820s.[1]

What becomes clear is that, to the Board at least, the construction of timekeepers was a secretive business over which it proved difficult to gain control, particularly in contrast to the production of astronomical data. Indeed, control

[1] Betts, *Marine Chronometers at Greenwich*, p. 59.

was at times wrested from the Board and only regained after decades of wrangling. Throughout, the management of timekeeper trials remained an area of often public dispute over who should conduct them, how they were executed and who had the competence and authority to interpret results.

A Change of Emphasis

While this chapter covers the period from around 1770, it was the Discovery of Longitude at Sea Act of 1774 that dominated relationships between the Board and those seeking rewards for timekeepers. The rewards available were halved to £5000 (for timekeepers proved to be accurate to within one degree of longitude), £7500 (to two-thirds of a degree) and £10,000 (to half a degree). More stringent conditions were also applied, a move intended to clarify matters underlying disagreements with the Harrisons and thus prevent similar disputes thereafter. The Act of 1774 not only specified that any timekeeper offered for reward must be based on principles not previously made public but also set out what trials might be used to assess performance. These changes made explicit the Board's control over the process.[2] Under the new legislation, two or more timekeepers of the same construction had to be submitted for testing, beginning with a trial of twelve months at the Royal Observatory. If successful, the watches would be sent on two sea voyages (in opposite directions) around Great Britain, with the possibility of 'such other Voyages to different Climates as the said Commissioners shall think fit to direct and appoint' and of additional testing at the Royal Observatory for up to twelve months. Should any timekeepers pass these trials, their maker had to make a full explanation of their construction and assign them to the Commissioners for public use before any reward was given.[3]

The reduced rewards and new conditions dominated later dealings with watchmakers. As early as 1775, Thomas Mudge noted his objections to both the Act and the Royal Observatory trials, having previously attempted to head off the legislation.[4] Two decades later, a parliamentary committee appointed to consider a petition on his behalf (discussed below) concluded that the Act imposed 'conditions so difficult, and so impossible to be surmounted … that it is to be feared that few artists will engage in an undertaking so discouraging and precarious'.[5] Indeed, no maker succeeded in fully meeting the new conditions.

[2] Maskelyne, *An Answer to a Pamphlet*, pp. ii–iv, confirms that the Commissioners pushed to make these aspects more explicit, particularly holding initial trials at the Royal Observatory.
[3] Discovery of Longitude at Sea Act 1774 (14 Geo 3 c 66), copy at CUL RGO 14/1, pp. 37r–42v.
[4] BoL, confirmed minutes, 27 May 1775, CUL RGO 14/5, pp. 279–280; Mudge, *A Reply to the Answer*, pp. 22–23.
[5] House of Commons Select Committee, *Report*, p. 35.

While timekeepers were not the Board's sole concern in subsequent decades, its financial outlay in this area was considerable due to substantial rewards to a small number of makers. Excluding the final payment to Harrison agreed by Parliament in 1773, £9250 was given in rewards between 1770 and 1821. This amounted to almost 40 per cent of all rewards in the period. An additional £2203 was spent on the purchase, maintenance and repair of timekeepers. Overall, 10 per cent of the Board's total expenditure (16 per cent if Harrison's final reward is included) related to timekeepers during a period in which the Board's publishing activities became its main expense.[6]

The surviving archive shows that the Board dealt with just under fifty individuals in relation to mechanical timekeeping in this period (Table 6.1).[7] Of these, twenty can be identified as clock- or watchmakers, with schemes also coming from, among others, naval officers, a 'poor Seaman', a gunmaker, a licensed dockyard porter, and a civil servant based in India.[8] Some of the same petitioners, though not in general those identifiable as clock- or watchmakers, submitted schemes on other longitude methods as well.[9]

Over half the approaches relating to mechanical timekeepers were schemes only and were dismissed with little further consideration. In 1786, for example, the Commissioners quickly decided that Reverend Butler's proposal for finding longitude by hour glass or clockwork merited no further attention.[10] Thirteen years later, when John Dumbell asked for £1000 to enable him to make a timekeeper with improvements to render it immune to the effects of heat and cold, he was told that the Board did 'not give money to any person who has not produced something to entitle them to such encouragement'.[11]

[6] Figures based on Howse, 'The Board of Longitude Accounts'; see Appendix 2. Payments to expeditionary astronomers have not been included in the totals for timekeepers, although some of their duties did relate to timekeepers.

[7] These figures and Table 6.1 refer to schemes for mechanical timekeeping – broadly speaking, clocks and watches – but not for sundials, clepsydra or other methods and devices (some of which may have mechanical elements). Many of these other schemes are in 'Correspondence Regarding Methods and Instruments Used to Establish Longitude and the Use of Chronometers at Sea', 1783–1828, CUL RGO 14/38, 'Correspondence Regarding Miscellaneous Schemes and Inventions', 1784–1826, CUL RGO 14/44, and 'Correspondence on Schemes and Inventions', 1785–1829, CUL RGO 14/45; for example, Martin and Jordan's clockwork globe (Messrs Martin and Jordan to BoL, 18 December 1794, CUL RGO 14/38, pp. 204r–206v; BoL, confirmed minutes, 6 June 1795, CUL RGO 14/6, p. 2:245), and William Chavasse's 'Marine Transit' or mercurial timekeeper (William Chavasse, papers relating to the Marine Transit, 1812, CUL RGO 14/38, pp. 36r–40v; BoL, confirmed minutes, 4 March 1813, CUL RGO 14/7, p. 2:172), both quickly rejected.

[8] Peter Derness to Sir Harry Parker, 20 February 1790, CUL RGO 14/24, p. 447r.

[9] Austin, Boyle, Brazill, Hudson and Mallison, for example, submitted proposals relating to Jupiter's satellites, while Edward Massey approached the Board with a mechanical log and a sounding machine, receiving a reward for the latter; Treherne, 'Massey family'; Poskett, 'Sounding in Silence'.

[10] BoL, confirmed minutes, 4 February 1786, CUL RGO 14/6, p. 2:91.

[11] BoL, confirmed minutes, 7 December 1799, CUL RGO 14/6, p. 2:319.

Table 6.1 *Individuals presenting mechanical timekeepers and schemes to the Board of Longitude, 1770–June 1821*

Name	Occupation/status	From	Contact period	Trial	Reward
Larcum Kendall	watchmaker	London	1765–1787	Y	
John Arnold	watchmaker	London	1770–1792	Y	£1322
Thomas Mudge	watchmaker	London	1774–1793	Y	£3000
Thomas Germain (or Gorman)	unknown	Winchester	1777–1778		
Johann Georg Thiele	watchmaker	Bremen	1778	Y	
William Coombe	watchmaker	London	1778–1799	Y	£200
Robert Blackburne	unknown	unknown	1781–1782		
Robert Warter	unknown	Salisbury	1783		
Ezekiel Walker	watchmaker/inventor	Lynn	1783–1825		
Revd Butler	priest	London and Carlow, Ireland	1785–1787		
Peter Derness	seaman	London	1790		
Thomas Earnshaw	watchmaker	London	1789–1809	Y	£3000
George Margetts	watchmaker	London	1790–1796		
Josiah Emery	watchmaker	London	1791–1794	Y	
William Pleadwell	millwright	London	1791–1796		
John Russell	watchmaker	Falkirk (?)	1796		
Mr Martin	unknown (possibly watchmaker)	unknown	1797–1798		
John Boyle	Captain, RN	London	1798		
John Dumbell	unknown	London	1799–1803		
John Roger Arnold	watchmaker	London	1801–1806		£1678
John Charles Sander	unknown	Leipzig	1803		
John Fisher	unknown	London	1805		
Joseph Hardy	unknown	London	1805		
William Hardy	clockmaker	London	1805–1822	Y	£50
Edward Massey	watchmaker; nautical instrument maker	London	1805–1823		
Joseph Manton	gunmaker	London	1807–1813	Y	
John Jacob (Johannes Jakob) Schmidt	watchmaker	Karlsruhe	1807–1808		
William Forder	former purser, RN	London	1810		
John Henry Harris	fellowship porter	London	1810–1814		
W. H. Mallison	unknown	London	1810–1815		
Samuel Grimaldi	watchmaker	Clapham	1812–1813		
Edward Owen	commodore, RN	HMS *Cornwall*	1812–1813		
Sigismund Rentzsch	watchmaker	London	1813		
Robert Gillespie	unknown	Leith	1814–1815		
John Tibbot	watchmaker	Wales	1816		
Joseph Wilkinson	unknown	Stockport and Liverpool	1816–1820		
Charles Young	watchmaker	London	1817		
J. Brazill	unknown	London	1817–1819		

Table 6.1 (*cont.*)

Name	Occupation/status	From	Contact period	Trial	Reward
W. Carr	master, RN	Gillingham	1818		
Edward Elliot	unknown	West Wickham, Kent	1818–1819		
James Peck	journeyman tradesman	London	1818–1822		
Seth Hunt	watchmaker/inventor	London	1819		
David Thomas	unknown	Bridgwater	1819–1821		
Charles Hudson	civil servant	Calcutta	1819–1825		
Thomas Cumming	watchmaker	London	1819–1823	Y	
Benjamin Austin	unknown	London	1821		
William Probert	unknown	Staffordshire	1821		

The whole period of contact with the Board is indicated, although discussions about mechanical timekeeping may have been for a shorter period.
Sources: BoL, confirmed minutes, 1737–1823, CUL RGO 14/5 to 14/7; BoL, 'Papers Regarding the Public Trials and Improvements of Clocks and Chronometers, 1779–1826', CUL RGO 14/23; BoL, 'Papers Regarding the Public Trials and Improvements of Clocks and Chronometers', 1784–1828, CUL RGO 14/24; BoL, 'Correspondence Regarding Methods and Instruments Used to Establish Longitude and the Use of Chronometers at Sea', 1783–1828, CUL RGO 14/38.

Without an instrument and proof that it worked, further discussion or reward proved impossible. Those who said they had a working timekeeper were pressed for evidence, as were both William Pleadwell and John Russell in June 1796.[12] There is no record that proof was forthcoming. Even those who came in person with a watch might get no further. London watchmaker Sigismund Rentzsch presented his timekeeper in 1813 but was sent away because he had no corroboration of its performance.[13]

What the Board required was a working timekeeper and trustworthy testimony that it offered an improvement in performance.[14] In other words, the proposer (and thus the Board) had to show their understanding of the current state of horology – something necessary to demonstrate the novelty of a specific timekeeper design – and of the market for these devices. While evidence might initially come from self-reported results or from some reliable witness, ultimately it meant testing at the Royal Observatory. Few proposals got to this stage, however, and, while the Act of 1774 required a trial of twelve months at the Royal Observatory, a shorter test might eliminate candidates (see Tables 6.1 and 6.2). In March 1778, Johann Georg Thiele, a maker from Bremen, attended

[12] BoL, confirmed minutes, 11 June 1796, CUL RGO 14/6, p. 2:269.
[13] BoL, confirmed minutes, 3 June 1813, CUL RGO 14/7, p. 2:176.
[14] A supporting certificate 'from person or persons of Science (whose Character or Characters are known)' was required from 1772; BoL, confirmed minutes, 13 June 1772, CUL RGO 14/5, p. 228.

with a timekeeper that went to Greenwich for trial. Nevil Maskelyne reported in June that he had concluded after a few days that it went no better than 'a common Watch, and had therefore caused it to be taken away, it being unnecessary to make any further trial'.[15] This seems to belie the claim by historian and former Prussian army officer, Johann Wilhelm von Archenholz, that Thiele might have received a reward had he submitted his ideas before Harrison, since 'in the opinion of the English themselves, his mechanism was more ingeniously constructed, and much more likely than Harrison's to obtain the end proposed'.[16]

Thirty years later, the Board agreed to test a timekeeper held in an evacuated vessel, an idea proposed by London gunmaker Joseph Manton.[17] Following tests from December 1808 to February 1809 (Figure 6.1), the Board questioned Manton, recalling schemes published long before to probe the originality of his idea:

> Ques. Who was your Watch made by
> Answ. By Mr Pennington
> Ques. Are you aware that the idea of keeping the Watch in Vacuo is not new, but was published in the year 1714 by Jeremiah Thacker of Beverly in Yorkshire. London, by John Roberts at the Oxford Arms, in Warwick Lane 1714. Price /6d
> Answ. No I did not know it

Unusually, the Board demonstrated its long institutional memory, albeit of a publication that may have had satirical intent (see Chapter 3). Manton explained, however, that he was preparing timekeepers on 'a new principle', two of which would be ready for trial within six months.[18] The Board agreed to hold trials but it is not clear that these took place. Manton received no reward.[19]

The rewards available naturally attracted the attention of well-established watchmakers. Having settled in London, Swiss-born Josiah Emery had a fine reputation by the time he approached the Commissioners in 1791.[20] As usual, they insisted he produce evidence of the performance of his timekeepers.[21]

[15] BoL, confirmed minutes, 7 March and 6 June 1778, CUL RGO 14/5, pp. 329, 336.
[16] Von Archenholz, *A Picture of England*, pp. 101–102. For Thiele, see Oestmann, *Auf dem Weg*, pp. 15–21.
[17] BoL, confirmed minutes, 5 March 1807 (initially rejected) and 1 December 1808 (trial agreed), CUL RGO 14/7, pp. 2:104, 2:124.
[18] BoL, confirmed minutes, 2 March 1809, CUL RGO 14/7, p. 2:126; Nevil Maskelyne, 'Report of the Astronomer Royal about the Going of Mr Manton's Watch in Vacuo, at the Royal Observatory, from Decr 5th 1808 to Feby 27th 1809', 2 March 1809, CUL RGO 14/10, p. 235r and RGO 14/24, p. 452r.
[19] BoL, confirmed minutes, 5 March 1812, CUL RGO 14/7, pp. 2:158–2:159.
[20] Betts, 'Emery, Josiah'.
[21] BoL, confirmed minutes, 5 March 1791, CUL RGO 14/6, p. 2:166.

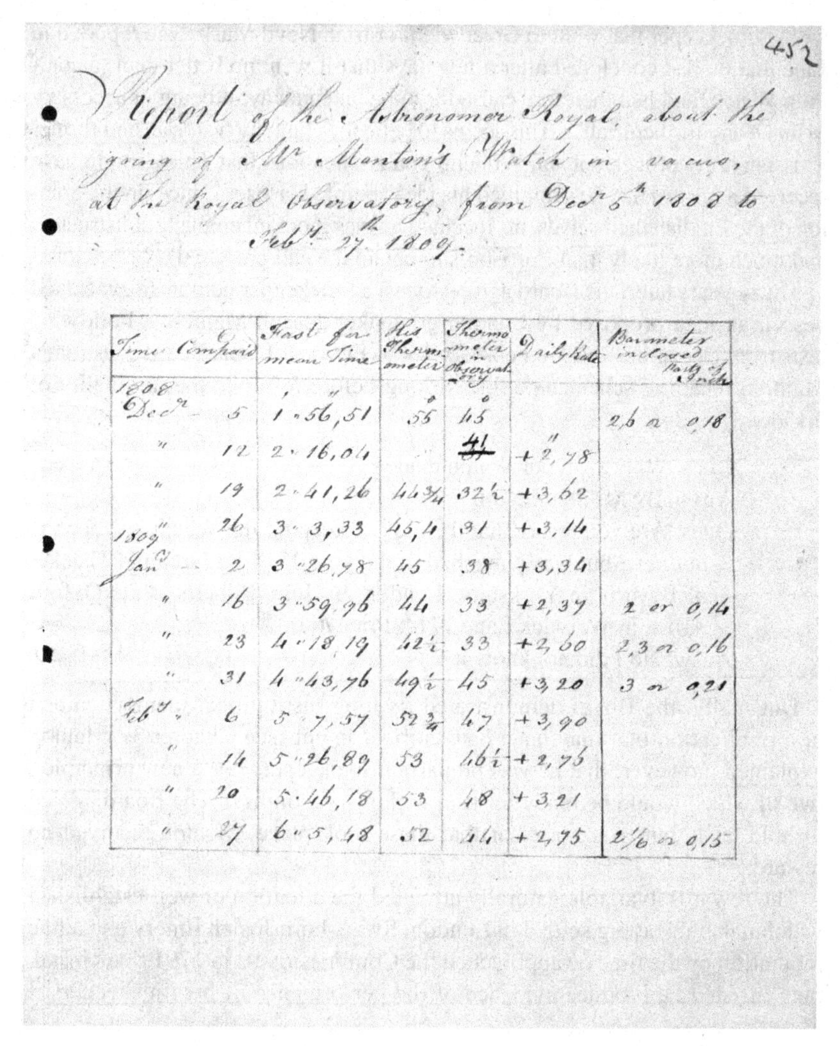

452

Report of the Astronomer Royal, about the going of Mr Manton's Watch in vacuo, at the Royal Observatory, from Decr 5th 1808 to Feby 27th 1809.

Time Compared		Fast for mean Time	His Thermometer	Thermometer in Observatory	Daily Rate	Barometer inclosed Parts of Inch
1808 Decr	5	1' 56,51"	55°	45°		2,6 or 0,18
"	12	2 16,04		41	+2,78"	
"	19	2 41,26	44¾	32½	+3,62	
"	26	3 3,33	45,4	31	+3,14	
1809 Jany	2	3 26,78	45	38	+3,34	
"	16	3 59,96	44	33	+2,37	2 or 0,14
"	23	4 18,19	42½	33	+2,60	2,3 or 0,16
"	31	4 43,76	49½	45	+3,20	3 or 0,21
Feby	6	5 7,57	52¾	47	+3,90	
"	14	5 26,89	53	46½	+2,75	
"	20	5 46,18	53	48	+3,21	
"	27	6 5,48	52	44	+2,75	2⅙ or 0,15

Figure 6.1 Results of the trials of Joseph Manton's 'watch in vacuo' at the Royal Observatory, December 1808 to February 1809, CUL RGO 14/24, p. 452r. Reproduced by kind permission of the Syndics of Cambridge University Library

After a preliminary trial by George Gilpin, former Royal Observatory assistant, the Board agreed a Greenwich trial, which took place in 1793–1794.[22] In light of the then recent enquiry into Thomas Mudge's petition (see below),

[22] George Gilpin, account of the rate of Emery's timekeepers, 1792, CUL RGO 14/23, p. 197r.

it is notable that the Commissioners called Emery into the meeting to agree the trial's conduct.[23] Neverthcless, neither watch performed sufficiently well to warrant reward.[24] Another pair of timekeepers, trialled in 1795–1796 (by which time Emery had died), also failed, and the Board subsequently declined an appeal from his widow, claimed on the basis of Emery's concentration on longitude matters, 'in preference to others of a more lucrative nature'.[25] Echoing Harrison's requests for financial support as he focused on perfecting longitude timekeepers to the exclusion of other business, this was an appeal other makers including John Arnold and Thomas Earnshaw would repeat.

Of the ten makers whose timekeepers underwent trials at the Royal Observatory in the period, six received a payment or reward: Larcum Kendall, John Arnold, Thomas Earnshaw, Thomas Mudge, William Coombe and William Hardy.[26] In addition, Arnold's son, John Roger, received a reward for his late father's and his own work. Kendall, Arnold, Earnshaw and Mudge accounted for most of the Board of Longitude's business and financial outlay relating to timekeepers, as discussed below. Coombe, an otherwise unknown figure, came to the Board's attention after sending a promising account of a watch, which went on trial in 1779.[27] It performed well enough to gain £200 to enable further work.[28] Requests for additional funds in 1783, 1785 and 1799 were declined.[29] Hardy approached the Board in 1807 with a new clock escapement rather than a marine timekeeper. It was tested between May and August that year and Hardy was awarded £50 to encourage further work.[30]

[23] BoL, confirmed minutes, 1 December 1792, 2 March and 1 June 1793, CUL RGO 14/6, pp. 2:189, 2:197, 2:201.

[24] BoL, confirmed minutes, 1 March 1794, CUL RGO 14/6, p. 2:213; 'An Account of the Going of Mr Emery's Watch No. 1 at the Royal Observatory', 1793–1794, CUL RGO 14/23, pp. 201r–202v; 'An Account of the Going of Mr Emery's Watch No. 2 at the Royal Observatory', 1793–1794, CUL RGO 14/23, pp. 203r–204v.

[25] BoL, confirmed minutes, 7 June 1794, 5 December 1795, 5 March and 11 June 1796, CUL RGO 14/6, pp. 2:220, 2:253, 2:261, 2:266–2:267.

[26] Howse, 'Britain's Board of Longitude'. Kendall's payments were not classified as rewards. Besides Thiele, Manton and Emery, Thomas Cumming was the only maker who had timekeepers tested but received no reward. A trial was recommended for a timekeeper by Seth Hunt in 1819, but there is no record that it took place.

[27] BoL, confirmed minutes, 28 November 1778 and 10 July 1779, CUL RGO 14/5, pp. 346, 350–351; McEvoy, 'Maskelyne's Time', pp. 179–182.

[28] BoL, confirmed minutes, 27 November 1779, CUL RGO 14/5, p. 356; trial results at CUL RGO 4/312, pp. 86r–89v.

[29] BoL, confirmed minutes, 22 November 1783, 17 August 1785 and 7 December 1799, CUL RGO 14/6, pp. 2:55, 2:86, 2:319–2:320.

[30] BoL, confirmed minutes, 5 March and 3 June 1807, 3 March 1808, CUL RGO 14/7, pp. 2:104, 2:107, 2:115; McEvoy, 'Maskelyne's Time', pp. 188–192. As the trial was for a land-based clock, it is not included in Table 6.2.

Testing and Deployment at Sea

In its dealings with Harrison from the mid-1760s, the Board of Longitude focused increasingly on the question of whether affordable timekeepers could be made to work reliably at sea. As the arguments with Harrison began to be settled, the Board focused on this matter more completely, above all through its relationship with Larcum Kendall, John Arnold and, later, Thomas Earnshaw. These dealings foregrounded questions around proper procedures for the issue and management of marine timekeepers. They also allowed some testing to move from land to sea, through the Board's involvement in high-profile expeditions, on which it sent instruments and astronomers from 1772. Significantly, all except one were to the Pacific and all were voyages in which leading Commissioners had an interest.

Larcum Kendall (1719–1790) was a well-respected London watchmaker when he was selected as one of the expert witnesses at Harrison's demonstration of H4's construction in August 1765.[31] Having possibly assisted in the making of H4, he seems to have been the obvious person to make a faithful copy, which the Board resolved to do within a month of Harrison's demonstration.[32] Completed in 1769, the watch (now called K1) was certified by the Board and William Harrison in March 1770.[33] Kendall was duly paid the remainder of the £450 agreed as the fee, plus a further £50 for his 'extraordinary trouble & loss of time' in adjusting it, taking it and H4 apart and reassembling both for the purpose of comparison. Kendall was also asked to 'report to the next Board the terms upon which he will undertake to instruct other workmen to make Watch Machines with Mr. Harrison's improvements'.[34] The Board wished to focus on the issue of replicability that had been called into question by some witnesses at Harrison's explanations.

The Board's relationship with Kendall differed, therefore, from their dealings with other makers. Rather than coming to the Commissioners as a projector or petitioner, Kendall was a trusted maker whom they approached and commissioned (rather than rewarded) over the next two decades to make, clean and repair timekeepers.[35] The Board trusted Kendall when he stated in May 1770 that training workmen to make copies of H4 would not be practicable but that he could make a watch of a simpler and more affordable design.[36]

[31] Betts, 'Kendall, Larcum'.
[32] BoL, confirmed minutes, 12 September 1765, CUL RGO 14/5, pp. 109–110.
[33] Larcum Kendall, one-day marine timekeeper (K1), c. 1769, NMM ZAA0038; Betts, *Marine Chronometers at Greenwich*, pp. 187–190; additional papers relating to the construction of K1 (and K2) at BL Add MS 39822, ff. 33–87.
[34] BoL, confirmed minutes, 3 March 1770, CUL RGO 14/5, pp. 191–192.
[35] BoL, payments to Larcum Kendall, 1776–1787, CUL RGO 14/17, pp. 393–398.
[36] BoL, confirmed minutes, 26 May 1770, CUL RGO 14/5, pp. 193–195.

Kendall presented his second watch (K2), for which he was paid £200, in 1772 and suggested that he could make 'others of the same kind for half the said sum'.[37] Again, the Commissioners readily agreed and were delivered a third timekeeper (K3) for £100 in 1774.[38]

Kendall's work therefore allowed the Board to consider how marine timekeepers could be made affordably in large numbers. The associated question was whether such timekeepers would prove reliable at sea. This required testing at the Royal Observatory, as had been established with H4, and then on long voyages. With K1 having undergone a comparison with H4 at the Royal Observatory by the end of 1771 (Table 6.2), Maskelyne suggested that the second of the Pacific voyages commanded by James Cook presented the perfect opportunity for such a trial. The planned expedition, he proposed, might be 'rendered more serviceable to the improvement of Geography & Navigation' if the Board and the Royal Society were to provide it with astronomical instruments 'and also some of the Longitude Watches', and if 'a proper person' could also be sent to deploy the instruments and teach the ships' officers 'the method of finding the Longitude'.[39]

In referring to 'Longitude Watches', Maskelyne was suggesting that at least one timekeeper by London maker John Arnold (1735/6–1799) be sent as well. Arnold, to whom Maskelyne had previously sent a copy of the printed description of Harrison's timekeeper, first approached the Board in May 1770, offering 'a Timekeeper of a new construction'. This was at the meeting at which Kendall reported that teaching other artisans to make timekeepers like Harrison's was not viable.[40] Arnold had already shown that his timekeepers might be seaworthy: one was aboard an East India Company ship, the *Egmont*, with Captain Mears. Indeed, this was the beginning of a profitable relationship between Arnold, the Company and its officers, helped by the support of Company hydrographer, Alexander Dalrymple.[41]

Arnold again attended the Board in March 1771, showing a timekeeper that Maskelyne had tested with promising results and indicating that the price of his watches, when perfected, would not exceed sixty guineas (£63). Another watch was then on board the *Northumberland* with Rear Admiral Sir Robert Harland on route to the East Indies, following orders by the Earl of Sandwich

[37] BoL, confirmed minutes, 7 March 1772, CUL RGO 14/5, p. 221; Larcum Kendall, one-day marine timekeeper (K2), c. 1771, NMM ZAA0078; Betts, *Marine Chronometers at Greenwich*, pp. 190–193; papers relating to the construction K2 (and K1) at BL Add MS 39822, ff. 33–87.

[38] BoL, confirmed minutes, 28 November 1772 and 25 June 1774, CUL RGO 14/5, pp. 232, 263; Larcum Kendall, one-day marine timekeeper (K3), c. 1774, NMM ZAA0111; Betts, *Marine Chronometers at Greenwich*, pp. 194–196.

[39] BoL, confirmed minutes, 28 November 1771, CUL RGO 14/5, p. 207.

[40] BoL, confirmed minutes, 26 May 1770, CUL RGO 14/5, p. 197.

[41] Cook, 'Alexander Dalrymple and John Arnold'; Davidson, 'The Use of Chronometers'; Davidson, 'Marine Chronometers', pp. 76, 78–79, 88.

Table 6.2 *Marine timekeepers trialled at the Royal Observatory, 1770–1821*

Maker	Timekeeper	Trial dates	Notes
Larcum Kendall	K1	March 1770–May 1772	§ Tested alongside H4; K1 not working Jan to March 1772 but repaired by Kendall
John Arnold	watch	by March 1771–Dec 1771	§ BoL, confirmed minutes, 2 March 1771, record a watch on trial. Results for Nov–Dec reported at BoL, confirmed minutes, 7 March 1772
John Arnold	1, 2 and 3	April 1772	§ Rates calculated before issue to *Adventure* and *Resolution*
Larcum Kendall	K2	May 1772–May 1773	§ Tested alongside H4, which was still running May-Oct 1773
		Oct 1773–Feb 1774	K2 issued to Captain Vandeput in 1774 but H4 kept running
John Arnold	1	July 1774–Dec 1774	§ Tested alongside H4 (still running from trials with Kendall watches), which was kept running to Sept 1775
Larcum Kendall Thomas Mudge	K3 watch	Dec 1774–March 1775	§ Two watches run together for comparison; BoL, confirmed minutes, 4 March 1775. As a result, Kendall was requested to make adjustments to K3
Larcum Kendall	K3	c. June 1775	§ BoL, confirmed minutes, 27 May 1775: K3 can go on trial in Transit Room
Larcum Kendall	K1	Aug–Sept 1775	§ One-month trial (with H4) after return from second Cook voyage
Larcum Kendall	K1 and K3	April–June 1776	§
Thomas Mudge	watch	Nov 1776–Feb 1778	§
Johann Georg Thiele	watch	1778	§ Trialled for a few days between March and June meetings
John Arnold	36	Feb 1779–July 1780	§ Results of first 13 months in Arnold, *An Account Kept During Thirteen Months*. BoL, confirmed minutes, 4 March 1780, refer to 'No. 39', presumably in error
Thomas Mudge	Green and Blue	April 1779–July 1780	§
William Coombe	watch	June 1779–Jan 1780	§
Larcum Kendall	K2	March 1780–Jan 1781	§
John Arnold	58	March–April 1782	
John Arnold	pocket 69	Oct 1782–June 1783	Two trial periods: Oct 1782–March 1783; April–June 1783
John Arnold	34	Dec 1782–Dec 1783	

Table 6.2 (*cont.*)

Maker	Timekeeper	Trial dates	Notes
John Arnold	55 and 64	Dec 1782–June 1783	
John Arnold	box 12	Jan–March 1783	
John Arnold	47	July–Sept 1783	
Thomas Mudge	Green and Blue	July 1783–Sept 1784	§
John Arnold	pocket 4/305	Oct 1783–March 1784	Trialled over two periods: Oct–Dec 1783; Jan–March 1784
John Arnold	88	Jan–Feb 1784	
John Arnold	47 and 54	March–April 1784	
John Arnold	59	May–July 1784	
John Arnold	pocket 61	May–Aug 1784	
John Arnold	15 and box timekeeper	July–Oct 1784	§ No. 15 stated by Arnold as one of the timekeepers to be considered for reward
John Arnold	85 and 11	Aug–Sept 1786	
John Arnold	80	Sept 1786	
Larcum Kendall	K1	Nov 1786–April 1787	§
John Arnold	no number	Jan–Feb 1788	
Thomas Mudge	Green and Blue	June 1789–June 1790	§
Thomas Earnshaw	watch	July 1789–April 1790	§ Presumably the timekeeper compared to Mudge's watches July–Aug 1789 (BoL, confirmed minutes, 15 August 1789). BoL, confirmed minutes, 4 December 1790, note results as Jan 1789–April 1790, probably in error. Earnshaw, *Longitude*, p. 28, calls it a watch
John Arnold	82	April–July 1790	§ Probably box timekeeper purchased by Captain Roberts, for which BoL reimbursed him
Thomas Earnshaw	watch (for use for Bligh)	Feb–July 1791	§ Earnshaw, *Longitude*, pp. 23–25, 59, describes trial of timekeepers by Earnshaw (5), Arnold (3) and Brockbank (1) for Bligh to purchase one for *Providence* voyage. Bligh identified winning watch as Earnshaw 1503, had one purchased for him and bought another for himself; BoL took 1514 for Gooch; Brinkley took another (pp. 24–25). Earnshaw indicates two of the Arnolds tested went with Gooch (p. 25). CUL RGO 14/25, pp. 333r–333v appears to be results for Earnshaw watches
Thomas Earnshaw	watch E (for use by Gooch)	April–August 1791	
Thomas Earnshaw	watches 1 and 2	May–July 1791	
Thomas Earnshaw	watch 3 (for use by Gooch)	May–July 1791	
John Arnold	14, 176 and one other	May–July 1791	
John and Myles Brockbank	timekeeper	May–July 1791	

Table 6.2 (*cont.*)

Maker	Timekeeper	Trial dates	Notes
Thomas Earnshaw	4259 and pocket E (for W. Brinkley)	Oct–Dec 1791	§ CUL RGO 14/25, p. 334r. BoL, confirmed minutes, 1 December 1792, notes rates of seven Earnshaw timekeepers reported, perhaps this and the trials above
John Arnold	18	July–August 1791	Possibly timekeeper mentioned above with Arnold 14 and 176
John Arnold	38	Dec 1792–April 1793	§
John Arnold	box timekeeper	by April 1793	Earnshaw, *Longitude*, p. 60, notes this trial and states that the Arnold was bought by Captain Cheap of East India Company ship *Britannia*, which departed 5 April 1793. Arnold timekeeper may be 38 (above)
Thomas Earnshaw	pocket timekeeper	by April 1793	
Josiah Emery	1 and 2	July 1793–Feb 1794	§
Josiah Emery	3 and 4	July 1795–Feb 1796	§
Thomas Earnshaw	265	Jan 1796–Jan 1797	§
Thomas Earnshaw	1 and 2	Jan 1798–Jan 1799	§ Box timekeepers
Thomas Earnshaw	1 and 2	Oct 1799–Dec 1800	§
Thomas Earnshaw	309	Feb 1800–April 1811	Watch belonging to Nevil Maskelyne
Thomas Earnshaw	1 and 2	July 1801–Nov 1802	§
Joseph Manton	chronometer in vacuo	Dec 1808–Feb 1809	§ Chronometer made by Pennington
Thomas Cumming	19	June 1820–Oct 1821	§ Trial begun at Royal Society (Sept 1819–June 1820). Performance compared with Molyneux 403 in 1821. 1819 results compared with other chronometers in 1822

Dates are those during which timekeepers are recorded as running (if known), which may be longer than assessment periods. Where sources vary, the longer period has been given.
§ = trial mentioned and/or results given in Board of Longitude archive.
Sources: William Bayly, 'Rating of Longitude Watches on HMS *Adventure*', 1772, CUL RGO 4/322; BoL, confirmed minutes, 1737–1823, CUL RGO 14/5, RGO 14/6, RGO 14/7; BoL, 'Papers Regarding the Public Trials and Improvements of Clocks and Chronometers, 1779–1826', CUL RGO 14/23; BoL, 'Papers Regarding the Public Trials and Improvements of Clocks and Chronometers', 1784–1828, CUL RGO 14/24; BoL, 'Papers Regarding Arnold and Earnshaw's Chronometers and Claims for Reward', 1783–1808, CUL RGO 14/25; Maskelyne, 'An Account of the Going of Dr Maskelyne's New Chronometer N.309 Made by Mr T. Earnshaw', 1799–1811, CUL RGO 4/153; Maskelyne, 'Books on Chronometer Trials', 1799–1811, CUL RGO 4/312; Arnold, *An Account Kept During Thirteen Months*; Arnold, *Certificates and Circumstances*; Banks, *Protest Against a Vote*; House of Commons Select Committee, *Report*; Maskelyne, *Arguments for Giving a Reward*; Betts, *Marine Chronometers at Greenwich*; Earnshaw, *Longitude*; Gould, *The Marine Chronometer*; Howse, 'Captain Cook's Marine Timekeepers. Part I'; Howse, 'Captain Cook's Marine Timekeepers. Part II'.

to have Arnold's work tried at sea.[42] With the promise of affordable timekeepers, the Commissioners offered Arnold £200 to perfect his improvements. By December, observatory trials warranted the inclusion of the Arnold watch in a list of instruments to be sent with Cook.[43] When Cook's ships departed, Arnold had in fact supplied three watches, allowing each vessel to carry two marine timekeepers. During the three-year voyage, however, none of Arnold's performed as well as K1, which Cook hailed as 'our faithful guide through all the vicissitudes of climates'.[44] Nevertheless, Arnold's work had shown sufficient promise that he remained a trusted supplier to the Board, the Navy and East India Company, and someone to whom the Board entrusted repair and maintenance work.[45] As discussed below, this relationship came under strain with the emergence of Thomas Earnshaw, when the Board's affiliations became divided: Banks remained a supporter of Arnold; Maskelyne increasingly favoured Earnshaw.

From Cook's second Pacific voyage of 1772–1775 to Matthew Flinders's circumnavigation of New Holland (Australia) in 1801–1804, the Board attempted to ensure the reliability of long-distance sea trials through the appointment of expeditionary astronomers issued with numerous instruments including timekeepers (Table 6.3).[46] While the management and testing of the timekeepers were not the astronomers' only tasks, they were among the first things discussed in the largely unchanging instructions issued.[47] The Board's astronomers had to ensure the timekeepers were wound daily, use them to determine longitude (to be compared with longitude from dead reckoning and lunar distances), and keep records of these results and of the going of the timekeepers.[48] Results were to be returned to

[42] BoL, confirmed minutes, 7 March 1772, CUL RGO 14/5, p. 220; Howse, 'Captain Cook's Marine Timekeepers. Part I ', pp. 190–191.

[43] BoL, confirmed minutes, 2 March and 14 December 1771, CUL RGO 14/5, pp. 203–204, 211; BoL to Navy Board, 2 March 1771 (£200 reward to Arnold), TNA ADM 49/65, pp. 103–104.

[44] James Cook to the Secretary of the Admiralty, 22 March 1775, quoted in Beaglehole (ed.), *Journals*, II, p. 692. Howse, 'Captain Cook's Marine Timekeepers. Part I', pp. 192–195, details the performance of the watches.

[45] BoL, payments to John Arnold, 1771–1791, CUL RGO 14/17, pp. 25–31. For Arnold's career, see Mercer, *John Arnold & Son*. Ishibashi, 'A Place for Managing Government Chronometers', p. 63, notes that chronometers by the Arnolds dominated the Navy's stock to 1821; see also Stewart, 'The British Naval Chronometers of 1821'.

[46] Higgitt, 'Equipping Expeditionary Astronomers'. [47] Phillips, 'Instrumenting Order'.

[48] The following sources discuss the deployment of timekeepers. Cook's voyages: Beaglehole (ed.), *Journals*, II, III and IV; Dunn, 'James Cook and the New Navigation'; Howse, 'Captain Cook's Marine Timekeepers. Part I'; Howse, 'Captain Cook's Marine Timekeepers. Part II'. Phipps's voyage: Bendall, *Kendall's Longitude*, pp. 41–51; Savours, 'A Very Interesting Point'. Vancouver's expedition (including Gooch): David, 'Vancouver's Survey Methods'; Davies, 'Horology and Navigation'; Davies, 'Vancouver's Chronometers'; Davies, 'Arnold's Chronometer No. 176'; Dening, *The Death of William Gooch*; Dunn, 'Heaving a Little Ballast'. Flinders's circumnavigation of New Holland: Barritt, 'Matthew Flinders's Survey Practices'; Dunn and Phillips, 'Of Clocks and Cats'; Morgan, 'Finding Longitude'; Morgan, *Australia Circumnavigated*; Phillips, 'Remembering Matthew Flinders'.

Table 6.3 *Expeditions to which the Board of Longitude supplied astronomers and marine timekeepers, 1772–1804*

Dates	Commander and expedition	Astronomers	Vessel BoL timekeepers	Notes
1772–1775	James Cook Second Pacific voyage	William Wales William Bayly	*Resolution* K1 Arnold 3 *Adventure* Arnold 1 and 2	
1773	Constantine Phipps Expedition towards the North Pole	Israel Lyons	*Racehorse* K2 One Arnold *Carcass* One Arnold	Phipps took his own Arnold pocket timekeeper
1776–1780	James Cook Third Pacific voyage	James Cook James King William Bayly	*Resolution* K1 *Discovery* K3	Cook and King shared astronomical duties. Maskelyne lent Cook a watch by Ellicott, for which the Board gave him 20 guineas (£21)
1791–1795	George Vancouver North-west coast of America	William Gooch	*Discovery* K3 *Chatham* Arnold 82 Earnshaw 1514 (pocket) Arnold 14 and 176	Arnold 96 said to have been with Vancouver Additional timekeepers issued to Gooch, sent in 1791 but killed before joining Vancouver
1794–1798	William Broughton Vancouver expedition	John Crosley	*Providence* Earnshaw 1 (pocket) Earnshaw 2 (pocket) Earnshaw 248 Arnold 56	Sent to join Vancouver expedition, with Crosley to replace Gooch, but instead charted coast of East Asia. Broughton purchased Arnold 45 from East Indiaman *Henry Addington* in Feb 1797 and saved it from *Providence* wreck in May
1801–1804	Matthew Flinders Survey of New Holland (Australia)	John Crosley; replaced by James Inman	*Investigator* Arnold 82 and 176 (to 1802) Earnshaw 520 and 543 K3 (from 1803)	Navy Board issued Arnold pocket 1736 to Flinders. Crosley had Earnshaw pocket 465, but left with it at Cape of Good Hope. Arnold 82 and 176 were sent back to England in 1802 after they stopped. K3 was issued to Inman, who joined in June 1803

Sources: BoL, confirmed minutes, 1737–1823, CUL RGO 14/5, RGO 14/6, RGO 14/7; BoL, 'Papers on the Loan of Instruments', 1766–1829, CUL RGO 14/13; BoL, 'Papers Regarding Arnold and Earnshaw's Chronometers and Claims for Reward', 1783–1808, CUL RGO 14/25; BoL, 'Observations on Voyages of Discovery', 1763–1805, CUL RGO 14/67; Betts, *Marine Chronometers at Greenwich*; Broughton, *A Voyage of Discovery*; Howse, 'Captain Cook's Marine Timekeepers. Part I'; Howse, 'Captain Cook's Marine Timekeepers. Part II'; May, 'How the Chronometer'; Savours, 'A Very Interesting Point'.

Maskelyne, who as Astronomer Royal was responsible for appointing and training the astronomers.[49] Maskelyne was also keen to receive intermediate reports on the timekeepers' performance and often passed these on to the Board.[50]

In the early expeditions, the fact that the timekeepers were still under test was emphasised by the protocols for their management. The timekeepers remained under lock and key and could only be checked or handled in the presence of trustworthy witnesses.[51] As Cook recorded from Plymouth in July 1772,

> before we departed the Watches were put in motion in the presence of my self Captain Furneaux, the first Lieutenant of each of the Sloops, the two astronomers and Mr Arnold and afterward put on board: Mr Kendals and one of Mr Arnolds on board the Resolution and the other two of Mr Arnolds on board the Adventure: the Commander, First Lieutenant and Astronomer on board each of the Sloops had each of them Keys of the Boxes which contained the Watches and were allways to be present at the winding them up and comparing the one with the other.[52]

Similar arrangements applied to timekeepers by Kendall (K2) and Arnold on Constantine Phipps's Arctic voyage the following year.[53]

Thus longitude timekeepers were initially subject to different standards of care, recording and reporting than other instruments taken on the voyages. For later expeditions, Maskelyne's training and disciplinary control of the expeditionary astronomers allowed the protocols around locks and witnessing to relax. A well-trained astronomer could be trusted to care for the timekeepers

[49] Cook's second voyage: William Bayly, 'Rating of Longitude Watches on HMS *Adventure*', 1772, CUL RGO 4/322; William Bayly, 'Log book of HMS *Adventure*' and 'Observations on HMS *Adventure*', 1772–1774, CUL RGO 14/56, RGO 14/57; William Wales, 'Log book of HMS *Resolution*' and 'Observations on HMS *Resolution*', 1772–1775, CUL RGO 14/58, RGO 14/59. Phipps's voyage: 'Comparison of Chronometers', 1773, CUL RGO 4/154. Cook's third voyage: James Cook, James King and William Bayly, 'Astronomical Observations Made During Captain Cook's Last Voyage', 1778–1780, CUL RGO 14/61. Vancouver's expedition: William Gooch, 'Observations in the Voyage of the *Daedalus*', 1791–1792, CUL RGO 14/62, RGO 14/63; see also 'Observations on Voyages of Discovery, 1763–1805, CUL RGO 14/67, which includes Crosley's observations on the *Providence*, and William Gooch and others, 'Letters, Memoranda and Journal Containing the History of Mr William Gooch', 1786–1835, CUL Mm.6.48. Flinders's voyage: John Crosley, 'Observations of John Crosley, 1801, RGO 4/186; Matthew Flinders and Samuel Ward Flinders, 'Log Book, Observations and Memoir of the HMS *Investigator*', 1801–1803, CUL RGO 14/64; Samuel Ward Flinders, 'Reduction of the Astronomical Observations Made During the Voyage of HMS *Investigator*', 1801–1803, CUL RGO 14/65; various, Observations on a Voyage of Discovery of the HMS *Investigator*', 1801–1803, CUL RGO 14/66. 'Correspondence and Related Papers Regarding Observations Made on Voyages of Discovery', 1787–1824, CUL RGO 14/68, includes correspondence to and from Wales, Crosley, Flinders and his brother, and Inman.

[50] For example, Nevil Maskelyne to Earl of Sandwich, 22 April 1773, NMM SAN/F/4, with results between Plymouth and Cape of Good Hope for timekeepers on Cook's second voyage, reported at BoL, confirmed minutes, 24 April 1773, CUL RGO 14/5, p. 239.

[51] BoL, confirmed minutes, 14 May 1772, CUL RGO 14/5, pp. 223–224.

[52] Beaglehole (ed.), *Journals*, II, pp. 16–17. Similar conditions applied on the third voyage; Beaglehole, *Journals*, III, p. 1504.

[53] Savours, 'A Very Interesting Point', p. 410.

and report their performance. Still, any timekeeper sent on an expedition first underwent testing at Greenwich to check its performance and establish an initial rate (see Tables 6.2 and 6.3).

As other authors have noted, the Board's interest in the long-distance deployment and trialling of marine timekeepers focused on a small number of exceptional expeditionary voyages rather than on more routine voyages.[54] In part, this was because the Board owned few timekeepers – George Gilpin reported in 1797 that there were only nine watches belonging to the Board then in his or Maskelyne's possession.[55] Occasionally, the Commissioners did consider results from other sea trials, notably in 1796, when the Admiralty ordered that timekeepers by Mudge, Earnshaw and Brockbank be compared on the *Sans Pareil*.[56] As Table 6.4 indicates, the Board also sometimes lent timekeepers for general use or for specific voyages. This seems to have been determined by who knew whom and the confidence that those borrowing timekeepers had the right skills and experience: Maskelyne recommended the loan of instruments including K1 to William Dawes when accompanying the First Fleet; Banks's sponsorship of the infamous *Bounty* voyage ensured the loan of K2 and other instruments to William Bligh; the First Lord of the Admiralty gave permission for the loan of timekeepers at least twice in the 1790s.[57] Only a few of these voyages returned reports to Maskelyne, usually from users knowledgeable about astronomical and surveying practices.[58] The picture is less clear after about 1805, however, with no records yet identified that detail the issue of the Board's timekeepers. Thomas Hurd, Hydrographer to the Royal Navy, seems to have assumed responsibility for the issue of chronometers belonging to the Navy Board in 1809, but it is unclear whether this also included the Board's, although he was its Secretary from 1810.[59]

[54] May, 'How the Chronometer', p. 638; Nockolds, 'Early Timekeepers at Sea', p. 112.

[55] George Gilpin to William Wales, 27 February 1797, CUL RGO 14/13, pp. 34r–34v.

[56] BoL, confirmed minutes, 3 December 1796, CUL RGO 14/6, p. 2:272; Earnshaw, *Longitude, an Appeal*, pp. 60, 68–75. Betts, *Marine Chronometers at Greenwich*, p. 209, and Nockolds, 'Early Timekeepers at Sea', p. 148, state that the trial was organised by or for the Board of Longitude. The Board's minutes indicate this was not the case. See also May, 'How the Chronometer', pp. 644–645.

[57] Regarding Dawes, see Nevil Maskelyne to Joseph Banks, 17 October 1786, NMM MSK/3/1/25.

[58] Home Riggs Popham to [Nevil Maskelyne], 23 July 1786, and Home Riggs Popham, 'Observations for Determining the Longitude by a Watch Made by Mr. Kendall on His Own Construction', 27 September 1785–22 July 1786, RGO 14/51, pp. 193r–207r, report on the performance of K2 on the *Grampus* and *Nautilus*. Popham had surveying duties. William Dawes to Nevil Maskelyne, 20 December 1786, 16 January 1787 and 3 September 1787, CUL RGO 14/48, pp. 239r–244v, 269r–275v, report on the going of K1, sent with the First Fleet. Deployed as an engineer and surveyor, Dawes established an observatory on what is now Dawes Point, New South Wales.

[59] For example, Thomas Hurd, reports of issue of chronometers, 1816–1817, CUL RGO 14/23, pp. 13r–24r. The timekeepers identified appear in a list of 'Government Time-keepers' compiled by Hurd in 1821, CUL RGO 5/229, pp. 5r–6r, and are not obviously Board-owned; see also May, 'How the Chronometer', p. 650. Webb, 'The Expansion of British Naval Hydrographic Administration', p. 301 n. 1421, notes that evidence for Hurd assuming responsibility for chronometers in 1809 remains uncertain.

Table 6.4 *Marine timekeepers supplied by the Board of Longitude for other purposes, 1771–1805*

Dates	Timekeeper	Issued to	Notes
1771	Arnold	Rear Admiral Sir Robert Harland, commander-in-chief, East Indies Station	Supplied by order of Earl of Sandwich. Harland bought a second Arnold and reported on both to Sandwich, who forwarded the remarks. BoL, confirmed minutes, 7 March 1772, confirms payment
1774–1777	K2	Captain George Vandeput, *Asia*	Used on North American Station
1777	H4	—	Intended for *Lyon*, going to the Arctic under Lieutenant Young; no evidence it went
1781–1784	K2	Rear Admiral Robert Digby, commander-in-chief, North American Station	Used during American Revolutionary War
1785–1786	K2	Lieutenant Home Riggs Popham, for Captain Edward Thompson, *Grampus*	Voyage to west coast of Africa to identify sites for settlements; K2 returned with Popham on *Nautilus*
1785	K3	Vice Admiral John Campbell, commander-in-chief, Newfoundland Station	BoL agreed in 1781 to lend K3 to Admiral Sir George Rodney and then in March 1782 to Captain Elliot; neither happened. Campbell took K3 on a voyage to Newfoundland with Mudge's first timekeeper
1786–1789	K3	Rear Admiral John Elliot, commander-in-chief, Newfoundland Station	K3 probably passed from Campbell to Elliot, who acknowledged possession from 1786, when he took over Newfoundland Station. Vice Admiral Mark Milbanke took possession in 1789 on relieving Elliot, but had to return it in 1790
1786–1792	K1	Captain Arthur Philip, *Sirius*	Taken on First Fleet voyage to New South Wales, departing 1787; survived *Sirius* wreck, March 1790
1787–1788	H4	Major General William Roy	Used while measuring distance between Royal Observatory and Observatoire de Paris
1787–	K2	Lieutenant William Bligh, *Bounty*	First breadfruit voyage K2 taken to Pitcairn following mutiny of 1789; bought by Captain Folger in 1808; returned to England about 1843
1790	Arnold	Admiral Mark Milbanke, governor of Newfoundland	Supplied in lieu of K3

Table 6.4 (*cont.*)

Dates	Timekeeper	Issued to	Notes
1790–1792	Arnold	Captain Edward Edwards, *Pandora*	Pursuit of *Bounty* mutineers
1790–1792	Arnold	Captain John Blankett, *Leopard*	In charge of convoy to China
1791–1793	Earnshaw 1503	Captain William Bligh, *Providence*	Second breadfruit voyage
1792–1794	Arnold	Erasmus Gower, *Lion*	Macartney Embassy to China Same timekeeper previously lent to Blankett
1793	Earnshaw	Lord Hood, commander-in-chief, Mediterranean Fleet	Permission given by Earl of Chatham (First Lord of the Admiralty); noted as cleaned by Earnshaw in BoL confirmed minutes, 1 June 1793, and as an Earnshaw at CUL RGO 14/13, p. 17r; listed as an Arnold at CUL RGO 14/13, p. 21r
1793–1802	K1	Vice Admiral Sir John Jervis, commander-in-chief, Leeward Islands Station	Permission given by Earl of Chatham (First Lord of the Admiralty). Jervis appointed Commander-in-Chief, Mediterranean in 1795; took over Channel Fleet in 1800; became First Lord of the Admiralty in 1801. K1 seems to have remained with him
1798	Earnshaw 1514	Whidby (possibly Joseph Whidbey)	Identified as 'once in the possession of Gooch' (CUL RGO 14/13, p. 17r)
1805	Arnold 176	William Bligh, appointed Governor of New South Wales	Bligh departed 1806, returned 1811. BoL requested 176 back in 1812; still missing in 1828
1805	Earnshaw	unknown	Identified as 'formerly with Flinders' (CUL RGO 14/13, p. 17r)

Sources: BoL, confirmed minutes, 1780–1801, CUL RGO 14/6; BoL, 'Papers on the Loan of Instruments', 1766–1829, CUL RGO 14/13; BoL, 'Papers Regarding Arnold and Earnshaw's Chronometers and Claims for Reward', 1783–1808, CUL RGO 14/25; Bendall, *Kendall's Longitude*; Betts, *Marine Chronometers at Greenwich*; Howse, 'Captain Cook's Marine Timekeepers. Part I'; May, 'How the Chronometer'; Mudge, *A Narrative of Facts*, pp. 10–12; Maskelyne, *An Answer to a Pamphlet*, p. 74; Phillips, 'Making Time Fit', pp. 94–99; Roy, 'An Account of the Trigonometrical Operation'.

Hurd was also made Superintendent of Timekeepers in 1819, following changes under the Longitude Act 1818.[60]

The third maker whose timekeepers the Board issued to voyages and expeditions was Thomas Earnshaw (1749–1829). Having established himself in

[60] Discovery of Longitude at Sea, etc. Act, 1818 (58 Geo 3 c 20), copy at CUL RGO 14/1, pp. 79r–82v. Hurd was replaced as Secretary of the Board in 1818.

London in the mid-1770s, Earnshaw turned his attention to marine timekeepers around 1780 and had one on trial at the Royal Observatory by July 1789.[61] As with Arnold, the good results of this trial brought Earnshaw into regular contact with the Commissioners. He sent several timekeepers for trial at the Royal Observatory and from 1791 supplied the Board with timekeepers for ships and expeditions (Tables 6.2, 6.3 and 6.4).[62] Like Arnold and Kendall, Earnshaw also undertook cleaning, repair and other work for the Board.[63]

Arnold and Earnshaw benefited from a productive relationship with the Board and its deployment of their timekeepers, not to mention the trials run at the Royal Observatory. Both published results of Greenwich trials and of the use of their timekeepers on Board-supported voyages, as well as of other users' trials and reports. Unsurprisingly, they chose not to publish the less favourable comments that became more common from the 1780s.[64]

Famously, it was as a result of the Royal Observatory trial of Arnold's pocket timekeeper number 36 in 1779–1780 that the word 'chronometer' began to be used for the new timekeepers. The idea seems to have come from Alexander Dalrymple, a close ally of Arnold and Banks, who wrote around this time that:

> The Machine used for measuring Time at SEA is here named CHRONOMETER, my friend Mr Banks agreeing with me in thinking so valuable a machine deserves to be known by a name, instead of a Definition. The name Time-Keeper is only proper to a perfect Chronometer.[65]

Arnold used the term throughout his account of the trial of number 36 and, although the word had been used previously, this marked the start of consistent usage in the modern sense.[66]

High-profile naval expeditions were thus important in allowing the Board to put timekeepers to the most arduous of sea trials, as well as ventures from which Arnold and Earnshaw gained the Board's trust and public recognition,

[61] Betts, 'Earnshaw, Thomas'; BoL, confirmed minutes, 4 December 1790, CUL RGO 14/6, p. 2:164.

[62] For Earnshaw timekeepers supplied elsewhere, see Arnott, 'Chronometers on East India Company Ships'; Davidson, 'Box Chronometers'; Davidson, 'The Use of Chronometers'; Davidson, 'Marine Chronometers'; May, 'How the Chronometer'; Miller, 'Longitude Networks'; Nockolds, 'Early Timekeepers at Sea'; Phillips, 'Making Time Fit', pp. 95–102; Webb, 'The Expansion of British Naval Hydrographic Administration', pp. 300–308.

[63] BoL, payments to Thomas Earnshaw, 1791–1806, CUL RGO 14/17, pp. 230–242.

[64] Arnold, *An Account Kept During Thirteen Months*; Arnold, *Certificates and Circumstances*; Earnshaw, *Longitude, an Appeal*; Phillips, 'Making Time Fit', pp. 58–59.

[65] Dalrymple, *Some Notes Useful*, p. 1, quoted in Cook, 'Alexander Dalrymple and John Arnold', p. 191; see also Betts, *Marine Chronometers at Greenwich*, p. 45.

[66] Arnold, *An Account Kept During Thirteen Months*; John Arnold, pocket chronometer 36, 1778, NMM ZBA1227. For earlier usage, see Köberer, 'On the First Use'.

in turn helping and helped by their work to build clientele directly with the Royal Navy and East India Company. Yet this did not guarantee that relationships with the watchmaking trade would remain harmonious and within the Commissioners' control. Arnold and Earnshaw met resistance from the Board as they made greater demands for financial support, culminating in a lengthy dispute in the early 1800s. Before then, however, the Board's control over its dealings with the trade was called to public account and threatened in hostile interactions with Thomas Mudge and his son.

'A Censure upon Their Conduct': The Mudges and the Board

Thomas Mudge (1715/16–1794) was, like Larcum Kendall, a well-established watchmaker on whom the Commissioners had called as an expert witness at the discovery of H4. He published some thoughts on timekeeping at sea the same year but seems only to have begun making marine timekeepers in the early 1770s, after ill health prompted his retirement from his London business.[67] By June 1774 he had asked the Earl of Sandwich for a formal trial of a watch, which ran alongside K3 at the Royal Observatory between December 1774 and March 1775 (Table 6.2).[68] It stopped twice, however, once with a broken mainspring. Following repairs, it was tested again from late 1776 and performed sufficiently well that the Commissioners awarded Mudge £500 to make two watches for trial and potential reward.[69] Referred to as Green and Blue, the two near-identical timekeepers underwent three trials between 1779 and 1790 (Figure 6.2 and Table 6.2).[70] Well before these began, however, Mudge had made clear his objections to the conditions of the Act of 1774, as well as to the conduct of the trial of his first timekeeper.[71] The trials went ahead but Green and Blue never performed well enough to warrant further reward. The Board halted testing in December 1790.[72]

The matter did not end there. In June 1791, Mudge sent the Board a memorial asserting that, although the watches had not succeeded, they were 'superior

[67] Mudge, *Thoughts on the Means*; Penney, 'Thomas Mudge and the Longitude'; Seccombe and Penney, 'Mudge, Thomas'.

[68] Thomas Mudge, marine timekeeper, 1774, British Museum, museum no. 1958,1006.2119; BoL, confirmed minutes, 6 June 1774, CUL RGO 14/5, pp. 259–260.

[69] BoL, confirmed minutes, 4 March 1775 and 1 March 1777, CUL RGO 14/5, pp. 273–274, 311; results recorded in 'Books on Chronometer Trials', 1770–1790, CUL RGO 4/312, pp. 49v–57r, 100v–105v.

[70] Thomas Mudge, marine timekeeper 'Green', 1776, private collection; Thomas Mudge, marine timekeeper 'Blue', 1776, Mathematisch-Physikalischer Salon, Dresden, inv. no. D IV b 11; results recorded in 'Books on Chronometer Trials', 1770–1790, CUL RGO 4/312, pp. 57v–83r, 91v–98v, 107v–125v.

[71] BoL, confirmed minutes, 27 May 1775, CUL RGO 14/5, pp. 279–280.

[72] BoL, confirmed minutes, 4 December 1790, CUL RGO 14/6, pp. 2:159–2:160.

Account of the going of Mr. Thos. Mudge's Time-keeper called Green at the Royal Observatory continued.

95

1784		Mean State of Ther.	Daily rate	Sum of daily rates	Compd. Varn.	Error of Watch
April	24	49	+ 2,98	+ 2.40,43	− 24,00	+ 3.4,4
	25	48½	+ 3,48	+ 2.43,91	− 24,09	+ 3.8,0
	26	48	+ 3,68	+ 2.47,59	− 24,17	+ 3.11,8
	27	54	+ 3,57	+ 2.51,16	− 24,26	+ 3.15,4
	28	48½	+ 3,36	+ 2.54,52	− 24,35	+ 3.18,9
	29	49½	+ 3,45	+ 2.57,97	− 24,43	+ 3.22,4
	30	50	+ 3,24	+ 3.1,21	− 24,52	+ 3.25,7
May	1	49	+ 3,13	+ 3.4,34	− 24,60	+ 3.28,9
	2	47½	+ 3,72	+ 3.8,06	− 24,69	+ 3.32,8
	3	50	+ 3,65	+ 3.11,71	− 24,78	+ 3.36,5
	4	51	+ 3,07	+ 3.14,78	− 24,86	+ 3.39,6
	5	54	+ 2,93	+ 3.17,71	− 24,95	+ 3.42,7
	6	55	+ 2,41	+ 3.20,12	− 25,04	+ 3.45,2
	7	58	+ 2,46	+ 3.22,58	− 25,12	+ 3.47,7
	8	58	+ 2,19	+ 3.24,77	− 25,21	+ 3.50,0
	9	60	+ 2,52	+ 3.27,29	− 25,30	+ 3.52,6
	10	63	+ 2,32	+ 3.29,61	− 25,38	+ 3.55,0
	11	62	+ 2,36	+ 3.31,97	− 25,47	+ 3.57,4
	12	58	+ 2,22	+ 3.34,19	− 25,55	+ 4.0,7
	13	56½	+ 2,38	+ 3.36,57	− 25,64	+ 4.2,2
	14	56	+ 2,45	+ 3.39,02	− 25,72	+ 4.4,7
	15	58	+ 2,45	+ 3.41,47	− 25,81	+ 4.7,3
	16	60	+ 2,41	+ 3.43,88	− 25,90	+ 4.9,8
	17	62½	+ 2,37	+ 3.46,25	− 26,00	+ 4.12,3
	18	62½	+ 2,19	+ 3.48,44	− 26,07	+ 4.14,5
		Watch forgotten to be wound up was found stopped at 0.59.36 and took 7¼ turns in winding up. Set it going again				
	20	63	+ 2,42	+ 3.50,86	− 26,16	+ 4.17,0
	21	62½	+ 2,35	+ 3.53,21	− 26,24	+ 4.19,4
	22	65	+ 2,45	+ 3.55,66	− 26,33	+ 4.22,0
	23	60½	+ 2,22	+ 3.57,88	− 26,42	+ 4.24,3
	24	68	+ 2,09	+ 3.59,97	− 26,51	+ 4.26,5
	25	69	+ 2,15	+ 4.2,12	− 26,59	+ 4.28,7
	26	69	+ 1,78	+ 4.3,90	− 26,68	+ 4.30,6
	27	64	+ 2,08	+ 4.5,98	− 26,76	+ 4.32,7
	28	62	+ 2,25	+ 4.8,23	− 26,85	+ 4.35,1
	29	58	+ 2,39	+ 4.10,62	− 26,94	+ 4.37,6

1784		Mean State of Ther.	Daily rate	Sum of daily rates	Compd. Varn.	Error of Watch
May	30	60½	+ 2,37	+ 4.13,19	− 27,02	+ 4.40,2
	31	60	+ 2,37	+ 4.15,56	− 27,11	+ 4.42,7
June	1	59	+ 2,51	+ 4.18,07	− 27,19	+ 4.45,3
	2	60	+ 2,53	+ 4.20,60	− 27,28	+ 4.47,9
	3	56	+ 2,74	+ 4.23,34	− 27,37	+ 4.50,7
	4	61½	+ 2,88	+ 4.26,22	− 27,45	+ 4.53,7
	5	63½	+ 2,56	+ 4.28,78	− 27,54	+ 4.56,3
	6	61	+ 2,60	+ 4.31,38	− 27,63	+ 4.59,0
	7	59	+ 2,92	+ 4.34,30	− 27,71	+ 5.2,0
	8	58	+ 2,74	+ 4.37,04	− 27,80	+ 5.4,8
	9	61	+ 2,83	+ 4.39,87	− 27,89	+ 5.7,8
	10	60½	+ 2,57	+ 4.42,44	− 27,97	+ 5.10,4
	11	59½	+ 2,76	+ 4.45,20	− 28,06	+ 5.13,3
	12	61	+ 2,74	+ 4.47,94	− 28,14	+ 5.16,1
	13	61	+ 2,74	+ 4.50,68	− 28,23	+ 5.18,9
	14	60½	+ 2,72	+ 4.53,40	− 28,32	+ 5.21,7
	15	61½	+ 2,61	+ 4.56,01	− 28,40	+ 5.24,4
	16	63½	+ 2,57	+ 4.58,58	− 28,49	+ 5.27,1
	17	61½	+ 2,23	+ 5.0,81	− 28,58	+ 5.29,4
	18	60	+ 2,74	+ 5.3,55	− 28,66	+ 5.32,2
	19	60	+ 2,73	+ 5.6,28	− 28,75	+ 5.35,0
	20	60	+ 2,70	+ 5.8,98	− 28,84	+ 5.37,8
	21	59½	+ 2,80	+ 5.11,78	− 28,92	+ 5.40,7
	22	58½	+ 2,74	+ 5.14,52	− 29,01	+ 5.43,5
	23	57½	+ 3,07	+ 5.17,59	− 29,09	+ 5.47,7
	24	56½	+ 2,74	+ 5.20,33	− 29,18	+ 5.49,5
	25	56½	+ 3,07	+ 5.23,40	− 29,27	+ 5.52,7
	26	58	+ 2,98	+ 5.26,38	− 29,35	+ 5.55,7
	27	58	+ 2,93	+ 5.29,31	− 29,44	+ 5.58,8
	28	58	+ 2,98	+ 5.32,29	− 29,53	+ 6.1,8
	29	57	+ 2,93	+ 5.35,22	− 29,61	+ 6.4,8
	30	58	+ 2,98	+ 5.38,20	− 29,70	+ 6.7,9
July	1	58	+ 3,01	+ 5.41,21	− 29,78	+ 6.11,0
	2	57½	+ 2,93	+ 5.44,14	− 29,87	+ 6.14,0
	3	59½	+ 3,13	+ 5.47,27	− 29,96	+ 6.17,2
	4	60½	+ 2,62	+ 5.49,89	− 30,04	+ 6.19,9
	5	64	+ 2,65	+ 5.52,54	− 30,13	+ 6.22,7
	6	68	+ 2,71	+ 5.55,25	− 30,22	+ 6.25,5
	7	70	+ 2,24	+ 5.57,49	− 30,30	+ 6.27,8

Figure 6.2 Some of the results of trials of Mudge 'Green' at the Royal Observatory, 1784, CUL RGO 4/312, p. 95r. Reproduced by kind permission of the Syndics of Cambridge University Library

to any that have hitherto been invented, and are constructed upon such principles as will render them permanently useful'. Having devoted almost twenty years to them, he hoped the Commissioners would offer him a reward, 'upon his making a discovery of the principles upon which his Timekeepers are constructed', principles he considered 'permanent in their nature'.[73] The Commissioners seem to have made no response until the matter was taken up by Mudge's son, lawyer Thomas Mudge junior, who was already acting on his father's behalf before the Greenwich trials ended.[74] Bypassing the Commissioners, Mudge junior petitioned Parliament in March 1792, having also published an account of his father's work and dealings with the Board of Longitude. His account accused Maskelyne of mistreating his father's timekeepers, withholding results, applying a poor method for calculating rate, carrying out and assessing the trials in ways that were inconsistent with those of Harrison's timekeeper, and deliberate deceit.[75] The parliamentary petition reiterated Mudge's long devotion to marine timekeepers and his dealings with the Board. It also placed the question of rate at the heart of the debate, suggesting that 'if the most proper Method of Calculation had been adopted … it would have been evident that they had gone much within the Limits which the Act prescribes'. Lastly, Mudge reminded MPs of Harrison's successful appeal to them and suggested that his father's timekeepers were superior.[76]

The House of Commons agreed to hear the petition and an investigating committee met for eight days in May 1793, publishing its report the following month.[77] By then, the Commissioners had moved to head off the attacks on its conduct. Early in 1792, Maskelyne had published a rebuttal of Mudge's allegations (a move he had contemplated in the face of Harrison's attacks in the 1760s but never executed). He had the support of the Board, which paid for the publication and advertisements in the *Morning Post* and other periodicals.[78] Mudge junior's even lengthier counter-response appeared later that year and heightened the attack, claiming that Maskelyne:

> has been a confirmed enemy to the success of the Time-keepers, and has supported his opposition not only by the most uncandid, and disingenuous conduct, but frequently by round and bold assertions, which he must have known at the time not to be true.[79]

In April 1793, with the parliamentary committee imminent, the Board nominated Banks, Maskelyne and Anthony Shepherd, Plumian professor at Cambridge,

[73] Thomas Mudge to BoL, 1791, CUL RGO 14/23, pp. 159r–159v; BoL, confirmed minutes, 11 June 1791, CUL RGO 14/6, p. 2:170.

[74] Thomas Mudge junior to BoL, 24 November 1790, CUL RGO 14/23, pp. 157r–157v.

[75] Mudge junior, *A Narrative of Facts*. The address to the Earl of Chatham, First Lord of the Admiralty, is dated 3 December 1791.

[76] *Journals of the House of Commons*, 47, p. 521.

[77] The committee comprised Sir Gilbert Elliot (chair), William Pitt (for two meetings), Sir George Shuckburgh, William Windham, Dudley Ryder and Charles Bragge. Pitt was Master of Trinity House at this time.

[78] BoL, confirmed minutes, 1 December 1792, CUL RGO 14/6, p. 2:188; bills for printing and advertising Maskelyne's *Answer*, February–May 1792, CUL RGO 14/15, pp. 234r–237r; Maskelyne, *An Answer to a Pamphlet*.

[79] Mudge junior, *A Reply to the Answer*, p. 133.

'to watch the progress of the said Petition through both Houses of Parliament', giving them 'access to all such Books and Papers belonging to the Board as they may judge necessary to disprove any of the Allegations contained in the said Petition'.[80] This subcommittee published its objections, while Banks circulated a pamphlet attacking Mudge's claim and wrote to MPs to ensure that Mudge's supporters did not pack the investigating committee.[81] Rewarding Mudge, he argued to William Windham, one of the leading MPs supporting the watchmaker, would be to reward 'an inferior artist, manifestly to the injury & discouragement of those who are superior to him in the Same Line'. The Board would be 'sorely humiliated', with its future decisions 'liable to be superseded by the jurisdiction of an incompetent Tribunal, & its Functions … thereby renderd useless'.[82]

The conduct of the observatory trials and the proper methods for assessing performance formed the core of Mudge junior's criticisms and of the evidence heard by the parliamentary committee. The principal issue was setting a rate as the benchmark for a watch's performance during the trial. Mudge junior argued that a rating period equivalent to that of the trial was required. Six months would be needed to assign an initial rate, he suggested, since the Act of 1774 expected a watch worthy of reward to be able to determine longitude within the specified limits 'in all Voyages for the Space of six Months'.[83] Maskelyne argued that this would never be practicable for typical voyages and that one month was pragmatic and sufficient. But there was further disagreement as to whether the month immediately preceding the formal trial period was the appropriate benchmark for every subsequent month in a trial lasting one year, with other options mooted.[84] The dispute repeated issues raised in Harrison's condemnation of H4's trial at the Royal Observatory. Maskelyne now argued that 'the more enlightened of the present age' should judge the merits of each rating method for themselves, adding the caveat that future timekeepers might require new modes of trial and calculation.[85] This would remain a contested matter but was crucial: although every timekeeper would not undergo testing at the Royal Observatory, each would need to have its rate determined before going to sea and, ideally, during any long voyage.

Faced with technical arguments about the merits of specific mechanical constructions, the parliamentary committee appointed a subcommittee: Samuel Horsley, mathematician and Bishop of St David's; mathematician George Atwood; natural philosopher Jean-André Deluc; instrument makers Jesse

[80] BoL, confirmed minutes, 2 March 1793, CUL RGO 14/6, p. 2:192.

[81] Joseph Banks to several members of the House of Commons, 22 April 1793, in Mudge junior, *A Description, with Plates*, p. v (the book dedicated to Windham); Anonymous, *Objections to Mr. Mudge's Pretensions*; Banks, *Observations on Mr. Mudge's Application*, dated 27 March 1793.

[82] Joseph Banks to William Windham, 20 April 1793, in Chambers (ed.), *The Letters of Sir Joseph Banks*, p. 151. Mudge junior, *A Reply to the Answer*, p. 170, commends Windham's conduct.

[83] Discovery of Longitude at Sea Act, 1774 (14 Geo 3 c 66).

[84] Mudge, *A Narrative of Facts*, pp. 19–31; Maskelyne, *An Answer to a Pamphlet*, pp. 83–106.

[85] Maskelyne, *An Answer to a Pamphlet*, p. 106.

Ramsden and Edward Troughton; and watchmakers John Holmes, Charles Haley and William Howells.[86] Meeting on 14 May, the subcommittee witnessed the dismantling of one of Mudge's timekeepers by watchmaker Matthew Dutton in order to assess its construction.

The witnesses called before the main committee included William Wales, Josiah Emery, John Arnold, Mudge junior, George Gilpin, Joseph Lindley, John Crosley and Aaron Graham. They offered often contradictory evidence on how to rate a watch and assess performance, as well as the relative quality of timekeepers by Mudge, Arnold, Emery and Haley. Maskelyne also gave evidence about the Royal Observatory trials, while Banks reiterated the Board's position. Appearing on 17 May, Banks restated the concerns he had expressed previously,

> that the majority, if not the whole of the Board, will consider a reward given by the House of Commons, in direct opposition to their opinion, as a censure upon their Conduct in the execution of the trust devolved on them by Parliament, and that they had far rather the House of Commons would make an enquiry into their management of public money, if they suspect them to have done wrong, than to thus cast an oblique reflection upon them.[87]

Banks's objections and Maskelyne's defence of the trials notwithstanding, the committee argued in Mudge's favour, citing his long dedication to making marine timekeepers. It found his watches to be 'of great and indubitable excellence' and believed that they contained 'an important improvement in the art of constructing timekeepers'.[88] As already noted, the committee also criticised the extreme difficulties imposed by the Act of 1774. It demurred, however, on questions around the conduct of the trials and the best method for establishing rate, noting only in the words of its subcommittee that:

> no judgment can be formed of the exactness of any timekeeper by theoretical reasoning upon the principles of its construction, with such certainty as with safety to be relied upon, except it be confirmed by experiments of the actual performance of the machine.[89]

Nevertheless, the committee seemed unconvinced that it would ever be possible to find a satisfactory method for assigning rate, noting the ambiguity in what different experts meant by the term.[90]

The Board of Longitude met shortly after the parliamentary committee submitted its report and resolved that its nominated subcommittee should pursue matters. Later in June, however, Parliament confirmed a further reward to Mudge of £2500 (bringing the total reward to £3000), a testament to his

[86] Nevil Maskelyne to George Atwood, 14 May 1793, NMM AGC/M/16, which discusses temperature compensation and isochronism, appears to show that Maskelyne used personal contacts to make the Board's case and steer the assessment. Atwood was a close friend of the prime minister, William Pitt.

[87] House of Commons Select Committee, *Report*, pp. 52–53. [88] Ibid., pp. 29, 35.

[89] Ibid., p. 32. [90] Ibid., pp. 8–9.

son's successful lobbying.[91] Mudge junior later sought to capitalise on the affair by establishing a manufactory for timekeepers of his father's design.[92] Approached in 1796 about a proposal to supply the Admiralty with these timekeepers, the Board of Longitude replied:

> We cannot but applaud the liberality manifested by the Board of Admiralty in shewing a disposition to encourage the making of good time-keepers for the use of the navy at the public expence. As such those of Mr. Mudge's construction deserve their attention. But we should not do justice to the trust reposed in us, and should, at the same time, pay a bad compliment to the liberality of the Board of Admiralty, if we concealed from them that other artists have invented time-keepers of a more simple construction, and sold them for about half the price required by Mr. Mudge, which from the trials of them that have been made at the Royal Observatory, and else where, may be depended on for keeping time full as well as Mr. Mudge's. These Artists are Mr. John Arnold and Mr. Thomas Earnshaw.[93]

'Persons Better Qualified': The Arnold–Earnshaw Dispute

Between them, John Arnold and Thomas Earnshaw are credited with providing the bridge between Harrison's ideas embodied in H4 and the commercial production of marine chronometers (with credit also to French makers such as Pierre Le Roy).[94] For the Board of Longitude, the two makers were long-standing suppliers and sometimes troublesome petitioners. Even after Arnold's death, their rivalry drew the Board into a protracted debate that ended only when another parliamentary committee acknowledged the Commissioners' right to make lasting judgements on matters entrusted to them.

Both makers' initial interactions with the Commissioners showed promise but friction with each arose within a few years. As Tables 6.2–6.4 indicate, Arnold was a trusted supplier to the Board and regularly had timekeepers tested at the Royal Observatory. These trials seem not to have all had equal status, however, with only a few relating to timekeepers owned or considered for purchase by the Board. Arnold's good relations with Maskelyne allowed him to bring other timekeepers to Greenwich for what George Gilpin described as 'private trials'.[95]

[91] *Journals of the House of Commons*, 48, p. 946; BoL, confirmed minutes, 1 June 1793, CUL RGO 14/6, pp. 2:200–2:201.

[92] Betts, *Marine Chronometers at Greenwich*, pp. 205–207.

[93] BoL, confirmed minutes, 11 June 1796, CUL RGO 14/6, p. 2:264; Betts, *Marine Chronometers at Greenwich*, p. 48. The sea trials on the *Sans Pareil* seem to have arisen from Thomas Mudge junior's dealings with the Admiralty.

[94] Betts, 'Arnold and Earnshaw'; Betts, *Marine Chronometers at Greenwich*, pp. 37–54; Betts, 'The Quest for Precision', pp. 330–333; Davies, 'The Life and Death', p. 511; Gould, *The Marine Chronometer*, pp. 105, 114–116, 127–128. For the development of marine timekeepers in France, see Fauque, 'Testing Longitude Methods'.

[95] House of Commons Select Committee, *Report*, p. 90. McEvoy, 'Maskelyne's Time', pp. 176–178, discusses Arnold's other work for the Royal Observatory.

Having awarded Arnold £700 by 1773 to allow him to pursue his work, the Board turned down a request for further support in 1774, insisting that he first provide proof either 'of the merits of the Watches, which they have already had of him, or … that he has already made some very considerable improvements'.[96] Only in 1779 was another award of £500 agreed.[97] In each case, the rewards were considered advances to cover the purchase of timekeepers and to be set against any future reward under the relevant Act.[98]

Additional payments followed until the end of 1782, when Arnold presented a memorial offering terms upon which he might surrender his patent for a new escapement and make a full discovery of its principles before expert witnesses.[99] He asked that a lifetime annuity be granted to him, his wife and his son, that no one else be eligible to receive a reward on the basis of his invention, and that he be eligible for the full reward under the Act of 1774 once its further conditions were met. He also suggested that Larcum Kendall be commissioned to make a watch with his new escapement to show that the watches could be produced by other makers. Should these conditions not be acceptable, Arnold suggested instead a £1000 advance for him to continue making low-priced watches with the new escapement.[100]

The Board considered and rejected the memorial the following July, after which Arnold declared his intention to submit two watches for trial under the terms of the Act of 1774.[101] By March 1784, however, he had offended the Commissioners with a memorial in which 'Professor Hornsby's name was mentioned in a very disrespectful manner', leading them to refuse further dealings until he withdrew his accusation.[102] The contentious charge was that Hornsby had 'used his utmost endeavours to depreciate' Arnold's professional character by seeking to use the performance of two unfinished watches (then with Alexander Aubert) to judge all his work, and that Hornsby had cast doubts upon Arnold's invention.[103] With the disagreement either resolved or dropped, Arnold had two timekeepers – number 15 and a box chronometer – under test in Greenwich by August 1784. He indicated, however, that only number 15 was intended for consideration for a reward; he had another nearly

[96] BoL, confirmed minutes, 25 June 1774, CUL RGO 14/5, p. 260.
[97] BoL, confirmed minutes, 6 March 1779, CUL RGO 14/5, p. 343.
[98] Arnold, *An Account Kept During Thirteen Months*, p. 4, shows the financial situation in 1780: of £1400 advanced to Arnold to date, £378 was for the supply of six timekeepers (itemised), leaving £1022 against any future reward.
[99] BoL, confirmed minutes, 7 December 1782, CUL RGO 14/6, p. 2:41. Howse, 'Britain's Board of Longitude', p. 406.
[100] John Arnold, 'The Memorial of John Arnold', 1782, RGO 14/25, pp. 86r–86v.
[101] BoL, confirmed minutes, 19 July and 22 November 1783, CUL RGO 14/6, pp. 2:50, 2:54.
[102] BoL, confirmed minutes, 6 March 1784, CUL RGO 14/6, p. 2:61.
[103] John Arnold to Lord Howe, 6 March 1784, CUL RGO 14/25, p. 99r.

ready for formal trial.[104] Nothing further is recorded, however, so it seems that no definitive trial took place.[105] Thereafter, until his death in 1799, Arnold's further dealings with the Board (besides appearing as a witness at the hearings for Mudge's petition) seem only to have been maintenance and supply work.

Thomas Earnshaw came on the scene some years after Arnold's near attempt at the longitude rewards. In 1789, he had a watch on trial at the Royal Observatory and seems to have forged a productive relationship with Maskelyne.[106] Late in 1791, he declared his intention to submit two timekeepers for trial under the terms of the Act of 1774 and six months later requested assistance to purchase a transit instrument and build an ice house, presumably for testing timekeepers. The Board deferred a decision on the latter request.[107] It was not until 1797 that Earnshaw petitioned again, requesting some reward in recognition of the good performance of his timekeepers.[108] The Commissioners declined but agreed to advance £300, the price of two watches for formal trial.[109] Identified as box timekeepers numbers 1 and 2, they travelled to Greenwich for three trials between January 1798 and November 1802 (see Table 6.2).[110] In December 1802, however, the Board ended the trials: the timekeepers' performance had been good but had not met the conditions of the Act and the Commissioners declined Earnshaw's suggestion of sea trials.[111]

The following March, Earnshaw petitioned again, requesting a reward 'if the Board are of opinion that his time-keepers exceed in accuracy of going all that have hitherto been made'. The Commissioners were willing to offer a reward equal to the £3000 given to Mudge (if Parliament supplied the additional funds), providing Earnshaw disclose the construction details.[112] Just two weeks later, however, their resolution had been challenged on behalf of the late John Arnold, with the suggestion that Earnshaw's timekeepers performed no better than those by other makers. While an extraordinary meeting a fortnight later upheld the decision in favour of Earnshaw, on the basis of a comparison of

[104] BoL, confirmed minutes, 6 March 1784, CUL RGO 14/6, p. 2:66–2:67, 2:71; results for the box timekeeper, July–August 1784, CUL RGO 14/25, p. 134r; results for Arnold's watch no. 15, July–August 1784, CUL RGO 14/25, p. 135r.

[105] Maskelyne, *An Answer to a Pamphlet*, p. 116, states that no Arnold watches were put on trial for the maximum rewards available under the Act of 1774.

[106] McEvoy, 'Maskelyne's Time', pp. 183–187.

[107] Thomas Earnshaw to BoL, 23 November 1791, CUL RGO 14/25, pp. 194r–195r; BoL, confirmed minutes, 3 December 1791 and 2 June 1792, CUL RGO 14/6, p. 2:177, 2:185.

[108] Thomas Earnshaw to BoL, 27 November 1797, CUL RGO 14/25, pp. 196r–197v.

[109] BoL, confirmed minutes, 2 December 1802, CUL RGO 14/7, p. 2:19.

[110] Results and analyses of the three trials at CUL RGO 14/25, pp. 288r–331v.

[111] BoL, confirmed minutes, 3 December 1802, CUL RGO 14/6, p. 2:19.

[112] Thomas Earnshaw to BoL, 3 March 1803, CUL RGO 14/25, pp. 200r–201r; BoL, confirmed minutes, 3 March 1803, CUL RGO 14/6, p. 2:21–2:22. The Discovery of Longitude at Sea, etc. Act, 1803 (43 Geo 3. c. 118), copy at CUL RGO 14/1, pp. 58r–60v, authorised a further £5000 for the Board's use in July 1803.

land trials, this began a dispute pitching Earnshaw against John Roger Arnold as proxy for his late father.[113]

In its attempt to resolve matters, the Board questioned a number of watchmakers about their work and that of Earnshaw and Arnold in 1803 and the first half of 1804.[114] In March 1804, Earnshaw and John Roger Arnold were asked for models and drawings of their mechanisms with written explanations.[115] Ahead of the meeting, Banks had printed and distributed to the Commissioners his objections to rewarding Earnshaw, who for his part kept up a steady stream of petitions and memorials.[116] By the end of the year, Maskelyne had published (using Parliament's official printer) his arguments in favour of Earnshaw, countering Banks's claims.[117] Far from opposing the efforts of watchmakers in longitude matters, Maskelyne stressed the need for equitable treatment. What this might mean was still unresolved, however, particularly the question of setting a timekeeper's rate and assessing its performance.

Printed versions of the watchmakers' descriptions were delivered to the Board in March 1805. These were edited – requiring the 'leaving out of Mr Earnshaws all passages in which any offensive mention is made of Mr Arnold, of his time-keepers, or other Watch Makers' – and reformatted with broad margins for comments. In April and May 1805, Banks and Maskelyne sent printed copies (at least eleven and six, respectively) to 'persons able and willing to suggest such questions as may tend to the better and more perfect explanation of the Timekeepers described therein'.[118] Those from Banks included a handwritten request:

[113] BoL, confirmed minutes, 17 March 1803, CUL RGO 14/6, p. 2:26–2:27. It is assumed that Banks orchestrated the challenge.

[114] In addition to John Roger Arnold and Earnshaw, the Board questioned Robert Pennington (June 1803), John Brockbank (June 1803 and March 1804), Robert Best (December 1803 and March 1804), William Frodsham (March 1804) and Miles Brockbank (June 1804); see CUL RGO 14/25, pp. 248r–262r.

[115] BoL, confirmed minutes, 23 March 1804, CUL RGO 14/7, pp. 2:51–2:52; Thomas Earnshaw, model escapement, 1804, NMM ZAA0123 (the whereabouts of Arnold's model is now uncertain); Betts, *Marine Chronometers at Greenwich*, pp. 254–257; John Roger Arnold, 'Descriptions and Drawings of Mr. Arnold's escapement and Time-keeper', 1804, CUL RGO 14/25, pp. 4r–47r; Thomas Earnshaw, 'Description of Mr. Earnshaw's Chronometer', 1804, CUL RGO 14/25, pp. 139r–155r; Arnold, *Explanation of Timekeepers*, copy at CUL RGO 14/25, pp. 52r–62r; Earnshaw, *Explanation of Time-Keepers*, copy at CUL RGO 14/25, pp. 158r–167r.

[116] Banks, *Protest Against a Vote*, dated 19 March 1804 on p. 6; Earnshaw, petitions and memorials, 1791–1807, CUL RGO 14/25, pp. 194r–236v.

[117] Maskelyne, *Arguments for Giving a Reward*, dated 1 December on p. 14.

[118] BoL, confirmed minutes, 4 April 1805, CUL RGO 14/7, pp. 2:66–2:67; Arnold, *Explanation of Timekeepers*, copies annotated by responding makers in CUL RGO 14/26; Earnshaw, *Explanation of Time-Keepers*, copies annotated by responding makers in CUL RGO 14/27. Copies were returned from: Paul Philip Barraud, John Barwise, John and Miles Brockbank, John Grant, Peter Grimaldi, William Hardy, Charles Haley, William Howells, Robert Molyneux, William Nicholson, Robert Pennington, James Petto, Louis Recordon, Owen

to note particularly every article which you conceive to be a new invention or a thing unknown to the trade till it was communicated by this explanation; and how far in your opinion, each of these is an improvement likely to be advantageous to the public, either by rendering time-keepers more accurate in their performance, or less costly to the purchasers.[119]

Having considered the responses and questioned the two makers further, the Board resolved in December 1805 to give Arnold and Earnshaw equal rewards of £3000 (less any advances), once the two had sworn an oath that they had disclosed:

every secret, every discovery, and every new manner of constructing the parts of time-keepers, of putting them together and of finishing them, which we are possessed of, and also that we have fully disclosed and described in the said papers the best manner of adjusting time-keepers that we are acquainted with, and we moreover solemnly swear that we have not concealed from the Board of Longitude or retained to ourselves for our own advantage any matter or thing relative to the construction, the finishing, and the adjustment of time-keepers, which we have any reason to believe is not well understood and fully known by those persons who exercise the art of making time-keepers.[120]

The following year, the descriptions were published together in an edition that included the questions put to Arnold and Earnshaw and their answers, with an introduction by Maskelyne that briefly summarised the history of the dispute.[121]

In contrast to the Board's clashes with the Mudges, which primarily concerned the conduct and adjudication of trials, the investigation of the Arnold–Earnshaw dispute forced the Commissioners also to consider questions around priority of invention and the practices of the watchmaking trade. Notionally, the principal issues were whether Earnshaw's timekeepers performed better than those by other makers and, if so, whether this arose from a technical innovation attributable to Earnshaw. This latter point centred on the invention of

Robinson, Edward Troughton and Charles Young. Comments on both descriptions survive from all except John Brockbank (Arnold only), Miles Brockbank (Earnshaw only) and William Howells (Earnshaw only). Others were sent but not returned; for instance, copies of both descriptions sent by Banks to Alexander Dalrymple on 26 April 1805, but with no annotations by Dalrymple, in the library of the Institution of Civil Engineers, London. James Duncan sent comments on the explanations, but no annotated copies survive; James Duncan to BoL, 30 May 1805, CUL RGO 14/25, pp. 130r–131r; BoL, confirmed minutes, 6 March 1806, CUL RGO 14/7, pp. 2:91–2:92.

119 Joseph Banks, handwritten note in a copy sent to Barraud of Arnold, *Explanation of Timekeepers*, CUL RGO 14/26, p. 4r.

120 BoL, confirmed minutes, 12 December 1805, CUL RGO 14/7, pp. 2:81–2:82. The decision reiterated and confirmed a resolution a year earlier agreeing rewards of £3000; BoL, confirmed minutes, 6 December 1804, CUL RGO 14/7, pp. 2:58–2:59.

121 Earnshaw and Arnold, *Explanations of Time-Keepers*; Betts, 'Introduction & Technical Appraisal'.

the spring-detent escapement, which Earnshaw claimed but Arnold had patented in 1782, and of the compensation balance, the successful form of which was attributable to Earnshaw.[122] In other words, the innovations at the heart of the dispute were well known and Banks's request regarding things hitherto unknown was unlikely to bear fruit.

As Banks's request reveals, the affair allowed the Commissioners to consider how timekeepers might be produced affordably. This was already evident in the verbal questioning in 1803–1804, which took two lines. First, the watchmakers were asked whether they used Arnold's or Earnshaw's construction in their own work, what difference there might be between the two, who should be credited with the disputed invention, and whether it would be worth rewarding Earnshaw to reveal the principles of his timekeepers. Second, the Board asked how many timekeepers each respondent typically made, whether pocket or box timekeepers appeared to go better, and how best to deploy and maintain them.

When the descriptions were published together in 1806, Maskelyne's preface noted the Board's ambition of making the manufacture of marine timekeepers 'more easy, common, and exact'.[123] This proved challenging. Several respondents felt that neither Arnold nor Earnshaw had communicated their principles well enough to have revealed anything of consequence. John Barwise believed that Arnold offered 'a just statement … of which the Trade in general have long been in compleat Possession, there is not a new Idea suggested in it'.[124] William Nicholson concluded that neither description would 'enable an Artist to make Time pieces less subject to vary in their performance or less costly to purchasers than those they now make'.[125]

The interviews and comments on the two descriptions also exposed enmity within the trade, mostly directed at Earnshaw.[126] For his part, Earnshaw threw copious accusations at his fellow watchmakers, notably the sharing of secrets, and singled out seven 'declared enemies': John Arnold, John Brockbank, Paul Barraud, Robert Pennington, Charles Haley, Peter Grimaldi and James Petto.[127] Of those respondents who chose a side, most supported Arnold. John Brockbank considered John Harrison the 'Father of Chronometry', but praised Arnold for having 'increased their [chronometers'] perfection'; Earnshaw had added

[122] Betts, *Marine Chronometers at Greenwich*, pp. 51–52, outlines the controversy and competing narratives.
[123] Earnshaw and Arnold, *Explanations of Time-Keepers*, preface.
[124] John Barwise, remarks on Arnold's *Explanation of Time-Keepers*, 1805, CUL RGO 14/26, p. 33v.
[125] William Nicholson, remarks on Arnold's *Explanation of Time-Keepers*, 5 June 1805, CUL RGO 14/26, p. 163v.
[126] Betts, 'Arnold and Earnshaw', p. 327. Gould, *The Marine Chronometer*, p. 124, suggests that the Board's investigation arose from accusations against Earnshaw by London watchmakers.
[127] Thomas Earnshaw to BoL, 3 July 1805, CUL RGO 14/25, pp. 275r–280r (p. 275v).

nothing.[128] Only Robert Best and William Frodsham supported Earnshaw's claim to prior invention. Conversely, William Hardy considered him no more than a copyist who offered nothing new, while John Barwise concluded that:

> A great many more Remarks might be made but I have neither Time nor Inclination to follow Mr Earnshaw through all his absurdities … Shall only beg leave to observe that this Explanation if it may be so calld abounds with Misrepresentations, nor does it communicate any Information that will enable me to make Time Keepers better than I have done … I sincerely wish it did.[129]

Robert Molyneux, however, hinted that Earnshaw 'has not imparted all the Secrets of his Business to the Board'.[130]

In seeking to resolve the dispute, the Board drew on a broad range of evidence: timekeeper trials at the Royal Observatory and elsewhere; verbal, written and printed evidence from John Roger Arnold, Earnshaw and several London watchmakers and other experts; and reports from users. This allowed them to probe the secrets of the watchmaking trade in the hope of gaining control where they had none and to attempt to understand how increased production of accurate and reliable marine timekeepers could be made cheaper. Where the investigations failed was in seeking to settle the question of priority of invention of the detached escapement, seemingly because this was not the Commissioner's true intention.[131]

While the final decision of the Board satisfied John Roger Arnold, Earnshaw felt that he had suffered an injustice.[132] Advertisements in the *Morning Chronicle* and *The Times* in February 1806 (and subsequently in other periodicals) accused Banks and Gilpin of 'extraordinary exertions' in encouraging 'the selfish combination of Watchmakers against him'. Earnshaw also took exception to Banks's pamphlet against the proposed reward to him.[133] It was a public attack that widened divisions within the Board, notably between Banks as a supporter of the Arnolds and Maskelyne of Earnshaw. At the Board's next meeting, Banks complained of the 'libellous nature' of Earnshaw's remarks,

[128] John Brockbank, remarks on Arnold's *Explanation of Time-Keepers*, 31 May 1805, CUL RGO 14/26, p. 40r.

[129] William Hardy, remarks on Earnshaw's *Explanation of Time-Keepers*, 1805, CUL RGO 14/27, p. 111r; John Barwise, remarks on Earnshaw's *Explanation of Time-Keepers*, 1805, CUL RGO 14/27, p. 33v.

[130] Robert Molyneux, remarks on Earnshaw's *Explanation of Time-Keepers*, 1805, CUL RGO 14/27, pp. 159r–159v. Betts, *Marine Chronometers at Greenwich*, p. 255, suggests that one of the secrets Earnshaw kept from the Board was his method for rating chronometers using the vernier method.

[131] Betts, 'Arnold and Earnshaw', p. 328; Phillips, 'Making Time Fit', pp. 70–113 (esp. pp. 90, 107, 111).

[132] By April, Arnold was mentioning the reward in advertisements; for example, *Morning Post*, 3 April 1806.

[133] 'Longitude', *The Morning Chronicle*, 4 February 1806; Banks, *Protest Against a Vote*.

proposing that the Board issue a suit against him. While the Commissioners acknowledged Banks's long and 'disinterested' contributions and condemned Earnshaw's 'ill advised and indecent expressions', they baulked at further action. Banks therefore proclaimed that he would personally sue Earnshaw.[134] Although he drew short of this, his long-time ally Alexander Dalrymple published a substantial response to Earnshaw, and Banks attended no further meetings of the Board until immediately after Maskelyne's death in 1811.[135]

Earnshaw, meanwhile, sent further petitions, initially offering to sell for 400 guineas (£420) the two timekeepers tested at the Royal Observatory. The Board offered 300 guineas (£315), a sum he finally refused, deciding instead to appeal to Parliament.[136] This led to discussion within the Board and the decision to inform the First Lord of the Admiralty, the Chancellor of the Exchequer and the Speaker of the House of Commons of the Commissioners' opposition to any appeal.[137]

Earnshaw submitted his petition in 1808, when he also published a long, self-aggrandising account of his longitude work and his many grievances. As with previous diatribes, Earnshaw dwelt on the many trials at the Royal Observatory (and elsewhere), arguing that the correct assignment of rate would have shown that his watches performed within the limits of the Act of 1774. Unlike those earlier attacks, however, it was Banks rather than Maskelyne whom Earnshaw identified as his foe.[138]

While the 1808 petition failed, a parliamentary committee was granted the following year after Earnshaw resubmitted. The committee convened in May 1809, by which time Maskelyne had composed a paper detailing the Board's objections.[139] The outcome proved quite different from those of Harrison's and Mudge's appeals. Reporting on behalf of the parliamentary committee on 31 May, Davies Giddy (a Fellow of the Royal Society and one of Banks's circle) noted that, while its members believed that Earnshaw had made more timekeepers of merit than any other maker, they were unable to decide on the matter of priority in inventions,

[134] BoL, confirmed minutes, 6 March 1806, CUL RGO 14/7, pp. 2:88–2:89.

[135] Dalrymple, *Longitude*, preface dated 28 March 1806.

[136] Thomas Earnshaw to BoL, 6 March, 5 June and 4 December 1806, CUL RGO 14/25, pp. 228r–233v; BoL, confirmed minutes, 6 March, 5 June and 4 December 1806, CUL RGO 14/7, pp. 2:90, 2:95, 2:98–2:99.

[137] Thomas Earnshaw to BoL, 2 December 1807, CUL RGO 14/25, pp. 234r–236v; BoL, confirmed minutes, 5 March and 3 December 1807, CUL RGO 14/7, pp. 2:101–2:102, 2:110–2:112.

[138] Earnshaw, *Longitude, an Appeal*. A copy of Earnshaw's book in the British Library (shelfmark G.19221) includes a letter from Earnshaw to Thomas Grenville, 15 February 1809, thanking him for his support and hoping for a 'just reward'. A former MP (and one again in later years), Grenville had been First Lord of the Admiralty in 1806–1807. He was also William Windham's cousin.

[139] BoL, confirmed minutes, 8 May 1809, CUL RGO 14/7, pp. 2:129; Nevil Maskelyne, 'Remarks on Mr Earnshaw's Petition', 8 May 1809, CUL RGO 14/10, pp. 237r–242r; *Journals of the House of Commons*, 63, pp. 101–102, and 64, p. 22.

which seem in part to have been borrowed from foreign artists, and rather to have proceeded from one contrivance or suggestion to another, than to have started into excellence by the discovery of any one individual.

Moreover, the committee deferred to:

the Board of Longitude appointed by the Legislature for the express purpose of deciding on this subject, composed of persons better qualified for the purpose, by their learning and habits of life, than almost any others can pretend to be, and who appear to the Committee to have diligently, honourably, and liberally discharged the duties of the high trust reposed in them.[140]

Parliament now recognised the specific expertise of the Commissioners and the rights of the Board to make lasting judgements.

Conclusion

Between the 1770s and the early nineteenth century, the Board of Longitude focused on the question of practicability that had become central in its dealings with John Harrison. This saw trial at the Royal Observatory enshrined as a standard for the assessment of timekeepers and thus an area of disagreement with those unhappy with their conduct or outcome. It brought a handful of makers into initially productive relationships with the Board, which was willing to award sizeable sums to encourage those whose work showed promise. Nevertheless, most of those same makers came into conflict with the Board as they pressed for larger rewards, with the conditions of the Act of 1774 proving insurmountable. Harrison's name cast a lasting shadow over these debates, with makers including Mudge and Earnshaw asserting that their timekeepers outperformed Harrison's, yet he had received far greater rewards. Indeed, the Act of 1774 made this inevitable, even if it could be shown that his timekeepers had been improved upon.

While Earnshaw's failed petition to Parliament marked the end of serious public challenges to the Board over its role in the development and assessment of marine timekeepers, the culmination of this phase is best symbolised by events in 1821. The minutes of the Board's Committee for Examining Instruments and Proposals on 11 June 1821 note that, '[t]he Royal Observatory at Greenwich having been selected as the Depot for Chronometers, the committee recommends that they be placed under the care of the Astronomer Royal'.[141] John Pond, sixth Astronomer Royal, thus took over Thomas Hurd's duties as Superintendent of Timekeepers (although Hurd remained as

[140] House of Commons Committee, *Report*; *Journals of the House of Commons*, 64, p. 367.

[141] BoL Committee for Examining Instruments and Proposals, minutes, 11 June 1821, CUL RGO 14/10, pp. 35v–36r; John Wilson Croker to John Pond, 23 July 1821, CUL RGO 5/229, pp. 2r–2v (official confirmation of Pond's appointment).

Hydrographer of the Navy until his death in 1823). That same month, public notice was given that:

> The Lords Commissioners of the Admiralty, being desirous of increasing the number of chronometers for the use of his Majesty's Navy, and of encouraging the improved manufacture of that important article, do hereby give notice, that a depot for the reception of chronometers is opened at the Royal Observatory of Greenwich, where the makers will be permitted to deposit their chronometers, in order to their being tried, and ultimately purchased for the use of the navy, or of being disposed of by the proprietors to private purchasers. And, for further encouragement, their Lordships will purchase, at the end of each year, the chronometer which shall have kept the best time, at the price of 300*l.*, and the second best at the price of 200*l.*, provided that there have been above ten chronometers in the competition ... The other chronometers their Lordships may purchase, as they may think proper, at such sums as may be agreed upon with the makers, and their Lordships have reason to expect, that their annual rate of purchase, for some years to come, will be not less than ten chronometers in each year.[142]

The possibility of chronometer trials had been discussed previously, sometimes at Board meetings, but had not been put into practice in such a public way.[143] Running between 1822 and 1835, the new premium trials enshrined the Royal Observatory as the site of instrumental accreditation of chronometers, with the Board retaining an involvement in their conduct.[144] The Observatory also continued testing and rating chronometers already in the Navy stock. These were still relatively few in number, with still fewer regularly issued to ships, but numbers grew rapidly thereafter.[145]

With this move, the authorities made clear that assessing the quality of a chronometer meant determining its rate at an official site distant from a maker's workshop or some other location. Challenges from the trade would no longer be countenanced, nor would methods for applying rate be brought into question.[146] With the Royal Observatory as the new repository for chronometers, the Board of Longitude also began to relinquish control over its stock of instruments, although its interest in geodesy and pendulum work continued through the 1820s.[147]

[142] *London Gazette*, 26 June 1821.

[143] William Marsden to George Gilpin, 7 March 1805, CUL RGO 14/23, pp. 5r–10r; BoL, confirmed minutes, 7 March 1805 and 4 April 1805, CUL RGO 14/7, pp. 2:63, 2:68; Phillips, 'Making Time Fit', pp. 64–65. Marsden was First Secretary of the Admiralty. Other proposals include Thomas Firminger, 'Memoir on the Usefulness of Time-keepers in the Service of the Navy; and a Plan for Introducing Them with the Best Prospect of Success to Ensure Their Accuracy at the Least Expense', *Philosophical Magazine*, Series 1, 42 (1813), pp. 241–246, and Basil Hall to the Admiralty, 7 April 1820, TNA ADM 1/1956 (supply and rating of Royal Navy chronometers).

[144] For the premium trials and chronometers in the Royal Navy, see Akkermans, 'Chronometers and Chronometry', esp. pp. 47–99; Ishibashi, 'A Place for Managing Government Chronometers'.

[145] Ishibashi, 'A Place for Managing Government Chronometers', pp. 56–57.

[146] Phillips, 'Making Time Fit', p. 66.

[147] Waring, 'Thomas Young', p. 62.

7 Manufacturing the *Nautical Almanac*

Richard Dunn, Simon Schaffer and Sophie Waring

Introduction

The Board of Longitude's most enduring legacy has been the *Nautical Almanac*, still produced annually by HM Nautical Almanac Office. The Board's publishing activities began in the wake of the 1763–1764 sea trials to Barbados. Overseen by Astronomer Royal Nevil Maskelyne, production of the *Nautical Almanac* came to dominate the Board's business. Just as the disputes with John Harrison resulted in the creation of the Board of Longitude, so the requirements of managing a demanding publishing venture shaped its later trajectory.

This chapter gives an overview of the production of the *Nautical Almanac* and related publications from the mid-1760s to the dissolution of the Board of Longitude. The first sections explore the creation and management of the publication under Maskelyne, who instituted a system for computing and checking the data needed for each edition, and oversaw printing and publication. These activities transformed the Board as it became entangled in the world of London publishing, resulting in administrative change and considerable financial outlay. They also allowed the Board to tap into and in turn influence astronomical and other work in Europe and elsewhere, although the Commissioners' piecemeal attempts to encourage adoption of new longitude methods met with mixed results. The later part of the chapter focuses on debates and changes following Maskelyne's death in 1811, with first John Pond and then Thomas Young overseeing production. These disputes were bound up in the demise of the Board in 1828 and eventual establishment of the Nautical Almanac Office.

An Almanac and an Ephemeris

Nevil Maskelyne took up the post of Astronomer Royal the day before the Commissioners of Longitude met on 9 February 1765 to discuss the Barbados trials described in Chapter 5. At his first meeting as an *ex officio* Commissioner, Maskelyne at once presented a memorial that set out the feasibility of the lunar-distance method based on reliable lunar tables used in conjunction with an instrument for observing the Moon's position from a ship. The memorial was backed with an account of deploying the method on a voyage to St Helena in 1761–1762, after which he had published a description in the *British Mariner's Guide*. His experiences had, his published account emphasised, convinced him of the value of Nicolas-Louis de Lacaille's proposal for an almanac containing pre-computed tables that would reduce the number of calculations required while at sea.[1] Backed by testimony from four East India Company (EIC) officers invited to the meeting, Maskelyne concluded that 'nothing is wanting to make the same [the lunar-distance method] generally practicable at sea, but a Nautical Ephemeris, an Assistance which they and many more hope for from this Board'.[2] The Commissioners resolved to accept Maskelyne's recommendation and push for a new Longitude Act that in turn instructed and enabled the Board to establish the new publication.[3]

It would take almost two years for the first edition, for 1767, to be published as *The Nautical Almanac and Astronomical Ephemeris*. As Bennett notes, titling it as both almanac and ephemeris was significant with regard to Maskelyne's ambitions. As an almanac, it was to be of practical use to mariners. As an ephemeris, it was to answer the needs of astronomers.[4] Satisfying two audiences proved to be far from straightforward, however, and would become a particular issue in the Board's final years.[5]

With only a minor change after the first edition, the *Nautical Almanac* had a consistent structure throughout the time the Commissioners oversaw publication (Box 7.1).[6] This comprised three main sections: front matter, including legally required material and preface; monthly tables for the year; and back matter, including instructions for using the tables and additional articles of interest as decided by Maskelyne and other Commissioners.

The monthly tables contained the bulk of the data needed to determine longitude by lunar distance or by timekeeper, as well as for finding latitude.[7] To make a lunar determination, it was necessary simultaneously to measure the angular distance between the Moon and Sun or another star, as well as the altitude of each body, preferably several times. The observations were then 'cleared' by means of calculations that drew data from the *Nautical Almanac* and other publications (see below) to derive a value for the lunar distance as

[1] Maskelyne, *British Mariner's Guide*, pp. iv–v; Bennett, 'Mathematicians'; Boistel, 'From Lacaille to Lalande', pp. 59–60.

[2] BoL, confirmed minutes, 9 February 1765, CUL RGO 14/5, pp. 78–81. See also John Horsley, 'Observations and Calculations', CUL RGO 14/67, p. 5v; Horsley, 'An Account'.

[3] Discovery of Longitude at Sea Act 1765, 5 Geo 3 c 20; copy at CUL RGO 14/1, pp. 29–33.

[4] Bennett, 'The First Nautical Almanac', pp. 145–146; see also Perkins, *Visions*, p. 22.

[5] The *Connoissance des Temps* wrestled with the same challenge; see Boistel, *'Pour la gloire'*.

[6] Bennett, 'The First Nautical Almanac', p. 153, notes that, although the arrangement of the tables changed slightly from the 1768 volume, the explanation of their use incorporated the change only from the 1781 volume.

[7] Cotter, 'A History'. p. 102; Howse, *Nevil Maskelyne*, pp. 95–96.

BOX 7.1 STRUCTURE OF THE *NAUTICAL ALMANAC* UNDER THE BOARD OF LONGITUDE

a) Front matter

- Title page
- Extracts of relevant Longitude Acts
- Warrants from the Commissioners to print and sell the *Nautical Almanac*
- Preface(s)
- Explanation of characters
- Obliquity &c./Obliquity of the Ecliptic
- Eclipses for the year
- 'Principle Articles' (title from 1794), including chronological cycles, ember days, obliquity, eclipses, terms, moveable feasts
- Errata and/or corrections (where needed)

b) Monthly tables, comprising twelve pages for each month*

- I: Sundays, festivals and other remarkable days; phases of the Moon, occultations, entry of the Sun into zodiacal signs and other phenomena
- II: Sun's longitude, right ascension and declination, equation of time for noon at Greenwich each day
- III: additional information for the Sun; eclipses of Jupiter's satellites
- IV: celestial latitude and longitude, declination and time of meridian passage of Mercury, Venus, Mars, Jupiter, Saturn
- V: celestial latitude and longitude of the Moon for each day at noon and midnight
- VI: age, time of meridian passage, right ascension and declination of the Moon for each day at noon and midnight
- VII: Moon's semi-diameter and equatorial horizontal parallax for each day at noon and midnight
- VIII–XI: lunar distances from the Sun and appropriate stars every three hours
- XII: configurations of Jupiter's satellites

c) Back matter

- 'Explanation and Use of the Articles Contained in the Astronomical and Nautical Ephemeris'
- Works published by the Commissioners of Longitude (from 1769 edition); advertisements for *Greenwich Observations* and other publications
- Additional articles and other material

* Arrangement from the *Nautical Almanac* for 1768, with lunar distances as two double-page spreads. The volume for 1767 had the configurations of Jupiter's satellites on page V.

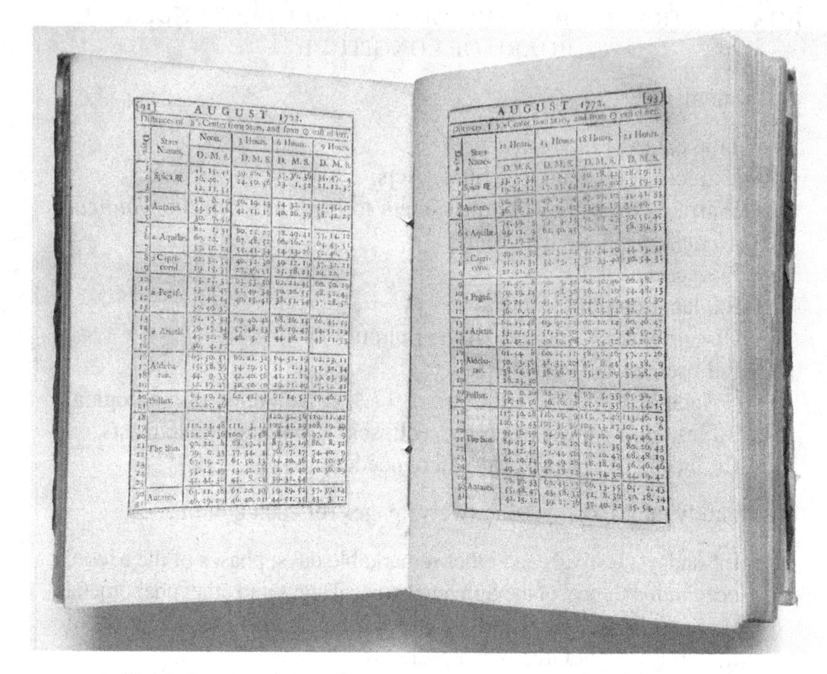

Figure 7.1 The *Nautical Almanac* for August 1772, showing the angular distances of the Moon from the Sun and stars at three-hour intervals.

if viewed from Earth's centre. Clearing was never a single method: at least forty variations had been proposed by the end of the century.[8] The final four pages of each month, which tabulated the Moon's distance from the Sun and selected stars at three-hour intervals, then allowed the navigator to find the equivalent time at Greenwich and, from the difference between that and the ship's local time (typically determined from Sun observations morning and afternoon), the longitude.[9] It was these tables in particular that relieved the mariner, in Maskelyne's words, 'from the Necessity of a Calculation, which he might think prolix and troublesome' (Figure 7.1).[10]

Although much of the data in the *Nautical Almanac* was for mariners, not all of it was. The tables relating to the planets and to Jupiter's satellites, for example, were suited instead to land-based astronomers and surveyors.[11] The same was true of the periodic inclusion of additional articles on a range of topics, often following rewards to their authors. These additions did include articles

[8] Cotter, 'A History', p. 305; Cotter, *A History of Nautical Astronomy*, pp. 208–237.

[9] Dunn and Higgitt, *Finding Longitude*, pp. 212–213; Howse, *Nevil Maskelyne*, pp. 92–93.

[10] *Nautical Almanac* for 1767, p. 164. See also Dunkin, 'Notes', p. 51; Grier, *When Computers*, p. 29; Wardhaugh, *Poor Robin's Prophecies*, p. 173.

[11] Bennett, 'The First Nautical Almanac', p. 153; Cotter, 'A History', p. 104.

of navigational use, such as alternative methods for finding latitude (in the editions for 1771, 1797 and others) and for clearing the lunar distance (1772), as well as articles relating to the Hadley quadrant (1774, 1788). Other articles were of purely astronomical interest, for example observations of the transit of Venus (1769), a forthcoming cometary return (1791), notes on Saturn's ring (1791) and the construction and care of reflecting telescopes (1787, 1788), as well as what were called 'Astronomical problems' (1778, 1781). Many of these additions were subsequently published together in a separate volume in 1813, following Board discussions that began in 1800.[12]

The main aim of the *Nautical Almanac* – to reduce the number of calculations needed to turn astronomical observations into navigational positions – and much of the tabulated data distinguished it from other serial publications. Some information echoed that in regularly updated navigational works, but the *Nautical Almanac* lacked data to support other procedures, such as traverse tables for dead reckoning, tables to determine the tides or detailed information about British coastal waters, as might be found in Colson's *Mariner's New Calendar* (editions from 1677 to 1785), Wakeley's *The Mariner's Compass Rectified* (1677–1796) or Haselden's *Seaman's Daily Assistant* (1757–1790).[13] Some of its tables likewise duplicated those in annually published almanacs, such as *The British Telescope*, *Merlinus Liberatus*, White's *Coelestial Atlas* and the immensely successful *Vox Stellarum*, notably the conventional calendar, positions of the planets, Moon and Sun, and the places and eclipses of Jupiter's satellites. But there were significant differences. Popular almanacs, for example, offered astrological and other information useful in daily life, not the data required for determining longitude. The *Nautical Almanac* did not, moreover, lie within the Stationers' Company monopoly control of almanacs, which was being challenged from the 1770s.[14]

Furthermore, while popular almanacs were typically published around the beginning of November for the following year, the Board's aspiration was that the *Nautical Almanac* would be published well in advance of when ships going on lengthy voyages might need its data.[15] Indeed, this desideratum was soon felt on James Cook's first Pacific voyage (1768–1771), since only the volumes for 1768 and 1769 were available when the *Endeavour* departed. From 1770, therefore, Cook and astronomer Charles Green had to refer to Maskelyne's *British Mariner's Guide* to perform the additional calculations the *Nautical Almanac*

[12] BoL, confirmed minutes, 7 June 1800, CUL RGO 14/6, pp. 2:327–2:328 (first proposed); Anonymous, *Selections*. Boistel, '*Pour la gloire*', discusses similar material in the *Connoissance des Temps*.

[13] Bolker, 'Lost at Sea', pp. 599–600; Cotter, 'A History', p. 81.

[14] Blagden, 'Thomas Carnan'.

[15] Capp, *Astrology*, pp. 238–269; Feist, 'The Stationers' Voice', p. 16; Perkins, *Visions*, pp. 13–16, 35.

tables avoided.[16] The initial aspiration was for publication two or three years in advance, but volumes were being published up to five years ahead by the late 1780s, and work on future volumes was paused in the 1790s when computing work had run even further ahead.[17]

Managing the Process

As Croarken and others have shown, regular and timely publication relied on the work of a geographically distributed network of mathematical practitioners known as computers and comparers. Under the system he created to ensure a steady flow of verified data, Maskelyne sent each computer instructions for the step-by-step calculations they were to undertake, as well as astronomical data, tables and other publications needed for the calculations. He assigned two computers to each month, one calculating for noon, the other for midnight. Once the computers returned their results, Maskelyne sent them to a comparer, who checked the calculations and made any corrections before returning them to Greenwich for assembly into the relevant edition.[18] Aided by the Board's Secretary, Maskelyne thus became manager of a network of pieceworkers, with whom he maintained regular postal communication to ensure they had the correct instructions and materials and returned data on time. Historians have plausibly seen Maskelyne's system as analogous to cottage-based forms of domesticated labour, eventually to be reorganised and managed under regimes of manufacture and large-scale computation (Figure 7.2).[19]

The arrangement suited the Commissioners, who limited their oversight to the authorisation of payments to computers, comparers, printers and publishers. Indeed, the minutes of Board meetings suggest that they were told few details.[20] Thus, although the Board was informed about the first appointments – of Israel Lyons, George Witchell, John Mapson and William Wales as computers and Richard Dunthorne as comparer – others were only mentioned if notable issues arose.[21] In January 1770, for instance, Maskelyne revealed that Joseph Keech and Reuben Robbins had been copying each other's work and were to be dismissed.[22] Thirty years later, he reported that Francis Simmonds had been found

[16] Dunn, 'James Cook', p. 95; Howse, 'The Principal Scientific Instruments', p. 133–134; see also Beaglehole, *Journals*, I, p. 392.

[17] BoL, confirmed minutes, 12 December 1767, CUL RGO 14/5, p. 165.

[18] Bennett, 'The First Nautical Almanac', p. 150; Croarken, 'Providing Longitude'; Croarken, 'Tabulating'; Croarken, 'Nevil Maskelyne', pp. 134–141.

[19] Daston, 'Calculation', pp. 14–15; Grier, *When Computers*, pp. 27–33.

[20] For evidence of the work involved, see Nevil Maskelyne, 'Diary of "Nautical Almanac" work', 1767–1811, CUL RGO 4/324; Nevil Maskelyne, 'Book of Accounts', 1765–1811, NMM MSK/2.

[21] BoL, confirmed minutes, 13 June (Lyons, Witchell, Mapson and Wales) and 18 July 1765 (Dunthorne), CUL RGO 14/5, pp. 101–102, 105.

[22] BoL, confirmed minutes, 13 January 1770, CUL RGO 14/5, p. 188. Croarken, 'Providing Longitude', p. 121, notes that both were re-employed but never allocated the same months.

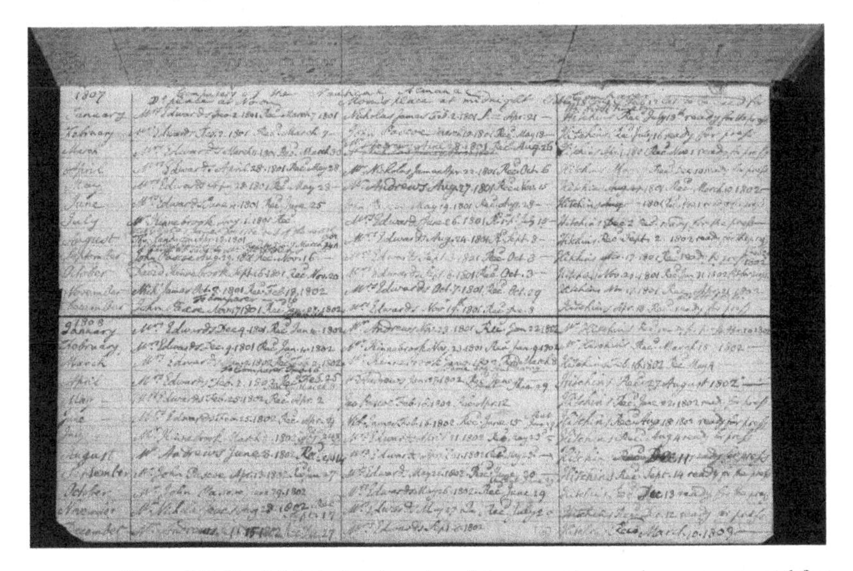

Figure 7.2 Nevil Maskelyne's notes of the computers and comparers used for the 1807 and 1808 editions of the *Nautical Almanac*, CUL RGO 4/324, p. 5r. Reproduced by kind permission of the Syndics of Cambridge University Library

'inadequate to the calculations'.[23] Even in these instances, Maskelyne simply informed the Board of decisions already made.

As Table 7.1 indicates, the computers and comparers were generally supplementing income from other professions, including clergyman, schoolmaster or practical mathematical occupations such as astronomical assistant, surveyor and table compiler. While the *Nautical Almanac* had different purposes to the popular almanacs, there were personnel in common. Most prominent was Henry Andrews, computer and comparer for forty-seven years. Besides working as a schoolmaster, Andrews edited the bestselling almanac, *Vox Stellarum*, from 1789 to 1820, by which time annual sales had risen to 500,000. William Wales was both computer for the *Nautical Almanac* and a regular contributor to the *Ladies' Diary*. The latter was one of several almanacs edited for the Stationers' Company by Charles Hutton, *Nautical Almanac* comparer from the mid-1770s alongside his role as professor of mathematics at the Royal Military Academy, Woolwich.[24]

Pay was initially £70 for computing or comparing twelve months of the *Nautical Almanac*, but had risen to £225 for computing and £250 for comparing

[23] BoL, confirmed minutes, 1 March 1800, CUL RGO 14/6, p. 2:324.

[24] Blagden, 'Thomas Carnan', p. 33; Capp, *Astrology*, pp. 242–243, 247, 264–267; Croarken, 'Nevil Maskelyne', pp. 141, 152; Curry, 'Andrews, Henry'; Feist, 'The Stationers' Voice', p. 43; Guicciardini, 'Hutton, Charles'; Perkins, *Visions*, pp. 20, 29–31, 105–106, 112; Wardhaugh, *Poor Robin's Prophecies*, pp. 190, 230.

Table 7.1 *Computers, comparers and others employed to produce the* Nautical Almanac *and perform related work for the Board of Longitude*

Name	Role	Dates employed	Lived	Other occupations
Richard Dunthorne	comparer; other	1765–1775	Cambridge	astronomer; surveyor; superintendent
†Israel Lyons	computer; other	1765–1769	Cambridge	botanist; astronomer
John Mapson	computer	1765–1770	Tetbury, Gloucestershire	unknown
†William Wales	computer; comparer	1765–1795	London	astronomer; table compiler; teacher
George Witchell	computer; other	1765–1768	London and Portsmouth	headmaster; computer
*Malachy Hitchins	computer; comparer; other	1767–1809	St Hilary, Cornwall	clergyman
Henry Andrews	computer; comparer	1768–1815	Royston, Hertfordshire	schoolmaster; bookseller; table compiler
Charles Barton	computer; comparer	1768–1778	Greenwich	schoolmaster
Joseph Keech	computer	1768–1778	London	clerk
Reuben Robbins	computer	1768–1778	London	mathematics teacher
Walter Steel(e)	computer	1768–1772	unknown	computer
*†William Bayly	computer	1771–1780	Greenwich	astronomer; schoolmaster
James Stephens	computer	1772	unknown	ship's master
John Edwards	computer; other	1773–1784	Ludlow	clergyman; instrument maker
*George Gilpin	computer; other	1775–1800	London	astronomical assistant; clerk
*John Hellins	computer	1776	Devon	clergyman
Charles Hutton	comparer	1776–1782	Woolwich	professor of mathematics
John Wales	computer	1776–1786	Warmfield, Yorkshire	weaver; cloth manufacturer
Richard Coffin	computer	1778	Devon (probably)	unknown
Michael Taylor	computer; other	1778–1789	London	table compiler
Mary Edwards	computer; comparer	1784–1815 (unofficially from 1773)	Ludlow	none identified
*Joshua Moore	computer	1787–1793	Cambridge	unknown
Francis Simmonds	computer	1794–1799	Hampshire	computer

Table 7.1 (*cont.*)

Name	Role	Dates employed	Lived	Other occupations
Philip Turner	computer	1794–1800	London	computer
*†John Crosley	computer	1799–1817	Spitalfields, London	astronomer; computer
Nicholas James	computer; comparer	1799–1828	St Hilary, Cornwall	schoolmaster
John Pascoe	computer	1799–1808	Devon	surveyor
John Williams	computer	1800–1803	London	unknown
*David Kinnebrook	computer	1801–1802	Norwich	astronomical assistant; schoolmaster
Thomas Sanderson	computer	1801–1802	London	unknown
§William Shires	computer	1804	London	mathematical and navigational teacher
William Dunkin	computer; comparer	1808–1828+	St Hilary, Cornwall	miner; private teacher
Richard Martyn	computer	1808–1828	St Mabyn, Cornwall	surveyor
Benjamin Workman	computer	1808–1809	London	teacher
Thomas Brown	comparer	1809–1828+	Tideswell, Derbyshire	clergyman; headmaster
Eliza Edwards	computer	1809–1828+	Ludlow	none identified
Second daughter of Mary Edwards	computer	c. 1813	Ludlow	none identified
*Thomas Taylor	other	1813–1815	Greenwich	astronomical assistant
*James Gordon Bradley	other	1816–1817	London	artist
Henry Meikle	computer; other	1818–1822	London	computer
Henry Jenkins	computer	1819–1828	Isleworth, Middlesex	computer
*Thomas Glanville Taylor	other	1819–1828+	Greenwich	astronomical assistant
George Carey	computer	1820–1828	London; Arbroath	mathematics teacher
Zerah Colburn	computer	1824	London	schoolteacher
Son of T. Robertson Young	computer	1827–1828	unknown	unknown

* Employed as astronomical assistant and/or undertook other mathematical or astronomical work for the Royal Observatory, Greenwich.
† Employed by the Board of Longitude as expeditionary astronomer.
§ Sent tables and other materials for becoming a computer but not fully employed.
1828+ Continued work for the *Nautical Almanac* after 1828.

Notes

- For role, 'other' includes writing additional material published in the *Nautical Almanac*, correcting and revising proofs, lunar computations and other calculations.
- Dates employed are approximate periods in which work was done for the *Nautical Almanac*, based on relevant notes and correspondence, dates of payments and known dates of death. Work may not have been continuous between those dates.
- Individuals may have had contact with the Board of Longitude on other matters outside the dates of *Nautical Almanac* work.
- Those indicated as astronomical assistants at the Royal Observatory or as expeditionary astronomers may have done so outside the dates of *Nautical Almanac* work.

Sources: BoL, confirmed minutes, 1737–1829, CUL RGO 14/5 to 14/8; BoL, 'Petitions and Memorials', CUL RGO 14/11 and 14/12; BoL, 'Printers' and Publishers' Accounts', CUL RGO 14/15 and 14/16; BoL, 'Papers on Payments for Board Work', CUL RGO 14/17 and 14/18; BoL, 'Vouchers', CUL RGO 14/19 and 14/20; BoL, 'Papers on the *Nautical Almanac* and Errata in Works of the Board of Longitude', 1793–1829, CUL RGO 14/22; Nevil Maskelyne, 'Diary of *Nautical Almanac* work', 1767–1811, CUL RGO 4/324; Nevil Maskelyne, 'Book of Accounts', 1765–1811, NMM MSK/2; Bryden, 'William Shires'; Colburn, *A Memoir*, pp. 136–137; Croarken, 'Astronomical Labourers', p. 291; Croarken, 'Nevil Maskelyne'; Croarken, 'Providing Longitude'; Dolan, 'People: John Pond'; Howse, 'The Board of Longitude Accounts'; Kennett, 'An Eighteenth-Century Astronomical Hub'; Myrone, 'Drawing after the Antique'; Wales, *Captain Cook's Computer*, pp. 104–105, 343.

by the time of Maskelyne's death.[25] For some, the work provided additional income work that could fit around other duties. For others, it was their main income. When in 1793 the Board resolved that calculation be paused for five years, since calculations were complete to the end of 1804, Mary Edwards (who seems to have had no other source of income), Henry Andrews and Malachy Hitchins complained of the loss of 'an Employment from which they derive their Support and on which they have been accustomed to depend'. It was therefore decided to find alternative work, notably recalculating data based on 'new improved French Tables' published the previous year.[26]

Table 7.1 also indicates where computers lived, showing them to be as far afield as Yorkshire, East Anglia, Shropshire and Cornwall, as well as in and relatively near London. This spread took on new significance following the revelations about Keech and Robbins. Thereafter, Maskelyne sent work for the same month to computers who lived relatively far apart.[27]

As Maskelyne controlled the appointment of computers and comparers, he relied on personal acquaintance and recommendations for new appointments. It was probably Edward Waring, Lucasian professor of mathematics and fellow Commissioner, for instance, who introduced John Edwards to the Astronomer

[25] Nevil Maskelyne, 'Diary of "Nautical Almanac" Work', 1767–1811, CUL RGO 4/324, p. 1r.

[26] BoL, confirmed minutes, 7 December 1793, CUL RGO 14/6, pp. 2:207–2:209; see also Nevil Maskelyne to Henry Andrews, 15 August and 14 December 1793, CUL RGO 4/149, pp. 12, 15.

[27] Croarken, 'Providing Longitude', p. 114; Croarken, 'Tabulating', pp. 53–54; Croarken, 'Nevil Maskelyne'.

Royal. Others, such as Benjamin Workman, approached Maskelyne directly or, like Michael Taylor and George Witchell, were known from previous dealings with the Board.[28] Malachy Hitchins proved to be an influential intermediary. As a former astronomical assistant at Greenwich and then long-time comparer, Hitchins introduced Maskelyne to several Cornish acquaintances who went on to become computers. He also played a significant part in the compilation of the *Nautical Almanac*, directing and teaching computers, overseeing their work and acting as comparer. It was Hitchins, for instance, who revealed Keech's and Robbins's malpractice.[29]

Maskelyne also decided which other agents were appointed. In April 1766, on his say-so, the printers William Richardson and Samuel Clarke were licensed to produce the *Nautical Almanac*. The partners printed other works for the Board, including lunar tables, the account of Harrison's marine timekeeper and works by instrument makers including John Bird.[30] In the next two decades the firm received over £2000 from the Board. Distribution was entrusted to the Tower Hill stationers John Mount and Thomas Page, sellers of nautical books, and to the eminent Strand publisher John Nourse, bookseller to the king and specialist vendor of works in mathematics and the sciences.[31]

In January 1767, on the point of publication, Maskelyne told Nourse to prepare for the first batch of 100 copies to arrive from Salisbury Court, advised on using cheap blue paper covers, and explained that the stationers would keep a fifth of the cover price. Several hundred copies were to be distributed freely to the Royal Navy, with other copies 'bound handsomely' for the Commissioners. Maskelyne retained a dozen 'for his own private use as an acknowledgment for his trouble in preparing them for the Press'.[32] Eventually 1000 copies were printed, a small number compared to other almanacs: contrast the 15,000 copies of the *Ladies' Diary* and 82,000 of *Vox Stellarum* for 1761.[33] Priced initially at 5*s* but reduced to 2*s* 6*d* for the 1768 edition, the *Nautical Almanac* was also considerably more expensive than the cost of between 6*d* and 9*d* of the more numerous almanacs.[34]

28 BoL, confirmed minutes, 18 September 1764 and 18 January 1765, CUL RGO 14/5, pp. 64, 67, 73; Croarken, 'Nevil Maskelyne', pp. 148, 152–153.

29 Croarken, 'Nevil Maskelyne', pp. 146–147; Kennett, 'An Eighteenth-Century Astronomical Hub'.

30 Mayer, *Tabulæ motuum solis et lunæ*; Harrison, *The Principles of Mr. Harrison's Time-Keeper*; Maskelyne, *An Account of the Going*; Bird, *The Method of Constructing Mural Quadrants*.

31 The Board subsequently used John Nichol, Christopher Buckton and Thomas Bensley as printers and Peter Elmsly, Payne and Mackinlay, and John Murray as distributors.

32 BoL, confirmed minutes, 26 April 1766, CUL RGO 14/5, pp. 125–126; Nevil Maskelyne to John Nourse, 3 January 1767, NMM AGC/8/29; Howse, *Nevil Maskelyne*, p. 86.

33 Blagden, 'Thomas Carnan', p. 24 n. 4. See also Capp, *Astrology*, pp. 239–240; Perkins, *Visions*, pp. 20–30.

34 The price rose to 3*s* 6*d* from the 1771 edition and 5*s* in 1801 (beginning with the volume for 1805 and for second/third editions of volumes for earlier years published from 1802); BoL, confirmed minutes, 6 June 1801, CUL RGO 14/6, pp. 2:347–2:348.

While the *Nautical Almanac* lay at the heart of its publishing activities, the Board also had to produce works that supported both the computation of the almanac and the adoption of new longitude methods. Seamen in particular required not just the annually changing data tabulated in each year's volume but also data that did not change or did so only slowly. These were supplied in the *Tables Requisite to Be Used with the Astronomical and Nautical Ephemeris*, of which 10,000 copies were printed in 1766 for sale the following year. The volume included, among other things, tables for correcting shipboard observations for refraction, parallax and the dip of the horizon, as well as star positions and proportional logarithms.[35] Updated editions, revised by William Wales and published in 1781 and 1802, included trigonometrical and logarithmic tables absent from the first edition.[36] Other volumes of tables and related data published by the Board were primarily of assistance to the creation of the *Nautical Almanac*. They included lunar tables and tables of logarithms and of the products and powers of numbers. Maskelyne sent copies to computers along with the *Tables Requisite*, previous years of the *Nautical Almanac* and various manuscript tables.[37]

Publishing the *Nautical Almanac* and other titles transformed the Board into an agent in the stormy world of London's print trade and an important state patron of skilled labour in mathematical practice, editorial control and long-range commerce. This had financial and operational implications. Before 1765, when almost all its budget was devoted to dealings with Harrison, the Board's accounts record no expenditure on computation or publications. During 1766, the Commissioners ordered £450 to be paid for computing and checking the forthcoming *Nautical Almanac*. The following year, £540 was spent on calculation and £325 on paper and printing for the 1767 edition. In the next three decades, almost £11,000 was spent on paper and print for the *Nautical Almanac*, about as much again on computation (see also Appendix 2). By contrast, it was possible to establish a London print shop with an initial investment of a few hundred pounds. The cost of Kendall's copy of Harrison's fourth marine timekeeper (H4) was the same as that for computing the first edition of the *Nautical Almanac*.[38]

[35] Anonymous, *Tables Requisite*, 1st ed.; Howse, *Nevil Maskelyne*, p. 91.

[36] BoL, confirmed minutes, RGO 14/6, 15 July 1780 and 7 June 1794, pp. 2:8, 2:220–2:221; Anonymous, *Tables Requisite*, 2nd and 3rd eds; Howse, *Nevil Maskelyne*, p. 157. In 1821, the *Tables Requisite* were replaced by Lax, *Tables*.

[37] Board publications sent to computers included: Bernoulli, *Sexcentenary Table*; Hutton, *Tables of the Products*; Mayer and Mason, *Mayer's Lunar Tables*; Taylor, *Sexagesimal Table*; Taylor, *Tables of Logarithms*; see Nevil Maskelyne, 'Books Necessary for the Use of the Computers and Comparer of the Nautical Almanac', 18 September 1799, CUL RGO 4/324, p. 21v; Croarken, 'Tabulating', p. 55.

[38] Howse, 'The Board of Longitude Accounts'; Howse, 'Britain's Board of Longitude', p. 404. Publication costs of works the Board sponsored by Ramsden (1774), Bayly (1782) and Wales (1777 and 1788), as well as Maskelyne's description of Harrison's sea watch (1767) and his reply to Mudge (1792), are excluded from these estimates. For example costs, see BoL,

Such expenses warranted rigorous fiscal and administrative attention. Although the Commissioners trusted Maskelyne with the day-to-day running of the *Nautical Almanac*, its production and management nevertheless began to take up a large part of what were now regular meetings. Routine *Nautical Almanac* business included issuing warrants to printers and publishers, and imprests and payments to computers, comparers, printers, publishers and paper suppliers. These expanded activities furthered the Board's transformation into a formal organisation and began to dictate the structure and timing of meetings. The salary of its Secretary, Admiralty clerk John Ibbetson, was doubled to £80 in 1765, recognising that he and Maskelyne were to administer a complicated accountancy and management system.[39] In 1780, the Board formed a separate standing committee to keep an eye on the expenses relating to printing and other matters, thus streamlining business at Board meetings.[40] More extensive discussions were needed only when trickier matters arose, for instance problems with printer Christopher Buckton arising from his bankruptcy, which led to his dismissal in 1798 and replacement by Thomas Bensley.[41] The resulting discussions exposed the Board's legal obligations as a licenser of publication. As Maskelyne explained to Joseph Banks in September 1797, circumventing Buckton's refusal to surrender copies he had printed would require a meeting of the Board to formally agree a licence to an alternative printer.[42]

In 1802, the Board looked to enforce its sole right to publish the *Nautical Almanac* when the publisher and bookseller David Steel began selling a pirate copy. The title page of *The Seaman's Almanac* stated that it had been 'carefully computed' by Thomas Kirby, whom Steel employed as a teacher at his Navigation Warehouse near the Tower of London.[43] Yet it included tables of lunar distances with data identical to that in the *Nautical Almanac*, except that they used civil time, with each new day starting at midnight rather than noon (as used in the *Nautical Almanac*). At 2*s* 6*d*, it was 1*s* cheaper than the

'Account of Works Printed, Published and Sold', 1784, CUL RGO 14/15, pp. 241v–253v; BoL, accounts relating to Charles Nourse, 1782–1789, CUL RGO 14/15, pp. 12r–16r. For comparative costs, see Kernan, *Printing Technology*, pp. 55–56.

39 Baker, 'Humble Servants', p. 211.

40 BoL, confirmed minutes, 15 July 1780, pp. 2:8–2:9.

41 BoL, confirmed minutes, 3 June 1797, 2 June and 1 December 1798, CUL RGO 14/6, pp. 2:279–2:280, 2:289–2:290, 2:294–2:296.

42 Nevil Maskelyne to Joseph Banks, 9 September 1797, NMM MSK/3/1/22. See also Nevil Maskelyne to Joseph Banks, 25 June and 28 December 1798, 4 February 1799, RS MM/8/10, MM/8/14, MM/8/17; William Wales to Joseph Banks, 10 December 1798, RS MM/8/12; Pearson and Loggon to Joseph Banks, 21 December 1798, 11 January and 2 February 1799, RS MM/8/13, MM/8/16, MM/8/19; T. Bygrave to Nevil Maskelyne, 25 January 1799, RS MM/8/15.

43 Kirby, *The Seaman's Almanac* (copy at University College London); Cotter, 'John Hamilton Moore', p. 324.

Nautical Almanac and Steel was already advertising future editions. In April 1802, Joseph Banks alerted his fellow Commissioner and First Lord of the Admiralty John Jervis, to Steel's infringement of the Board's monopoly. Jervis agreed that the case should be pursued, though he was worried that the maximum fine of £20 was an insufficient deterrent.[44] The Board resolved to prosecute Steel, but 'difficulties that had not been foreseen' delayed progress and in the end no action was taken.[45]

The Astronomical Networks of the *Nautical Almanac*

As noted above, the production of the *Nautical Almanac* drew on a distributed network of British mathematicians and astronomers. It also tapped into wider European networks and in turn became an important element in them.

The almanac's credibility lay above all in the accuracy of its data. Maskelyne's system of computers and comparers was a trustworthy foundation, but their computations inevitably relied on existing astronomical tables for the motions of the Sun, Moon and planets, and eclipses of Jupiter's satellites. The preface of each edition thus set out which tables had been used. For the first editions, these included Tobias Mayer's manuscript tables for the Sun and Moon (which the Commissioners finally published in 1770), Edmond Halley's for the planets, and Swedish astronomer Pehr Wilhelm Wargentin's for Jupiter's satellites.[46] These were updated as better tables became available: Mason's revised version of Mayer's lunar tables from the *Nautical Almanac* for 1777 and planetary tables by Wargentin from the volume for 1780.[47] Wargentin's tables of the planets and of Jupiter's satellites were later replaced by those of French astronomer Jérôme Lalande – those of the planets for the *Nautical Almanac* for 1784 onwards (using tables in the second edition of Lalande's *Astronomie*), and of both the planets and Jupiter's satellites from the volume for 1805 (tables from the third edition of *Astronomie* published in 1792).[48] Indeed, part of the work allocated to the computers while computation of the *Nautical Almanac* was paused in the 1790s was to do the calculations to switch to Lalande's tables. These proved sufficiently onerous to warrant an increased salary.[49] The last changes during Maskelyne's time occurred for the *Nautical Almanac* for 1813,

[44] Joseph Banks to John Jervis, 24 April 1802, RS MM/8/35; John Jervis to Joseph Banks, 25 April 1802, RS MM/8/37.

[45] BoL, confirmed minutes, 5 June and 2 December 1802, CUL RGO 14/7, pp. 2:8, 2:13–2:14.

[46] Halley, *Astronomical Tables*; Mayer, *Tabulæ motuum solis et lunæ*; Wargentin's tables appeared in Lalande, *Tables Astronomiques*.

[47] Mason's tables first appeared in the *Nautical Almanac* for 1774. Wargentin's tables appeared in Lalande, *Astronomie*.

[48] BoL, confirmed minutes, 1 March 1794 (Lalande's tables sent to computers), CUL RGO 14/6, pp. 2:212–2:213.

[49] BoL, confirmed minutes, 7 December 1799, CUL RGO 14/6, pp. 2:316–2:317.

with the adoption of the solar tables of French astronomer Jean Baptiste Joseph Delambre and the lunar tables of Austrian astronomer Johann Tobias Bürg, both published by the Bureau des Longitudes in 1806, though Bürg's had been communicated to the Board of Longitude by 1805. The extra labour required the Board to raise the salaries of the computers and comparer for one almanac.[50]

Besides incorporating data from astronomers elsewhere in Europe, the *Nautical Almanac* and other Board publications were sent as gifts to contacts and institutions in Britain and across the rest of Europe and in America (Box 7.2). The Board's published works thus formed part of the correspondence network

BOX 7.2 RECIPIENTS OF GRATIS COPIES OF BOARD OF LONGITUDE PUBLICATIONS

Notes

- Lists compiled in the period 1784–1792
- One copy of each work published unless otherwise stated

Board of Longitude – Commissioners, Officers and Other

- Commissioners of Longitude (two copies)
- Secretary of the Board of Longitude
- Computers and comparers of the *Nautical Almanac* (one copy of each *Nautical Almanac*)
- Head Master and Mathematical Usher (Second Mathematical Master), Royal Naval Academy, Portsmouth (*Nautical Almanac* plus other publications as Astronomer Royal sees fit for instructing Royal Navy masters)

Britain and Ireland

- Board of Admiralty
- The King's Library
- The Queen's Library
- The Royal Society

[50] BoL, confirmed minutes, 11 July 1805, 4 December 1806, 3 December 1807, CUL RGO 14/7, pp. 2:78, 2:99, 2:109–2:110; Bureau des Longitudes, *Tables astronomiques*.

- British Museum
- Public Library Cambridge
- Bodleian Library Oxford
- Royal Observatory, Greenwich
- New Observatory, Oxford
- Dublin Observatory
- University of Edinburgh
- University of Aberdeen
- University of Glasgow
- University of St Andrews
- University of Dublin
- William Bayly
- Vice Admiral Campbell
- Vice Admiral Darby
- Rear Admiral Elliot
- William Herschel
- Earl Howe
- John Ibbetson
- Captain James King
- Lord Mulgrave
- Sir Harry Parker
- Sir Philip Stephens
- William Wales

Other Countries

- The Vatican
- Bibliothèque du Roi, Paris
- Académie Royale des Sciences
- Observatoire de Paris
- Königliche Akademie der Wissenschaften, Berlin
- Royal Society of Sciences, Göttingen
- Royal Swedish Academy of Sciences, Stockholm
- Royal Society of Sciences, Uppsala
- Imperial Academy of Arts, St Petersburg
- Royal Academy of Sciences, Turin
- Academy of Sciences of the Institute of Bologna
- Kurpfälzische Akademie der Wissenschaften, Mannheim
- Imperial and Royal Observatory, Vienna
- Royal Observatory, Cádiz

- Royal Observatory, Milan
- La Specola, Padua
- Royal Observatory, Copenhagen
- Royal Observatory, Vilnius
- Royal Observatory, Uppsala
- Imperial and Royal Academy, Brussels
- Philosophical Society, Philadelphia
- American Academy of Arts and Sciences, Boston
- Royal Academy of Sciences, Lisbon
- Lisbon Observatory
- Göttingen Observatory
- Saxe-Gotha Observatory
- Pisa Observatory
- Malta Observatory
- Turin Observatory
- Jean-Nicolas-Sébastien Allamand, Leiden
- Christiaan Hendrik Damen (Allamand's successor)
- Paolo Andreani, Milan
- Giovanni Battista Beccaria, Milan
- Johann Bernoulli, Berlin
- Johann Elert Bode, Berlin
- Jean-Charles de Borda, Paris
- Roger Joseph Boscovich, Ragusa
- Thomas Bugge, Copenhagen
- Joseph-Bernard de Chabert-Cogolin, Brest and Paris
- Nicolas de Condorcet, Paris
- Johann Albrecht Euler, St Petersburg
- Abbé Gasparo Ferdinando Felice, Florence
- Paolo Frisi, Milan
- Maximilian Hell, Vienna
- Edme-Sébastien Jeaurat, Paris
- Joseph-Louis Lagrange, Berlin
- Pierre-Simon de Laplace, Paris
- Jérôme Lalande, Paris
- Pierre Lévêque, Nantes
- Anders Johan Lexell, St Petersburg
- Antonio Maria Lorgna, Verona
- Marsilio Landriani, Milan
- Pierre Méchain (instead of Jeaurat from 1784)
- Joseph de Mendoza y Ríos, Cádiz

- Charles Messier, Paris
- Antoine Darquier de Pellepoix, Toulouse
- Abbé Giuseppe Piazzi
- Alexandre Guy Pingré, Paris
- Achille Pierre Dionis du Séjour, Paris
- Giuseppe Toaldo, Padua (*Nautical Almanac*, *Tables Requisite* and Taylor's *Tables*)
- Antonio d'Ulloa, Cádiz
- Henrik Nicander (Pehr Wargentin's successor)

Sources: BoL, 'List of Public Societies and Learned Men of Foreign Nations, to whom the Committee of the Board of Longitude Propose the Publications of the Board Should be Sent as Presents', 1784–1792, CUL RGO 14/6, pp. 2:358–2:361; BoL, 'Lists of Societies and Individuals Entitled to Copies of the Board's Publications', 1766–1795, CUL RGO 14/13, pp. 93r–96v.

linking European and American men of science, in much the same way as comparable publications including the *Connoissance des Temps*.[51] Attempts to continue correspondence with successors to specific posts (for instance, when Jean Allamand, professor of mathematics and philosophy at the University of Leiden, died in 1787) suggest that the Board operated within a network that could outlast the individuals who formed it.[52]

A significant influence came with the use of the *Nautical Almanac* as the basis for nautical and astronomical almanacs in other countries. From 1772 (for 1774), the *Connoissance des Temps* began to include lunar distances from the *Nautical Almanac*. The relationship between the French and British publications remained, however, subject to the vagaries of national rivalry in the decades that followed, with the tables in the *Connoissance des Temps* calculated entirely in France from 1788 (for 1790).[53] Nonetheless, Maskelyne and his counterparts remained in dialogue, and each made regular comparisons of the data in their almanacs. For the French, the *Nautical Almanac* stood as a touchstone. This was conspicuous in Abbé Henri-Baptiste Grégoire's speech to the Convention Nationale in 1795, which looked to Britain's Board of Longitude in proposing the foundation of a Bureau des Longitudes. Grégoire argued that

[51] Boistel, *'Pour la gloire'*, pp. 126–127, 164, 318–320.
[52] Margócsy, 'A Long History', pp. 313–315.
[53] Boistel, 'From Lacaille to Lalande', pp. 56–57; Croarken, 'Greenwich, Nevil Maskelyne', pp. 60–61; Howse, *Nevil Maskelyne*, pp. 120–121.

the *Nautical Almanac* was a striking instance of the Board's success, which in turn underpinned British maritime dominance.[54] Those responsible for the *Connoissance des Temps* aspired above all to emulate Maskelyne's success in ensuring publication several years ahead, as well as to surpass the *Nautical Almanac* in terms of accuracy. This was something Hungarian astronomer Franz Xaver von Zach felt would be possible, writing of his confidence that Lalande's *Connoissance des Temps* would:

> henceforth take a rank ahead of the Nautical Almanac of the English, as much for the correction of calculations as for its printing: the English employ for its production only two computers and Maskelyne as editor. You have three calculators and two editors ... What barbarism that the English gentlemen still make use of the tables of Mayer and Halley for the planets![55]

Von Zach's comments aside, the British system of computers and comparers was a matter on which Pierre Méchain later corresponded with Maskelyne in his efforts to improve the production of the French almanac.[56]

Other nations took advantage of the *Nautical Almanac* too. In 1788, a Dutch translation appeared, as did the *Almanach ten dienste der zeelieden*, which recalculated the British data for Tenerife (which Dutch navigators took as their prime meridian).[57] The same year saw the publication of the Portuguese *Ephemerides Nauticas ou Diario Astronomico*, for which *Nautical Almanac* data was reduced to the meridian of Lisbon.[58] Likewise, from 1791 the Spanish *Almanaque náutico y efemérides astronómicas* borrowed freely from its British and French counterparts, while American commercial reprints of the *Nautical Almanac* appeared from 1802.[59]

The Commissioners had, of course, long been linked into European and American astronomical networks by virtue of the rewards on offer since 1714, with astronomical schemes figuring prominently among proposals. These included projects for the production of almanacs akin to the *Nautical Almanac*. In September 1764, for instance, the London mathematics teacher George Witchell submitted a 'General Table & Ephemerides', soon examined by the virtuoso Matthew Raper and the mathematician Gael Morris. By the time the Board heard their evaluation, however, Maskelyne was in command and Witchell simply recruited to his programme as a computer.[60]

[54] Débarbat, 'L'évolution administrative', pp. 23–25; Feurtet, 'Le Bureau des longitudes', pp. 71–90; Howse, *Greenwich Time*, pp. 81–82.

[55] Franz Xaver von Zach to Jérôme Lalande, 4 August 1796, Observatoire de Paris MS 1090/19.

[56] Boistel, *'Pour la gloire'*, pp. 133–139, 161–162, 222, 262, 275.

[57] Davids, 'The Longitude Committee', pp. 34–35.

[58] Figueiredo, 'Astronomical and Nautical Ephemerides, pp. 161–166.

[59] Cotter, 'A History', pp. 96–97; Large, 'How Far West', p. 123.

[60] BoL, confirmed minutes, 18 September 1764 and 18 January 1765, CUL RGO 14/5, pp. 64, 67, 73.

The Board of Longitude's significant investment in the *Nautical Almanac* signalled unambiguously the ways in which astronomical methods for determining longitude would be supported thereafter. Nevertheless, the Commissioners remained open to suggestions for improvements. New mathematical methods for clearing, for instance, gained rewards of £50 for Israel Lyons and George Witchell in 1766 and were included in the first edition of the *Tables Requisite*.[61] Cornelis Douwes, teacher and examiner of sea officers and pilots in Amsterdam, likewise received a reward of £50 for new solar tables and a method for finding latitude from two observations of the Sun. These, with improvements by Captain John Campbell, were appended to the *Nautical Almanac* for 1771 and appeared in the second edition of the *Tables Requisite*.[62] Witchell's proposal for general tables of refraction and parallax took him longer than the Board felt appropriate and so required additional funds and the guiding hand of Anthony Shepherd, Plumian professor at Cambridge, before publication in 1772.[63]

The Board's focus on astronomical tables as one of the longitude solutions to be supported long-term, alongside the timekeeper method, was made explicit in the Longitude Act of 1774. As discussed in Chapter 6, the Act addressed issues that had arisen in disputes with the Harrisons by creating separate rewards for timekeepers and for improved solar and lunar tables. Whereas rewards of up to £10,000 were available for timekeepers, a single reward of just £5000 was available for astronomical tables. Submissions would only qualify for a reward if they showed the Moon's distance from the Sun or stars to within 15 seconds of arc, were obviously constructed upon Newtonian principles of gravitation, could be successfully compared with astronomical observations covering a period of eighteen and a half years (the assumed period of irregularities of the lunar motions), and were assigned to the Commissioners for public use.[64]

As with timekeepers, no large reward in line with the terms of the Act was ever granted. Nevertheless, more than £4000 was given in lesser rewards for astronomical tables between 1774 and the Board's dissolution, about half that

[61] BoL, confirmed minutes, 26 April 1766, CUL RGO 14/5, p. 128; BoL, papers on errors in lunar distances, 1765–1766, CUL RGO 4/292 (Witchell's and Lyons's methods); Anonymous, *Tables Requisite*, 1st ed., pp. 17–27, 51–67.

[62] BoL, confirmed minutes, 2 May 1767 and 18 June 1768, CUL RGO 14/5, pp. 161, 169; Anonymous, *Tables Requisite*, 2nd ed., pp. 57–80; Cotter, 'A History', pp. 239–241; Cotter, *A History of Nautical Astronomy*, pp. 148–151.

[63] BoL, confirmed minutes, 12 September 1765 (£100 to Witchell), 10 January 1767 (Shepherd to take charge), 18 June 1768 (3000 copies to be printed), CUL RGO 14/5, pp. 111–112, 139–140, 169; George Witchell to Harry Parker, 21 December 1782, CUL RGO 14/2, pp. 212r–212v (previous payments for tables published by Shepherd); Shepherd, *Tables*; Forbes, 'The Foundation', pp. 394–395.

[64] Discovery of Longitude at Sea Act 1774 (14 Geo 3 c 66), copy at CUL RGO 14/1, pp. 37r–42v.

given for timekeepers. Having approached the Board in 1791 about errors in the tables of refraction and parallax, for instance, London watchmaker George Margetts requested a reward for his 'Longitude, and Horary Tables' the following year. They warranted a 'temporary Reward' of £100, since 'the Tables might be of great Use to Navigation', with a further sum possible after further consideration. An additional £200 was awarded in 1794, following further correspondence and a petition from Margetts supported by several EIC captains.[65] The largest of the rewards given in this later period was to the Spanish naval officer and mathematician Joseph de Mendoza y Ríos. Mendoza approached the Commissioners in 1796 with a new double altitude method for latitude, then with two volumes of tables for reducing the lunar distance, with further correspondence the following decade. The tables gained the support of both the Board and the EIC, although it was not until 1814 that the Commissioners agreed a substantial reward of £1200.[66]

The *Nautical Almanac* in the Wild

The Board intended the *Nautical Almanac* to become part of shipboard routine, making the lunar-distance method quicker and easier as well as supporting the use of timekeepers. As the Commissioners came to realise, however, dictating the practices of those far from the metropolis was no easy matter. Simply publishing the *Nautical Almanac* would not guarantee adoption.

[65] BoL, confirmed minutes, 3 December 1791 (errors in existing tables), 2 June 1792 (reward requested; £100 agreed) and 1 March 1794 (further £200), CUL RGO 14/6, pp. 2:176–2:177, 2:186, 2:214; George Margetts to BoL, 5 March 1790, CUL RGO 14/32, pp. 37r–38v (including copy of paper signed by EIC officers); George Margetts, correspondence and memorials, 1790–1794, CUL RGO 14/46, pp. 40r–63r; George Margetts, petition, c. 1790–1800, CUL 14/12, p. 322r.

[66] BoL, confirmed minutes, 5 March 1796 (double altitude method), 11 June 1796 (tables), 1 December 1814 (reward), CUL RGO 14/6, pp. 2:259, 2:267, RGO 14/7, p. 194; Joseph de Mendoza y Ríos to BoL, CUL RGO 14/41, 2 December 1796, pp. 12r–12v (double altitude method); Joseph de Mendoza y Ríos, correspondence and memorials on tables for lunar observations, 1796–1814, CUL RGO 14/46, pp. 66r–87r; BoL, 'Report of a Committee of the Board of Longitude, into Assisting Mendoza Rios Publish Tables Facilitating the Practice of Astronomy at Sea', 1802, RS MM/8/39; copies and extracts of letters from Joseph de Mendoza y Ríos to Nevil Maskelyne, 1802–1803, RS MM/8/40; James Mortlock to Joseph de Mendoza y Ríos, 11 March 1803, RS MM/8/46. For relations with Banks, see Joseph de Mendoza y Ríos to Joseph Banks, 2 June 1801, 16 October and 24 November 1813, 26 February and 3 March 1814, RS MM/8/31, MM/8/71, MM/8/72, MM/8/74, MM/8/75; Joseph de Mendoza y Ríos, 'Comparative Analysis of Jose de Mendoza Rios's Method for Reducing the Lunar Distances', October 1813, RS MM/8/76; Joseph Banks, note on Mendoza y Ríos and his method of calculations, c. 1813, RS MM/8/73; BoL, correspondence concerning Joseph de Mendoza y Ríos, 1813–1814, CUL RGO 14/32, pp. 256–261; Joseph Mendoza y Ríos to Thomas Jefferson, 23 August 1805, US National Archives and Records Administration RG 45 https://founders.archives.gov/documents/Jefferson/99-01-02-2294 [accessed 18 May 2023].

As the first volume became publicly available at the beginning of 1767, the Commissioners began to consider strategies for encouraging take-up. In March, they recommended that the Admiralty:

cause the Ephemeris & Nautical Almanack for the present year and also the Tables of Parallax and Refraction when Printed to be circulated on board His Majesty's Ships in such manner as their Lordships may think most proper in order to make the same as public as possible.

Nourse and Mount were summoned and told to send copies to the main ports of Britain and British North America, 'a few to each of the Maritime Countries in Europe' and 'in short to endeavour to circulate them by every means in their power'.[67] At the same time, London journals began to puff the new almanac, urging 'that our navigators will no longer neglect to put in practice a method which cannot fail of enabling them to correct their journals and of determining the true place of the ship. Let them no longer call for a discovery of the longitude, as if no such thing was in being.'[68]

By the following year, further strategies were being considered. In November, the Board recommended to the Admiralty that masters (navigating officers) of any naval vessels in Portsmouth:

do perfect themselves at the Royal Academy there in the Knowledge & Use of the said Almanacs & also of Hadley's Quadrant in making Lunar observations; and produce Certificates that they have done so, from the Head Master or Mathematical Usher of the said Academy who will have directions to instruct them gratis.

The Commissioners would pay the instructors (George Witchell, headmaster, and John Bradley, usher) half a guinea for each master certificated. They further suggested that in future the Navy Board only appoint as masters those with certificates from the Royal Naval Academy or from London navigation teachers Samuel Dunn and Robert Bishop.[69] At the same time, Bishop was involved in a parallel initiative, designing printed forms that guided users through the calculations needed for longitude by lunar distances.[70]

The Admiralty issued a formal notice to all masters in January 1769 but the scheme proved difficult to implement.[71] The following month, the Navy Board suggested to Admiralty Secretary Philip Stephens that the requirement

[67] BoL, confirmed minutes, 21 March 1767, CUL RGO 14/5, pp. 149–150.

[68] 'The Nautical Almanac for the Year 1767', *Monthly Review*, 36 (1767), pp. 379–382 (p. 382).

[69] BoL, confirmed minutes, 12 November 1768, CUL RGO 14/5, pp. 173–174; Howse, *Nevil Maskelyne*, p. 94.

[70] Robert Bishop, lunar distance form, 1768, NMM G298/1/3; completed examples for 1772 at CUL RGO 14/67, pp. 41r–49v. See also Robert Bishop to Admiralty, 11 February 1772, TNA ADM 106/1207/133 (new logbook including longitude by lunar distance). Mendoza y Ríos produced similar forms in 1818; see John Franklin, lunar observations taken on board HM brig *Trent*, 1818, NMM FIS/3.

[71] Order of 6 January 1769, reported in *London Gazette*, 7–10 January 1769, p. 1.

for certificated masters be suspended 'for some time longer' since there were too few for the ships that would need them.[72] Some of those hoping to qualify found it difficult to access the training. Five masters wrote in March that they had applied to the Academy but been told that they would have to pay two guineas for instruction as they were not employed on ships then in Portsmouth harbour. They requested 'some encouragement' so that they could be taught 'without paying any reward'.[73] Just fourteen masters were certificated at Portsmouth over the next two years.[74] In 1771, William Wales was nominated to replace Samuel Dunn, and the Board's earlier request to the Navy Board was modified to suggest only that new masters (rather than all masters) should have a certificate of competence in the lunar method.[75] The following decade, the Board turned to the idea of offering financial 'encouragement' to ships' schoolmasters 'to make themselves Masters of the Lunar Tables' and wrote to the Admiralty with a proposal that schoolmasters might teach 'Longitude Observations and other practical parts of Navigation'.[76] In March 1785, with peace established following the American Revolutionary War, the Commissioners tried once more to encourage the Admiralty to enforce their previous suggestion about the training and appointment of masters. There was no response by the August meeting and no more was recorded on the matter.[77]

The long-distance expeditions to which the Board of Longitude gave its support provided a more conducive context in which to demonstrate the practicability of longitude determination supported by the *Nautical Almanac* and promote its adoption. As numerous sources have discussed, the lunar-distance method was successfully deployed by James Cook and his officers, under instruction from astronomer Charles Green, on the first Pacific voyage of 1768–1771. Cook found the method 'more than Sufficient for all Nautical purposes' and anticipated that officers would find it straightforward once they took the effort to learn it.[78] It was deployed on the two later Pacific voyages

[72] Navy Board to Philip Stephens, 24 February 1769, NMM ADM/B/182/44.

[73] Brice Black, Peter Austin, John Wilson, Thomas Shortland and John Covey to Commissioners of the Navy, 1 March 1769, TNA ADM 106/1180/250.

[74] BoL, confirmed minutes, 17 November 1770, CUL RGO 14/5, p. 201; 'A List of the Names of those Masters of His Majestys Navy who have been Instructed in the Use of the Nautical Almanac Quadrant &c.', 1772, CUL RGO 14/18, p. 449r.

[75] BoL, confirmed minutes, 2 March 1771, CUL RGO 14/5, pp. 202–203; Forbes, 'The Foundation', p. 395; Howse, *Nevil Maskelyne*, p. 121.

[76] BoL, confirmed minutes, 7 December 1782 and 1 March 1783, CUL RGO 14/6, pp. 2:41–2:42, 2:45–2:46; Harry Parker to Philip Stephens,17 December 1782, enclosing minute from Commissioners, CUL RGO 14/9, p. 6r; Commissioners of Longitude to Admiralty, 1 March 1783, CUL RGO 14/9, pp. 7v–8r; see also 'Memorandum on Naval Schoolmasters by the Board of Longitude', c. 1783, RS MM/7/34.

[77] BoL, confirmed minutes, 5 March and 17 August 1785, CUL RGO 14/6, pp. 2:80–2:83; Harry Parker to Philip Stephens, 10 March 1785, CUL RGO 14/9, pp. 22r–22v, 23v–24v.

[78] Beaglehole, *Journals*, I, p. 392.

Cook led and on other expeditions supported by the Board of Longitude (see Chapter 6). For each, the Commissioners included copies of the *Nautical Almanac* among the books and instruments supplied to the astronomers they appointed. These long voyages became an important training arena, therefore, with the Board's astronomers instructed to teach the lunar-distance method to officers.[79] The particular aims and expectations of the expeditions naturally favoured the deployment of new methods that promised increased positional accuracy that was of obvious benefit for, among other things, extending geographical knowledge. Used together, the new longitude methods, by lunar distance and by chronometer, supplemented older navigational and surveying techniques to great effect.[80]

Nonetheless, as Wess points out, the lunar-distance method involved 'tedious, complex, unintuitive and non-visual mathematics', something that may explain suggestions of poor take-up on more routine voyages.[81] Adoption seems to have been slower in the Royal Navy than in the EIC, the directors of which ordered in 1768 that candidates for officer and mate produce a certificate of competence in lunar distances.[82] Over half of EIC ships were using lunar distances by 1772/1773, with 80 per cent using chronometers by 1792, not long after Company logbooks incorporated spaces for longitude by both methods.[83]

The adoption of the new methods was a piecemeal process, subject both to institutional directives and initiatives and to individual predilections. Navigators continued to use older methods as well as gradually learning the newer chronometric and astronomical ones. While proponents on land made competing claims about the relative merits of lunar and chronometric methods as supposed rivals, for those at sea the issue was not about choice between competing methods but about the complementary use of old and new techniques to build as complete a navigational picture as possible. Moreover, each technique was not necessarily used at all times. Particular methods might be more applicable according to prevailing conditions and assumptions about the proximity of land, hazards or critical turning points.[84]

[79] Phillips, 'Instrumenting Order', pp. 40–43. See Schotte, *Sailing School*, pp. 157–159, on navigational skills learned during Cook's voyages.

[80] Higgitt, 'Equipping Expeditionary Astronomers'. See also references in Chapter 6, notes 48–49.

[81] Wess, 'Navigation and Mathematics', p. 216.

[82] Miller, 'Longitude Networks', pp. 232–233.

[83] Davidson, 'The Use of Chronometers'; May, 'The Log-Books', p. 118. Davidson, 'Marine Chronometers', implies that chronometers replaced other methods, but dead reckoning continued on all vessels, with lunar observations often made in addition.

[84] Davidson, 'Marine Chronometers', p. 79; Dunn, 'Longitude Found'; Miller, 'Longitude Networks'; Werrett, 'Perfectly Correct', p. 111.

Some measure of the adoption of the *Nautical Almanac* is offered by the data in Table 7.2. This indicates that for the first volumes up to 1783 (for which no further editions have been identified), at least half and often many more of the usually 1000 printed were sold for all years bar the first two. The production of second and sometimes third editions became commonplace thereafter, to satisfy rising demand and to correct errors discovered in the first edition.[85] Print runs of first editions rose to 1250 from the volume for 1788 and eventually to 4500 for the volumes for 1812 to 1814.[86] Anything from 250 (for 1784) to 3000 copies of second and third editions might be published, although the total number for any year never exceeded 5000.[87] As Figure A.2 in Appendix 2 suggests, however, profit was not a motive for the Board's publishing activities. Editions of the *Nautical Almanac* to 1784, for instance, generated modest returns of between £37 and £128 against outlays mostly in excess of £300.[88] The Board's publishing activities underwrote the new longitude methods by maintaining critical infrastructure that commercial publishers could not support.

The success or otherwise of the Board of Longitude in promoting new methods and the take-up of the *Nautical Almanac* relied not just on the efforts of the Commissioners but also on perceptions of the quality of its data and the practicability of new methods. Bennett rightly suggests that the data was 'the *vital* content', hence Maskelyne's vigorous defence in the face of criticism, for instance in the preface to the 1774 volume, following claims by astronomer Johann Bernoulli about errors in the solar data for 1769 and 1770. Maskelyne took pains to point out that Bernoulli's claims arose from false assumptions about the tables on which the data was based.[89] He similarly challenged von Zach's allegations in 1786 of errors in tables of the Sun's right ascension and equation of time. These had been checked at Joseph Banks's request by computer Michael Taylor and von Zach was mistaken.[90]

Perceived accuracy was sufficiently important that Maskelyne tasked his assistants with regularly comparing *Nautical Almanac* data with observations

[85] For example, the first edition for 1784, p. 98, printed 'Add' on the column for September's equation of time, corrected to 'Subtract' in the second edition.

[86] BoL, confirmed minutes, RGO 14/6, 15 July 1780, p. 2:8 (1250 copies printed from edition for 1788).

[87] Nevil Maskelyne, 'Diary of 'Nautical Almanac' Work', 1767–1811, CUL RGO 4/324, p. 1v; BoL, 'Papers on Payments for Board Work', 1766–1828; CUL RGO 14/17, pp. 117r–168v, and RGO 14/18, pp. 214r, 279r–297r; BoL, 'Account of Works Printed, Published and Sold', 1784, CUL RGO 14/15, pp. 241r–263r.

[88] BoL, 'Account of Works Printed, Published and Sold', 1784, CUL RGO 14/15, pp. 241r–263r.

[89] *Nautical Almanac* for 1774, Preface; Bennett, 'The First Nautical Almanac', p. 154. On Nathaniel Bowditch and errors in the *Tables Requisite*, see Bowditch, *New American*, p. vi; Thornton, *Nathaniel Bowditch*, pp. 65–67, 72, 76.

[90] Maskelyne, *An Answer*, pp. x–xxii and Appendix 3, pp. 16–20; Howse, *Nevil Maskelyne*, pp. 171, 174.

Table 7.2 *Publication and take-up of the* Nautical Almanac *and other Board of Longitude publications*

Title	Issued	Printed (1st ed.)	Printed (2nd ed.)	Printed (3rd ed.)	Gifts	Sold 1 Jan 1784	Remaining 1 Jan 1784	Remaining 28 June 1800
Nautical Almanac for 1767	Jan 1767	1000	—	—	100	459	441	236
Nautical Almanac for 1768	Dec 1767	1000	—	—	86	379	535	434
Nautical Almanac for 1769	Oct 1768	1000	—	—	82	582	336	135
Nautical Almanac for 1770	Oct 1769	1500	—	—	81	571	848	741
Nautical Almanac for 1771	Jan 1770	1500	—	—	109	672	719	417
Nautical Almanac for 1772	Dec 1770	1500	—	—	107	699	694	421
Nautical Almanac for 1773	June 1771	1000	—	—	104	835	61	
Nautical Almanac for 1774	July 1772	1000	—	—	93	798	109	
Nautical Almanac for 1775	Dec 1773	1000	—	—	110	680	210	140
Nautical Almanac for 1776	Feb 1775	1000	—	—	94	679	227	122
Nautical Almanac for 1777	Feb 1776	1000	—	—	102	513	385	46
Nautical Almanac for 1778	June 1776	1000	—	—	116	793	91	67
Nautical Almanac for 1779	April 1777	1000	—	—	105	735	160	
Nautical Almanac for 1780	May 1777	1000	—	—	103	849	48	
Nautical Almanac for 1781	July 1779	1000	—	—	85	915	0	
Nautical Almanac for 1782	Oct 1779	1000	—	—	82	909	9	
Nautical Almanac for 1783	Oct 1779	1000	—	—	82	881	37	
Nautical Almanac for 1784	Jan 1780	1000	250	—	80	872	48	161
Nautical Almanac for 1785	July 1780	1000	250	—	80	487	433	
Nautical Almanac for 1786	April 1781	1000	500	—	88	319	593	
Nautical Almanac for 1787	July 1783	1000	750	—	130	71	799	
Nautical Almanac for 1788	Aug 1783	1250	1000	—	130	43	1077	530
Nautical Almanac for 1789	Oct 1783	1250	1250	—	130	10	1110	539
Nautical Almanac for 1790	Nov 1783	1250	1250	—	130	8	1112	249
Nautical Almanac for 1791		1250	1250	—				357
Nautical Almanac for 1792		1250	1250	—				130

Nautical Almanac for 1793		1250	1250	—				
Nautical Almanac for 1794		1250	1250	300				97
Nautical Almanac for 1795		2000	1000	—				
Nautical Almanac for 1796		2000	1000	500				100
Nautical Almanac for 1797		2000	1500	—				578
Tables Requisite (1st ed.)	Jan 1767	10,000	—	—	57	2694	7249	6351
Tables Requisite (2nd ed.)	May 1781	10,000	—	—	88	836	9076	
Harrison, *Principles of Mr. Harrison's Timekeeper*	March 1767	1000	—	—	55	449	444	
Maskelyne, *An Account of the Going*	April 1767	500	—	—	55	284	161	
Bird, *The Method of Dividing*	April 1767	500	—	—	135	295	70	
Bird, *The Method of Constructing Mural Quadrants*	July 1768	500	—	—	75	244	181	
Mayer, *Tabulæ motuum solis et lunæ*	March 1770	2000	—	—	105	334	1561	1174
Mayer, *Theoria lunae*	March 1770	1000	—	—	85	166	749	
Shepherd, *Tables for Correcting the Apparent Distance*	June 1772	3024	—	—	100	235	2689	
Wales and Bayly, *Original Astronomical Observations*	May 1777	500	—	—	87	62	351	
Ramsden, *Description of an Engine*	Jan 1777	500	—	—	74	161	265	
Ramsden, *Description of an Engine for … Strait Lines*	Aug 1779	500	—	—	72	37	391	254
Bernoulli, *A Sexcentenary Table*	June 1779	500	—	—	91	41	368	
Taylor, *A Sexagesimal Table*	April 1781	1000	—	—	88	21	891	770
Hutton, *Tables of the Products*	Nov 1781	500	—	—	70	5	425	
Cook, King and Bayly, *Original Observations*		150	—	—				
Mayer and Mason, *Mayer's Lunar Tables*		500	—	—				

Sources: BoL, 'Books in the Warehouse Belonging to the Board of Longitude', 28 June 1800, CUL RGO 14/14, p. 357r; BoL, 'Account of Works Printed, Published and Sold', 1784, CUL RGO 14/15, pp. 241r–263r; BoL, Papers on payments for Board work, CUL RGO 14/17, pp. 117r–168v; BoL, Papers on payments for Board work, CUL RGO 14/18, pp. 214r, 279r–297r.

at the Royal Observatory.[91] It was also a subject of ongoing discussion between Maskelyne and counterparts in France, Lalande in particular, which related also to the astronomical tables on which their respective almanacs were based.[92] Mistakes called for action. In January 1809, Maskelyne noted an error found by the French in the lunar tables, remarking that it would affect the volume for 1813, already published, and the calculations for 1814 and 1815.[93]

The question of the practicability of the lunar method and thus the value of the *Nautical Almanac* remained a live topic throughout Maskelyne's tenure and had to be defended with equal conviction. Within a couple of years of the first volume, the issue was being pursued in print by the Wearside mathematician William Emerson, stern critic of metropolitan learned authority. Emerson was one of the age's most prolific textbook writers in mathematical sciences and navigation, principally through his 1763 publishing deal with Nourse (the Board's agent for the *Nautical Almanac*), from whom he obtained invaluable texts such as up-to-date lunar tables.[94] In his astronomy coursebook, Emerson briefly explained how the combination of a lunar ephemeris and a reflecting instrument could in principle give a ship's longitude, but 'it is hardly to be performed at the time when wanted, besides the danger of error', since no lunar theory was accurate to within two minutes.[95] He went further in an essay, 'Of Finding the Longitude at Sea', contending that 'no body can be so silly as to think that any methods by the Moon should be practised every day'. Though he accepted that a lunar almanac would be helpful 'in time of danger and distress, when the reckoning is lost and when a ship is rambling she knows not where', its frequent and regular use 'would be slavery, indeed, when the common known methods are sufficient to guide a ship in moderate weather. And therefore the seamen will never thank us for imposing such a task upon them; nay, they will curse us for it.'[96]

Emerson's comments did not go unchallenged. A contributor to the *Gentleman's Magazine* damned him as an 'astronomical foetus' ignorant of the development of astronomical navigation since Newton, of the workings of Hadley's quadrant and of lunar tables' accuracy: 'you are angry without cause at being referred to almanacs and observations for conviction, instead of theory'. Thanks to the *Nautical Almanac*, Emerson was told, computations 'may now very easily be performed in three quarters of an hour by a moderate computer and in less by an expeditious one'. Emerson dismissed these as the views of a 'fiery Zealot' and 'astronomical conjuror'. A few years later, however,

[91] Volumes of computations of the Moon's place compared with the *Nautical Almanac*, 1777–1811, CUL RGO 4/91 to RGO 4/102.
[92] Boistel, *'Pour la gloire'*, pp. 262–267.
[93] Nevil Maskelyne, 'Diary of 'Nautical Almanac' Work', 1767–1811, CUL RGO 4/324, p. 33r.
[94] Bowe, 'Short Account', p. vii; Gow, 'Letters', p. 44. [95] Emerson, *A System*, p. 365.
[96] Emerson, *Miscellanies*, pp. iv, 204–205.

former *Nautical Almanac* computer and expeditionary astronomer William Wales went further in defence of the method, claiming that calculation times had been cut from three or four hours to just fifteen minutes 'by a very moderate computer'.[97] It was incumbent on Maskelyne and his workforce to realise such zealous claims about the new almanac system's efficacy.

Another fierce critic was mathematician and army officer Robert Heath, former editor of the *Ladies' Diary* and author of works on navigation and the annually published *British Palladium*. A satirical ode in the 1768 edition lambasted the *Nautical Almanac* as 'a fam'd *Crack*', while a letter from 'A Sea Officer', presumably Heath, offered a more concerted attack in 1774.[98] It incorrectly asserted that the data for the *Nautical Almanac* for 1774 had been copied from the *Connoissance des Temps* and so the money spent on computing was wasted. In fact, this was the year in which the *Connoissance des Temps* began to include lunar-distance tables based on the British almanac. The letter went on to question the practicability of the *Nautical Almanac*, claiming that 'we, on-board the *Navy*, make the same Use of the *Nautical Ephemeris* as we do of a *Pack of Cards* or the *Back-gammon Tables*; to pass an idle Hour or to kill Time!' Heath's authorship is supported by the fact that he had previously compared the *Nautical Almanac* to 'the *Nostrums* of the *Quack-Doctors*'.[99]

Criticism of the practicability of lunar distances recurred in disputes about timekeepers between the Board, the Harrisons and the Mudges (discussed in Chapter 6). Published attacks on Maskelyne's timekeeper trials at the Royal Observatory in particular set out the challenges the lunar method faced: the number of days each month when Sun–Moon observations were impossible; the frequent lack of a distinct horizon at night; and the difficulty of making accurate astronomical measurements from a ship at sea.[100] Such attacks sought to create a rivalry between methods that, at sea, were complementary.

Attacks also made much of the close association between Maskelyne, the *Nautical Almanac* and the lunar method, accusing the astronomer of financial self-interest. While Maskelyne rejected such accusations, his close association with lunar distances was inevitable. He had promoted the method in print since

[97] ΑΣΤΡΟΝΟΜΙΚΟΣ, 'Refutation of Several of Mr Emerson's Hypotheses', *Gentleman's Magazine*, 41 (March 1771), pp. 113–114 (p. 114), and 'Reply to Mr Emerson's Vindication' (August 1771), pp. 398–400; Emerson, 'Vindication of Mr Emerson's Astronomy', *Gentleman's Magazine*, 41 (July 1771), pp. 349–350, and 'Second Vindication' (October 1771), pp. 490–493, 538–540, 589–593; citation from Emerson, *Miscellanies*, pp. vi–vii; Wales and Bayly, *Original Astronomical Observations*, pp. xxxviii, xliii.

[98] 'Longitude Ode. By Mr. Moonsby', *The British Palladium* for 1768, p. 72; 'To the Palladium-Author', *The British Palladium* for 1774, pp. 3–4.

[99] *The British Palladium* for 1774, p. 4; Heath, *The Seaman's Guide*, p. 42; Barrett, *Looking for Longitude*, pp. 164–165, 220.

[100] Harrison, *Remarks*, pp. 30–32; Mudge, *A Reply*, p. 83.

1763, he oversaw the computation of the *Nautical Almanac*, of which he conspicuously authored each preface, and the explanations of the almanac's use made explicit reference to his *British Mariner's Guide*. It is no surprise, then, that Heath considered Maskelyne 'the reverend *Superintendent* or *Commander in Chief* of *Longitude*' or that von Zach gave him the satirical sobriquet 'the lunatic Doctor'.[101] Nonetheless, the Astronomer Royal wrote with pride of 'a work to which every lover of Astronomy, Navigation & Commerce would readily say, with Emphasis, Esto Perpertua!'[102]

The *Nautical Almanac* under John Pond

Maskelyne's death in February 1811 and the resulting appointment of John Pond as Astronomer Royal has typically been presented as the moment when the *Nautical Almanac* went into decline. The historiography presents Pond as 'a poor manager' with 'a lack of interest in computational astronomy and the management of the *Nautical Almanac*'.[103] Yet it is worth noting that by then the direction of the publication had suffered the loss not just of Maskelyne but also of experienced comparer Malachy Hitchins in 1809 and Secretary of the Board George Gilpin in 1810, while veteran workers such as Henry Andrews and Mary Edwards were over 60.

Nonetheless, the Commissioners assumed that the handover would be straightforward, with the minutes of Pond's first meeting in post stating that he should:

> conduct the business of the Nautical Almanac by superintending & paying the Computers & Comparers in the manner that has been heretofore done & that the Secretary be requested to allow him access to the Minutes of the Board in order that he may collect from them the mode in which that business has been hitherto carried on.[104]

As already noted, because Maskelyne kept the management under his personal control, the Board was so little involved that the minutes reveal little of what was required to keep the *Nautical Almanac* on schedule and error-free. The Commissioners assumed Pond was inheriting a functioning system and simply had to maintain it, with the editions to 1816 and the first months for 1817 already complete and ready for publication.

[101] *The British Palladium* for 1774, p. 4; Franz Xaver von Zach to Jérôme Lalande, 22 March 1795, Observatoire de Paris, MS 1090/9.

[102] Nevil Maskelyne, autobiographical notes on his life up to about 1789, CUL RGO 4/320, pp. 8:1r–8:9v (p. 8:7r).

[103] Croarken, 'Nevil Maskelyne', p. 157; Peacock, *Life of Thomas Young*, p. 358; Perkins and Dick, 'The British and American Nautical Almanacs', pp. 165–167; Sadler, *Man Is Not Lost*, p. 13; Wilkins, 'The Expanding Role', p. 239; Wilkins, 'The History', p. 56. Dolan, 'People: John Pond', offers a detailed counterview.

[104] BoL, confirmed minutes, 7 March 1811, CUL RGO 14/7, p. 2:146.

As in Maskelyne's time, subsequent minutes offer little detail for the seven years Pond oversaw the *Nautical Almanac*, and nothing by way of explicit criticism. They mainly note the routine granting of licences and warrants, the issue of imprests and payments, the production of related publications and the consideration of table-related proposals, often rejected. In this sense at least, continuity was achieved.

There was also correspondence with existing and former personnel. In 1811 John Simeon, MP for Reading, petitioned Charles Yorke, First Lord of the Admiralty, on behalf of Mary Edwards, asking that she be confirmed in the role of comparer, which she had taken on after the death of Hitchins.[105] The Commissioners resolved instead to appoint Thomas Brown, Anglican minister and brother-in-law of the Cambridge professor William Lax, although they had reiterated that there was no intention of ending the work of any existing computers or comparers.[106] It is not clear, however, that the inexperienced Brown (employed only since 1809) was able to offer the quality of administrative support provided by Hitchins, whose 'unremitting care and attention' Davies Gilbert later commended to Parliament.[107]

Edwards persisted in her requests for the Board to honour promises she said Maskelyne had made before his death. The Board conceded in part, allowing her to be paid at a higher rate (until her death in 1815), and employing her daughters 'at an extra price'.[108] The Commissioners were concerned, however, that Andrews, who had given equally long service, might warrant similar favour. Owing to poor health, Andrews had requested 'some less laborious occupation' than *Nautical Almanac* work. The Board instructed Pond to find something suitable, and after another petition three years later agreed to make a request to the Treasury on Andrews' behalf.[109]

Insofar as one can discern a style of financial administration, Pond seems, like Maskelyne, often to have personally paid for anything needed on the understanding that the Board would reimburse relevant expenses.[110] This seems largely to have worked. There was, however, increasing ambiguity as to whether the work Pond commissioned was for the Royal Observatory,

[105] John Simeon to Charles Yorke, 27 February 1811, enclosing petition from Mary Edwards, RGO 14/11, pp. 134r–136v.

[106] BoL, confirmed minutes, 16 July 1811, CUL RGO 14/7, pp. 2:151–2:152; Thomas Hurd to John Simeon, 8 March 1811, CUL RGO 14/11, p. 138r.

[107] Longitude Discovery Bill, 6 March 1818, *Hansard*, 37, cc. 876–880; BoL, confirmed minutes, 7 December 1809, CUL RGO 14/7, p. 2:135 (Brown recommended by Lax).

[108] Correspondence concerning petitions of Mary Edwards, CUL RGO 14/11, pp. 133–168; BoL, confirmed minutes, 5 December 1811, 5 March 1812 and 9 June 1813, RGO 14/7, p. 2:153, 2:158, 2:163.

[109] BoL, confirmed minutes, 3 December 1812 and 2 March 1815, CUL RGO 14/7, pp. 2:168, 2:195; Henry Andrews to BoL, 27 February 1815, CUL RGO 14/11, p. 5.

[110] Dolan, 'People: John Pond', notes this also of Pond's administration of Royal Observatory finances.

the *Nautical Almanac* or other Board of Longitude activities. When noting in 1817 that his observatory assistants were 'unequal to make the required computations many of which are necessary to the Calculations & correctness of the Nautical Almanacs', Pond recognised the extent to which the two were intertwined.[111]

Errors in the *Nautical Almanac* remained an area of concern and one for which Pond's stewardship received particular criticism at the time and later. That the edition for 1820 (published in 1817) required four pages of errata, for example, was of grave concern for bookseller George Coleman, who claimed to have sold between 160 and 170 copies by 1819.[112] The Board received some correspondence from those who had found errors, but it was debate in the *Philosophical Magazine* that projected the matter into the public sphere.[113] Indeed, the issue became more heated in 1817–1818 in pieces by the pseudonymous 'Astronomicus', eliciting a defence from 'Manchestriensis', while other correspondents pressed for data such as occultations to be added.[114] These debates would outlast Pond's time in charge.

Thomas Young as Superintendent of the *Nautical Almanac*

The *Philosophical Magazine* debates set the scene for consideration of the Longitude Discovery Bill in March 1818. Admiralty Secretary and Commissioner of Longitude John Croker observed during the bill's parliamentary debate that the *Nautical Almanac* had recently been in decline, evidenced by the fact that the edition for 1818 'did not contain less than eighteen grave errors'. His hope was that 'the House should select a proper person, with a moderate but adequate salary, to superintend the publication of that work'.[115] The resulting Act therefore created a Superintendent of the Nautical Almanac,

[111] BoL, confirmed minutes, 4 December 1817, CUL RGO 14/7, p. 2:238. Airy, 'Report … 1839', p. 3, describes similar ambiguity.

[112] George Coleman to Everard Home, 17 April 1819, CUL RGO 14/22, pp. 69r–71v.

[113] For instance, Captain Christian Becker, a Danish prisoner of war, pointed out errors in the editions for 1814, 1815 and 1816; CUL RGO 14/22, pp. 72r–93r. Rear Admiral Home Riggs Popham to John Croker, 22 January 1818, TNA ADM 1/269, notes two minor errors absent from errata for the 1818 edition.

[114] Astronomicus, 'On the Errors in the Nautical Almanac, &c.', *The Philosophical Magazine and Journal*, 48 (1816), pp. 34–35; Astronomicus, 'Further Errors in the Nautical Almanac for 1820', *The Philosophical Magazine and Journal*, 49 (1817), pp. 305–306; Astronomicus, 'On the Nautical Almanac for 1820', *The Philosophical Magazine and Journal*, 50 (1817), pp. 440–443; Manchestriensis, 'Defence of Nautical Almanac', *The Philosophical Magazine and Journal*, 51 (1818), pp. 146–147; Astronomicus, 'On the Nautical Almanac', *The Philosophical Magazine and Journal*, 51 (1818), pp. 186–187 (published just after discussion of the Longitude Discovery Bill); Anonymous, 'Nautical Ephemeris', *The Philosophical Magazine and Journal*, 48 (1816), pp. 460–461.

[115] Longitude Discovery Bill, 6 March 1818, *Hansard*, 37, cc. 876–880.

with Thomas Young, an ally of Joseph Banks, appointed to the role and that of Secretary of the Board of Longitude as part of a radical reorganisation of the Board along Banksian lines (see Chapter 11).[116]

The Act also formalised administration of the Board's finances. The first meeting after its passage noted not only Young's appointments but also the creation of a Committee of Accounts to oversee financial matters.[117] Three months later the Board agreed that one of the Committee's duties would be to ensure speedier clearing of imprests.[118] Meanwhile, Young opened an account in his own name at Coutts and Co. to administer Board expenses.[119]

These changes allowed John Pond to concentrate on the increasingly demanding work of the Royal Observatory, although he remained a Commissioner to be consulted on proposals and other astronomical matters. He also kept contact with several *Nautical Almanac* computers, employing them for other observatory tasks. Conversely, the work of the Royal Observatory's assistants was at times intertwined with or funded by the Board. Payments to the assistant Thomas Glanville Taylor, for instance, detail work related to both the *Nautical Almanac* and the eventual publication of Stephen Groombridge's catalogue of circumpolar stars, a project into which the Board was drawn.[120]

Young left the management and content of the *Nautical Almanac* largely unchanged during his time as Superintendent, much to the frustration of those within the Board and externally who wished to see changes, above all to make it more useful to astronomers. While the tension in trying to serve the differing interests of mariners and astronomers was not new, it re-emerged as an area of fierce dispute from 1818 in debates that bound up control of the *Nautical Almanac* and its content with calls for scientific reform being spearheaded by members of the nascent Astronomical Society (founded in 1820).[121]

Initial calls came from within the Board, when in 1818 Davies Gilbert proposed five changes: calculations of mean times; publication of lunar right ascensions and declinations to seconds; inclusion of Moon–Jupiter distances; listing of more stellar occultations; and hiring a fifth computer.[122] The matter was deferred at the next meeting, however, and the issue does not seem to have

[116] Discovery of Longitude at Sea, etc. Act 1818 (58 Geo 3 c 20); copy at CUL RGO 14/1, pp. 79r–82v.

[117] BoL, confirmed minutes, 5 November 1818, CUL RGO 14/7, pp. 2:255, 2:257.

[118] BoL, confirmed minutes, 4 February 1819, CUL RGO 14/7, pp. 2:265.

[119] Thomas Young to Coutts and Co., 4 February 1819, CUL RGO 14/19, p. 153r. Vouchers in CUL RGO 14/19 and RGO 14/20 include payments to computers and comparers.

[120] For example, Thomas Glanville Taylor, invoice, February-April 1822, CUL RGO 14/22, p. 525r (*Nautical Almanac*); Thomas Glanville Taylor, invoice, December 1827, CUL RGO 14/22, p. 895r (Groombridge catalogue). Airy (ed.), *Catalogue*, pp. iii–iv, acknowledges the ambiguity. The Board initially declined to publish the catalogue; BoL, confirmed minutes, 16 July 1811, CUL RGO 14/7, pp. 2:152.

[121] Ashworth, 'The Calculating Eye', pp. 430–434; Miller, 'The Royal Society', pp. 313–329.

[122] BoL, confirmed minutes, 5 November 1818, CUL RGO 14/7, pp. 2:257–2:258.

returned for consideration until 1824, when Gilbert called once again for additions 'for the use of Astronomers and of Observatories'. The Commissioners therefore instructed Young to add new tables of the places of sixty principal stars (for every ten days of the year) and the Pole Star (for every five days).[123]

By this time, calls were growing from outside the Board for more sweeping changes, led by two founding members of the Astronomical Society, Francis Baily and James South. Baily began the debate at the same time as suggesting the foundation of the Astronomical Society.[124] In diatribes that attacked both Young and the Board, he and South subsequently argued that the *Nautical Almanac* had fallen behind foreign ephemerides since Maskelyne's time in quality and, importantly, utility to astronomers.[125] What was needed, they argued, was 'an Astronomical Ephemeris, worthy of the country', to replace an almanac primarily for seamen that had become 'lamentably deficient'.[126]

Further calls for change came at Board meetings with the arrival as Commissioners of Longitude of two other leading Astronomical Society members: John Herschel from 1821 and George Airy from 1826. In 1827, Airy proposed improvements to the solar tables but found his ideas opposed by Young.[127] Herschel presented a more radical suggestion earlier that year. Lamenting that the *Nautical Almanac* was now 'essentially defective, inasmuch as it neither affords all, nor nearly all, the information which Astronomers require', he offered a series of improvements. Yet although his suggestions were briefly trialled as a supplement to the *Nautical Almanac*, Young concluded that it would be 'a great prodigality' to continue such a publication purely for the benefit of astronomers.[128]

Baily's and South's crusade made much of what they saw as the rise in errors in the *Nautical Almanac*. This remained an area of correspondence and vigilance for Young.[129] He therefore worked to ensure the timely correction of errors discovered, whether through lists of errata or corrections incorporated

[123] BoL, confirmed minutes, 5 February and 1 April 1824, CUL RGO 14/8, pp. 16–18. The new tables appeared in the *Nautical Almanac* for 1827, replacing earlier tables with fewer stars.

[124] Cagnoli, *Memoir*, pp. 29–36.

[125] Baily, *Astronomical Tables*; Baily, *Remarks*; South, *Practical Observations*; [Thomas Young], 'A Reply'. Both also corresponded with the Board, e.g. Francis Baily to Humphry Davy, 31 March 1824, CUL RGO 14/35, pp. 429r–430v; BoL, confirmed minutes, 1 April 1824, CUL RGO 14/8, p. 19; James South to BoL, 2 February 1826, CUL RGO 14/22, pp. 13r–v.

[126] South, *Practical Observations*, pp. 48, 59. Baily, *Astronomical Tables*, was a privately printed model intended for astronomers.

[127] BoL, confirmed minutes, 1 November 1827, CUL RGO 14/8, p. 56; Airy, *Autobiography*, pp. 76–77.

[128] BoL, confirmed minutes, 5 April, 7 June, 1 November 1827 and 7 February 1828, CUL RGO 14/8, pp. 50, 52–53, 55, 58; Herschel, 'The Following Paper' (copy also at CUL RGO 16/50 244/2); Young, 'Report', p. 10.

[129] BoL, confirmed minutes, 6 April 1820, CUL RGO 14/7, p. 2:304 (thanks for reporting errors already spotted).

into second and third editions, to defend the publication's accuracy and to monitor the basis of its data.[130] In 1820, for example, he undertook a detailed comparison of lunar observations with predicted values from the sequence of lunar tables that had been used to calculate the *Nautical Almanac*. He concluded that, with the adoption of Burkhardt's tables just a few years earlier, 'the degree of accuracy ... desirable is, in all probability, fully attained'.[131]

Conclusion

With the dissolution of the Board of Longitude by Act of Parliament in 1828, the would-be reformers sensed an opportunity to seize the initiative. Young, however, retained his position as Superintendent of the Nautical Almanac and favoured continuity, as he had done during the final years of the Board. Writing to Croker early in 1829, he felt that:

> it is in the intention of the Board of Admiralty that the computers, hitherto employed should proceed with their labours as heretofore, and that I should make no essential changes in the work or the workmen without their express directions.[132]

A few days later, he confided to his friend Hudson Gurney that 'the "Radical" abuse of the Nautical Almanac is likely to continue', but now felt secure within Admiralty protection.[133] He proved correct in that Baily and South resumed their attacks and, with their fellow Astronomical Society reformers, began lobbying government.[134] What they proposed was a new Board of Longitude independent of any state department to manage a restructured *Nautical Almanac* and judge potential improvements in navigation and astronomy. Parliament took no action, however, although things came to a head following Young's death in May 1829. While oversight of the *Nautical Almanac* initially reverted to Pond, Admiralty Secretary John Barrow sought the Astronomical Society's advice on possible reforms the following year, conceding to more than a decade of concerted pressure. The resulting proposals underpinned the creation of the Nautical Almanac Office in 1831. This saw a significant reform of the content of the *Nautical Almanac* and an almost complete change in the computers employed, while also bringing to an end the turbulent arguments about the best way to advance scientific thinking in a post-Board world.[135]

[130] Thomas Young, regarding changes to *Nautical Almanac* for 1826, 2 November 1822, CUL RGO 14/22, pp. 231r–232v; Young, 'Report'.
[131] BoL, confirmed minutes, 2 November 1820, CUL RGO 14/7, pp. 316–319.
[132] Thomas Young to John Wilson Croker, 10 Jan 1829, ADM 1/3469.
[133] Thomas Young to Hudson Gurney, 11 January 1829, quoted in Miller, 'The Royal Society', pp. 323–324.
[134] Baily, *Further Remarks*.
[135] Miller, 'The Royal Society', pp. 324–328; Perkins and Dick, 'The British and American Nautical Almanacs', p. 174.

8 Managing, Communicating and Judging Longitude after Harrison, 1774–c. 1800

Rebekah Higgitt

With the Longitude Act 1765, an enduring role for the Commissioners of Longitude had been established. This had been found particularly in publishing and overseeing the *Nautical Almanac* but also in continuing to judge and trial submitted schemes for and improvements to longitude methods or 'other Discoveries or Improvements useful to Navigation'.[1] With the Harrison affair finally dispatched, in 1774 yet another Act took the opportunity to tidy up their future business. This revoked all previous Acts, except with regard to the appointment of Commissioners and the publication of the ephemerides, and, while retaining the broadened 1765 remit, reframed the amount and conditions attached to the rewards.[2] Over the course of the next three decades, the activity of the Commissioners – now decisively the Board of Longitude – became increasingly regular and bureaucratic. In this, they can be seen as part of a broader movement across the departments and functions of the British state, responding to calls for disinterest and bureaucratic efficiency. Nevertheless, like other branches of the state, the Board's modes of working still also straddled the old world of personal connection and private patronage. They were dominated by the networks and interests of individual Commissioners, particularly the Astronomer Royal, Nevil Maskelyne, and, after he was elected president of the Royal Society in late 1778, Joseph Banks.

The Bureaucratising Board

Looking at the finances of the Board, its changing nature becomes immediately apparent (see also Appendix 2). Following Derek Howse's breakdown of the period of the Board's active existence into roughly equal thirds (1737–1766, 1767–1796, 1797–1828), expenditure shows a 171 per cent increase between the first and second, and another 22 per cent in the final third.[3] The increase

[1] Discovery of Longitude at Sea Act 1765 (5 Geo 3 c 20), copy at CUL RGO 14/1, pp. 29r–33v.
[2] Discovery of Longitude at Sea Act 1774 (14 Geo 3 c 66), copy at CUL RGO 14/1, pp. 37r–42v.
[3] Howse, 'Britain's Board of Longitude'. The following analysis is based on this and Howse, 'The Board of Longitude Accounts'.

between the first and second periods is largely accounted for by the amount spent on publications, including the *Nautical Almanac*, but also by increased overheads (including costs of salaries and expenses of Board members and functionaries) and spending on expeditions and trials (including instruments and the salaries of expeditionary astronomers). All of these elements increased again in the final third of the Board's existence, with overheads more than doubling, although there was a reduction in the amount spent on rewards. The Board was clearly characterised by a pre- and post-*Nautical Almanac* existence, with its role as a publisher having a significant impact on its expenditure and the regularity of its business from 1766. Its role as a distributor of rewards also changed, here the crux being pre- and post-Harrison (see Figures A.3 and A.4 in Appendix 2). The total spent on rewards in the thirty-six years between 1737 and the final payment to Harrison in 1773 was £28,799. After the passing of the 1774 Act, in the fifty-four years to the Board's abolition in 1828, it was £24,216. The last quarter of the eighteenth century, the focus of this chapter, saw less than £10,000 spent on rewards. The balance of Board of Longitude activity had fundamentally shifted.

This shift was a matter of policy, encapsulated in the Act of 1774. This reflected the belief that the major breakthroughs in the astronomical and timekeeping methods had taken place, and that it was unlikely that anything new would usurp them. It also reflected the lessons learned from the dispute with Harrison, carefully stating all the hurdles that had to be cleared before the Commissioners could have confidence that they had access to a replicable and useful method. The Act is known for Thomas Mudge's claim that Maskelyne had been heard to say that it 'had given the Mechanics a bone to pick which would crack their teeth'.[4] While Mudge asserted that it was prejudiced against mechanics and in favour of the lunar-distance method, it in fact reduced the awards and increased the conditions for all methods and created separate processes that meant astronomical methods did not compete against mechanical.[5] The rewards for timekeepers and new ideas were halved (offering £5000, £7500 and £10,000 rather than £10,000, £15,000 and £20,000), while improvements to lunar and solar tables might achieve £5000 only. Chapter 6 described the stricter conditions timekeepers had to pass to gain rewards, while improved tables needed to be accurate to 15 arcseconds and tested against a whole lunar cycle of 18.5 years. Unsurprisingly, no such rewards were distributed. More important were the smaller sums that the Act allowed for the improvement, manufacture or trialling of instruments or methods, and as lesser rewards, given at the Commissioners' discretion. Those rewarded included watchmakers, instrument makers, astronomers and mathematicians. These last might also benefit from the money (some £11,704 between 1767

[4] Mudge, *A Narrative of Facts*, p. 5. [5] Maskelyne, *An Answer to a Pamphlet*, pp. 54–55.

and 1796, and £23,344 between 1797 and 1828) spent on computing for Board publications.

The increased spending on overheads reflects the increased amount and regularity of Board business discussed in Chapter 5. In some years in the late 1760s and early 1770s there had been multiple meetings, but from 1776 a regular pattern of three meetings was established until the passing of the transformational Act of 1818 (see Chapter 11), with occasional additional meetings if business required. From 1780, with a backdated payment, George Gilpin was paid for looking after the instruments belonging to the Board and for undertaking or assisting with trials, eventually receiving £10 p.a.[6] Thus, with a payment of £15 per meeting to the various professors (and, from 1775, the Astronomer Royal, in recognition of the large volume of Board business he dealt with), the Secretary, the custodian of instruments, a series of expeditionary observers and computers and calculators, the Board was responsible for regular payments of government money to many individuals, above and beyond those given one-off payments or rewards.

The first two secretaries of the Board, John Ibbetson (1764–1782) and Sir Harry Parker (1782–1795), were both clerks to the Secretaries of the Admiralty. This underlines the administrative as well as financial links of the Board of Longitude to its parent department of state. The various First Lords of the Admiralty remained major figures at Board meetings until around 1808, and then again after the end of the Napoleonic Wars in 1815. From 1782 Philip Stephens, First Secretary to the Admiralty, became a regular attendee, meaning that, when a 1790 Longitude Act added the Secretaries of the Admiralty as Commissioners of Longitude, it was acting retrospectively.[7] Other Admiralty or political figures, such as the Comptroller of the Navy, the Speaker of the House of Commons and Admirals of the Red, White and Blue, occasionally appeared, with some taking particular interest. Maskelyne was the most regular of all attendees, missing only a handful of meetings in his forty-six years as Astronomer Royal. The presidents of the Royal Society were key figures, usually taking the chair if the First Lord was absent, until Joseph Banks, who began attending meetings in 1779, stepped away between June 1805 and Maskelyne's death in 1811 after they took opposing sides in the dispute between Arnold and Earnshaw. Banks's subsequent reshaping of the Board, after a period of quiet business involving largely only Maskelyne and the five professors, will be dealt with in Chapter 11. In the pre-1818 period, despite the presence of Admiralty personnel, and perhaps partly because of the distractions of war with France, the activities of the Board were at a remove from more central business.

[6] BoL, confirmed minutes, 15 July 1780 and 5 December 1795, CUL RGO 14/6, pp. 2:7, 2:252.
[7] Discovery of Longitude at Sea Act 1790 (30 Geo 3 c 14), copy at CUL RGO 14/1, pp. 52r–55r.

In their minor way, the changes of the Board of Longitude reflected the wider growth of the state in the eighteenth century. This was the result of war but it was also manifest in an increasing desire for expert advice and non-partisan civil servants. From the mid-1760s, it has been argued, in salient cases such as maritime and colonial affairs, the state increasingly mobilised proficient practitioners both for intelligence and for technical projects.[8] This growth provoked frequent calls to rein back the size and expense of the state, encouraging awareness that the money and activity involved belonged to the 'public'. Parliament and lobbyists were increasingly involved in both tracking expenditure and demanding that government act to address public concerns.[9] Language about public benefit had, of course, been core to the Commissioners of Longitude since the beginning, and it recurred frequently in the minutes of the later Board, echoing the terms of the original and later Acts. In considering action regarding Jesse Ramsden's dividing engine, for example, the minutes record the need to 'secure to the public the benefit of his discovery' and that 'the Inventor deserves Encouragement from the Public'.[10] Things as specialised as Johann Bernoulli's sexcentenary tables were claimed as 'of public utility' and published 'for the use of the Public'.[11] The increased turn to publication represented a desire to demonstrate the fruits of publicly funded labour and to make accessible what might otherwise remain private, in ways that clashed with artisanal norms.[12] We may see this too in Maskelyne having been the first Astronomer Royal to publish the Royal Observatory's observations, and his and the Board's campaign to rescue and publish the papers of his predecessors.[13] Flamsteed's papers were acquired 'for the use of the Publick' in 1771 and the imperative behind the long process of extracting Bradley's observations from the University of Oxford was underscored by claims that 'the Public have reason to complain', since he, his assistants, the instruments and the buildings were paid for and maintained with 'Public Money'.[14]

With responsibility for public interest came the continuing possibility of public scrutiny. Although wide discussion of the merits of decisions had always been a significant issue for the Board, there was more concern for transparency

[8] See Brewer, *The Sinews of Power*; Drayton, *Nature's Government*, p. 66.

[9] Gascoigne, *Science in the Service*, pp. 1–2.

[10] BoL, confirmed minutes, 27 May and 1 June 1775, CUL RGO 14/5, pp. 278, 283.

[11] BoL, confirmed minutes, 22 November 1777, CUL RGO 14/5, pp. 320–321.

[12] See Brewer, *The Sinews of Power*, pp. 181–204.

[13] Howse, *Nevil Maskelyne*, pp. 102, 106–108. Maskelyne devoted considerable space in his brief 'autobiographical notes' to the matter of Bradley's papers, and also highlighted his role in ensuring future observations were published, both before and after he became Astronomer Royal: see Nevil Maskelyne, autobiographical notes, CUL RGO 4/320, pp. 8:5r–8:6r and esp. p. 8:7r.

[14] BoL, account of money issued by the Treasurer of the Navy, 19 April 1774, TNA ADM 49/65, fols. 32–33 (fol. 33); see also Howse, 'Britain's Board of Longitude', p. 412; BoL, confirmed minutes, 11 June 1791, CUL RGO 14/6, p. 2:173.

in regular financial business by the later eighteenth century. This sits alongside the political campaigns of the 1780s against 'Old Corruption', and government responses such as removal of sinecures, and enquiries and committees to examine and audit the finances of public offices.[15] The Board of Longitude must have been well aware of this movement, for Fletcher Norton, Speaker of the House of Commons, known for championing parliamentary control over expenditure, was a regular attendee to 1799 (the last Speaker to so be).[16] The Board itself set up a standing committee 'to examine and audit the Accounts of printing and other Expences of this Board' in 1780.[17] This streamlined the business with bills and accounts at Board meetings, helped keep a significantly closer eye on the whereabouts of assets such as instruments and publications, and inspected accounts and receipts from those who made significant expenditure on their behalf, particularly the Secretary and Maskelyne. The committee was chaired by the president of the Royal Society, Joseph Banks, and could be attended by anyone who wished, with three stated as quorum. In practice, it often appears to have simply been a meeting between Banks and Maskelyne, sometimes with Anthony Shepherd, the Plumian professor of astronomy at Cambridge. Such procedures and appearance of scrutiny were a means of protecting individuals from being charged with profiteering, although Maskelyne did still attract such criticism from Thomas Mudge and Robert Heath.[18] The casting of individual action as part of collective institutional activity was a second line of defence.[19]

Nevertheless, despite being an increasingly bureaucratic organisation within a developing state apparatus, the activity of the Board of Longitude was both at a remove from government and essentially the product of individual networks and connections. The fact that Harrison and Mudge could appeal to Parliament, which felt competent to come to a different decision to the experts of the Board, is perhaps unsurprising. Similarly, the Board's ability to encourage mathematical and astronomical teaching – and thus take up of the lunar-distance method – was limited. The request, promoted by Maskelyne, that the Admiralty consider 'some encouragement' to schoolmasters on board the Navy's ships, that 'so useful a body of Men' might 'make themselves Masters of the Lunar Tables', was not more obviously successful than earlier attempts to have masters become qualified.[20]

[15] Gascoigne, *Science in the Service*, pp. 9–11; see Harling, *The Waning of 'Old Corruption'*, ch. 2.

[16] Laundy, 'Norton, Fletcher'. [17] BoL, confirmed minutes, 15 July 1780, pp. 2:8–2:9.

[18] Mudge, *A Narrative of Facts*; Howse, *Nevil Maskelyne*, pp. 170–175. On Heath see Barrett, *Looking for Longitude*, pp. 163–166, 206–207.

[19] At Maskelyne's request, the Board minutes recorded that the only money he dealt with was 'regularly & punctually accounted for, without any benefit or advantage whatever to himself' (BoL, confirmed minutes, 3 March 1792, CUL RGO 14/6, p. 2:181).

[20] BoL, confirmed minutes, 7 December 1782, CUL RGO 14/6, p. 2:41; Harry Parker to Philip Stephens, Secretary of the Admiralty, 17 December 1782, RGO 14/9, p. 6r; BoL, confirmed minutes, 12 November 1768, CUL RGO 14/5, p. 174. On the lack of use of the lunar-distance method, see Wess, 'Navigation and Mathematics'.

The business that the Board dealt with at its now regular meetings came to it in the variety of ways we might expect for those operating in a time described by Gascoigne as in 'transition from the informal methods of patronage and connection' to the 'beginnings of a bureaucratic order based on career civil servants'.[21] He has traced the personal, political and institutional connections used at this period by Banks to serve national and class interests and to shape the relationship between science and government. Many of these connections – including key patrons like the Earl of Sandwich and significant civil servants like Stephens and Evan Nepean – can also be found within or around the Board of Longitude as a result of their positions in the Admiralty. Gascoigne largely ignores the Board, suggesting that Banks's influence on or ability to work through it was limited because of the greater part played by Maskelyne.[22] This choice does significant disservice to Banks's active work in reshaping the Board after Maskelyne's death, and also overlooks his activity through and on behalf of the Board from when he joined it as president of the Royal Society in 1779 to the end of the century.[23] First, however, we should turn to Maskelyne's primary role within and beyond the Board.

Nevil Maskelyne: Chief Commissioner or Servant of the Board?

It can be argued not only that the way Maskelyne conducted himself as a Commissioner of Longitude differed from that followed by previous Astronomers Royal but also that he virtually 'invented' the Board, doing 'more than any other single official to forge a standing Board of Longitude'. He also raised its national and international standing, at least among astronomers and mathematical practitioners.[24] Drawing on his existing network of correspondents, and those acquired and cultivated through his position as Astronomer Royal and chief authority on astronomy within the Royal Society, Maskelyne helped place the Board among the top rank of astronomical and mathematical institutions in Europe. This is evidenced particularly by the exchange of publications. The Board, rather than any individual Commissioner, received observations from Jean-Dominique Cassini at the Observatoire de Paris or charts and an atlas from Vicente Tofiño of the Real Observatorio de Cádiz.[25] It was probably Maskelyne who suggested the Board's 1777 resolution 'That copies of every Work which hath been, or may be, printed and published by this Board be sent to Foreign Academies and Professors of Mathematic's'. It was certainly he that the Board trusted to draw up

[21] Gascoigne, *Science in the Service*, p. 5. [22] Ibid., pp. 4, 29.

[23] On Banks's activity after Maskelyne's death, see Higgitt and Dunn, 'The Bureau and the Board'.

[24] Baker, 'Humble Servants', p. 203.

[25] BoL, confirmed minutes, 14 November 1786, 11 June 1791, and 29 November 1788, CUL RGO 14/6, pp. 99, 171, 132.

the relevant list, and who later suggested adding several more learned societies, libraries, observatories and individuals across Britain and Europe in 1781 (Box 7.2).[26] Maskelyne's own working copy of this list shows additions through to the 1800s, including the academies in Philadelphia and Boston in the United States, and deletions due to death or, in the case of Edme-Sébastien Jeaurat, ceasing to be editor of the *Connoissance des Temps*.[27]

By the 1770s, Maskelyne's efforts were central to the regular work of the Board. He not only oversaw the system for publishing the *Nautical Almanac* and near-constant trials of one set of timekeepers or another at the Royal Observatory but was also almost always the person to whom new schemes, tables and instruments were referred for comment. He reported back, generally by the following meeting, whether to indicate that they 'did not contain any thing that merited the Board's attention' or to report at length on ideas ingenious but flawed, or 'ingenious and likely to be useful'.[28] He was typically director of any publication of the Board, and was the individual trusted to select, instruct and supply expeditionary observers, and to receive and evaluate their observations.[29] In such work he was supported by a number of trusted individuals from the community of practical mathematicians among which he felt most comfortable. William Wales, who became the Board's Secretary in 1795, had, for example, acted as an observer on a voyage of exploration, a computer for the *Nautical Almanac*, an author or editor of publications, an observer for the trial of an instrument and, occasionally, an alternative or second opinion to Maskelyne's judgements of submitted schemes. As well as payments received as a computer, he was paid £543 19*s* in rewards, £1433 8*s* 5*d* for Cook's second voyage plus an additional £50 for completing the related drawings and calculations, £10 1*s* for trialling Ralph Walker's compass and a total of £240 as Secretary.[30] Thus the Board of Longitude connection also allowed Maskelyne to act as a patron within this British network: individuals who had been his assistants at the Royal Observatory and/or *Nautical Almanac* computers would become expeditionary astronomers, computers for one-off publications, or, in the case of Wales and Gilpin, salaried officials for the Board.[31]

[26] BoL, confirmed minutes, 7 June 1777, CUL RGO 14/5, p. 316; BoL, confirmed minutes, 3 March 1781, CUL RGO 14/6, p. 16.

[27] Nevil Maskelyne, 'A List of Public Libraries, Learned Societies, Observatories, and private persons, both at home and abroad, to whom presents of the publications of the Board of Longitude are to be made', 1781–c. 1805, RGO 4/324, fols. 26v–27r.

[28] For example, BoL, confirmed minutes, 8 December 1787 and 1 March 1788, CUL RGO 14/6, pp. 2:113, 2: 119–2:121 (on John Churchman); BoL, confirmed minutes, 7 March 1789, CUL RGO 14/6, p. 2:134 (on M. Grenier).

[29] Higgitt, 'Equipping Expeditionary Astronomers'; Eóin Phillips, 'Making Time Fit', pp. 114–163; Phillips, 'Instrumenting Order'.

[30] Sums taken from Howse, 'Board of Longitude Accounts'.

[31] Gilpin's receipts for sums received from the Board between 1781 and 1809 are at CUL RGO 14/17, pp. 295r–309Av. Gilpin, Maskelyne's Assistant at the Royal Observatory between 1776

As had historically been the case, those who hoped to offer schemes to the Board often wrote to individual Commissioners, and sometimes multiple individual or institutional recipients. Where he was impressed, as in the case of Andrew Mackay at Marischal College, Aberdeen, Maskelyne was happy to support ideas at the Board and to explore other options, such as an assistantship at the Royal Observatory, an expeditionary appointment, Fellowship of the Royal Society, or simply sharing mathematical information and allowing dedications or an advertisement in the *Nautical Almanac*.[32] Henry Andrews, who was a teacher and seller of mathematical instruments and books in Hertfordshire, computed for the *Nautical Almanac* between 1768 and 1815 after being introduced to Maskelyne, probably by the *Almanac*'s first comparer, Richard Dunthorne.[33] This personal recommendation became a professional tie, then, in turn, a long personal correspondence. Correspondence with computers John and Mary Edwards, comparer Malachy Hitchens and astronomers Wales and Crosley turned into warm friendships. Maskelyne's paternalism is clear in several cases, not least his extensive efforts to settle the affairs of William Gooch, who had been sent out as the Board's astronomer to join George Vancouver's expedition and was killed en route.[34]

Such ties not only crossed multiple scientific and institutional contexts; they also included family and friends. Andrews's wife was, for example, tasked with helping Maskelyne and his wife find a nursery maid for their daughter, and Maskelyne promised to help find a position for the Andrews's son.[35] Maskelyne's sense of personal responsibility towards this network is evidenced many times over.[36] In one of his most significant acts of patronage, he persuaded the Board of Longitude to ease the financial shock caused to the *Nautical Almanac*'s longest-serving computers and comparer when work was suspended because they had computed more than sufficient years ahead. One-off sums were granted to Mary Edwards, Malachy Hitchens (£34 each), Andrews and Joshua Moore (£17 each), and Maskelyne also got agreement

and 1781, also became Secretary to the Royal Society from 1785 to 1810. He had been Wales's assistant on Cook's second voyage, made money testing timekeepers while Assistant at the Royal Observatory, did some *Nautical Almanac* computing and was ultimately salaried as both Secretary to the Board of Longitude and its Keeper of Instruments. On Gilpin and other assistants under Maskelyne, see Croarken, 'Astronomical Labourers'.

[32] See Baker, 'Humble Servants', pp. 223–224; Nevil Maskelyne to Andrew Mackay, 1787–1805, NMM MKY/8. From 1788 it had been necessary to gain permission from the Board to advertise in the *Nautical Almanac*; BoL, confirmed minutes, 1 March 1788, CUL RGO 14/6, pp. 2:125–2:126.

[33] Mary Croarken, 'Nevil Maskelyne', p. 151.

[34] Maskelyne's letters to Gooch's father are among the Gooch papers at CUL Mm.6.48, fols. 94–127; see also Dunn, 'Heaving a little Ballast', p. 92.

[35] Baker, 'Humble Friends', p. 225; letters from Nevil Maskelyne to Henry Andrews, 1768–1811, CUL RGO 4/149.

[36] See Croarken, 'Nevil Maskelyne', for some examples.

from the Board to use and pay the computers for any other 'publicly useful' work he could come up with, whether to improve the *Almanac* or reduce the Royal Observatory's observations, nominally to prepare them for use in any future trials of tables submitted to the Board for reward.[37]

While Maskelyne's role and connections as Astronomer Royal shaped the work of the Board, he also used it to enhance the reach of the Royal Observatory. As well as the specific instance above, more broadly the work of the Board created audiences, networks of assistance and users for the work of the Observatory. Above all, this was through the publication of the *Nautical Almanac*, which served to meet the original aim of the Observatory's foundation. However, the possibility of rewards or paid work from the Board of Longitude also generated a need for the Royal Observatory's output, whether from computers, teachers of navigation, those attempting to create improved tables, or observers and surveyors on voyages. Maskelyne did what he could to create demand for this output, asking the Navy to enhance the role of schoolmaster, as mentioned above, or again pushing for masters of ships to be examined in use of the lunar-distance method.[38] Loaning instruments or giving publications to those who might use them properly, such as George Witchell at the Royal Naval Academy in Portsmouth, likewise developed or rewarded the community of users of the *Nautical Almanac* and *Greenwich Observations*.[39]

The Board undoubtedly created a great deal of work for Maskelyne personally. His friend, the mathematician and teacher of navigation Patrick Kelly, later wrote with sympathy of this 'important and laborious duty', describing it as 'arduous as well as unpleasant' to have to act as judge, whether of friends, skilled individuals or the 'numerous candidates of very slight pretensions, and even visionaries'.[40] It might have been even worse had the Board advertised its rewards more widely. In fact, the *Nautical Almanac*, which would be used only by those with some understanding of mathematics and existing longitude techniques, seems to have been the chief vehicle for publicity. The terms of the Acts were set out at the front of each volume as an invitation of sorts to this audience. Likewise, when it became clear, ten years on, that the terms of one reward were not well known, the Board decided to reprint the advertisement and include it in the *Nautical Almanac*, as well as to send a copy to the enquirer.[41] Choosing not to target general publications or trade institutions, the

[37] BoL, confirmed minutes, 7 June 1794, 7 December 1793 and 1 March 1794, CUL RGO 14/6, pp. 2:220, 2:208, 2:215–2:216.

[38] BoL, confirmed minutes, 5 March 1785, CUL RGO 14/6, pp. 2:80–2:81.

[39] BoL, confirmed minutes, 3 March 1781, CUL RGO 14/6, p. 2:22. Witchell had received £310.50 between 1765 and 1770 as rewards for tables and computing methods, plus half a guinea per master who passed examinations on using the lunar-distance method at Portsmouth (only seven seem to have done so), and was an early *Nautical Almanac* computer, 1767–1769.

[40] [Kelly], 'Maskelyne, Nevil', p. 700.

[41] BoL, confirmed minutes, 5 December 1789, CUL RGO 14/6, p. 2:145.

Board reached largely those with mathematical and astronomical skills, or who were known to individual Commissioners or within their networks.

Joseph Banks as Commissioner of Longitude

Banks's presence on the Board of Longitude undoubtedly helped develop its role in linking scientific advice and maritime exploration, although it was Maskelyne who had earlier initiated such activity within the Board. Having played a central role in recruiting for, instructing and supplying the Transit of Venus voyages of 1761 and 1769 on behalf of the Royal Society, Maskelyne was active in 1771 and 1776 in ensuring that the Board placed good instruments, including Kendall's watches, and well-instructed observers on Cook's subsequent voyages.[42] It was at his suggestion that the Kendall watches, returned from use on *Resolution* and *Discovery* in 1780, were cleaned and repaired 'for further Use', and he who selected instruments and/or observers for Constantine Phipps's 1773 Arctic voyage, George Vancouver's 1791 Northwest Pacific voyage and for the use of William Dawes on the 1787 First Fleet.[43] However, Banks's networks and interests also played a role: notably, he loaned Kendall's second watch (K2) and other instruments and publications to William Bligh, with the Board simply assenting to a *fait accompli*.[44] It is possible that at this period Banks used his role on the Board to assist with his relations with the Admiralty rather than vice versa. Gascoigne notes that Banks was in low favour with the Admiralty after 1772 because of his behaviour when he hoped to travel on Cook's second voyage, but that he was clearly back on good terms by the time of the 1787 *Bounty* breadfruit voyage.[45] Sitting on the Board of Longitude from 1779, and being in a position to hand over a valuable instrument, may have helped restore his favour.

[42] For the second two Cook voyages, Maskelyne wrote to Sandwich in the month before the Board meeting to suggest the inclusion of instruments and someone who could use them: Maskelyne's letter of 25 October 1771 is partly transcribed in BoL, confirmed minutes, 28 November 1771, CUL RGO 14/5, p. 207; BoL, confirmed minutes, 2 March 1776, CUL RGO 14/5, pp. 296–297. Maskelyne to Lord Sandwich, 21 February 1776, NMM SAN/F/36, expressed his happiness that he agreed with the idea of the Board sending an astronomer with Cook.

[43] BoL, confirmed minutes, 4 November 1780, CUL RGO 14/6, p. 2:12; BoL, confirmed minutes, 6 March 1773, CUL RGO 14/5, p. 234; BoL, confirmed minutes, 5 December 1789 and 14 November 1786, CUL RGO 14/6, p. 2:146, 2:101–2:102. In these cases Maskelyne's actions responded to the initiatives of John Pringle (president of the Royal Society), Lord Chatham (First Lord of the Admiralty) and Dawes respectively. In 1789 Chatham informed the Board of the planned voyage of discovery to the southern whale fisheries, initially asking for two Kendall timekeepers, and subsequently the desire to send an astronomer with instruments on the revised voyage to the Northwest Pacific (BoL, confirmed minutes, 5 March 1791, CUL RGO 14/6, pp. 2:167–2:168). On Maskelyne's role in voyages of exploration, see Higgitt, 'Equipping Expeditionary Astronomers'.

[44] BoL, confirmed minutes, 8 December 1787, CUL RGO 14/6, p. 2:117.

[45] Gascoigne, *Science in the Service*, p. 124.

As the one Commissioner explicitly identified as an *ex officio* member of the standing committee, Banks was central to the Board's inner workings, although this is not always explicit in the Board minutes. They do record, though, that he missed just eight of the seventy-five meetings between 1779 and 1803.[46] This was despite the disputes at the Royal Society, when Banks's presidency was attacked by those who saw themselves as representing real (often mathematical) learning and skill, rather than mere dilettanti, and he and Maskelyne found themselves in opposition.[47] Banks seems to have allowed himself at least one moment of revenge within the context of the Board, before each found ways to rebuild their working relationship, including through their mutual interest in placing men of science on voyages of exploration.[48] Banks described 'revenging myself on the R Astronomer at the board of Longitude' and reducing him 'to a compleat state of humiliation'.[49] Interestingly, given Maskelyne's anxiety about being accused of profiting from the Board, Banks chose to have his revenge by accusing him of paying bills without authority, and threatened to resign from the auditing committee. The hatchet seemed to have been buried by the later 1780s and their relationship remained steady throughout the 1790s.

The minutes record a number of clear interventions by Banks, and some occasions where he seems to have had more covert influence. He was certainly a more active Commissioner than the five Oxbridge professors. His first recorded intervention, at his third meeting on 27 November 1779, was markedly different to the typical business transacted. It suggested a new reward for something beyond timekeepers or tables and was couched in significantly less specialist language. His suggestion was that the Board should address the recent 'impossibility of obtaining Flint Glass' for telescopes, apparently owing to a lack of understanding of how to produce glass of the best quality and recent use of cheaper but lower-quality techniques.[50] While the other Commissioners may have been interested in the potential to improve sextants and octants, Banks rather more vaguely referred to telescopes' 'great Utility to Navigation by enabling Comm~drs of Ships to decry their Enemies at a distance, to get an early knowledge of their strength', as well as 'to obtain, by observations of the Coelestial Bodies, the real situation of places which they are to visit'.[51] This initiative is in line with the neo-mercantilism that Gascoigne associates

[46] Schaffer, 'Joseph Banks'. [47] See Heilbron, 'A Mathematicians' Mutiny'.

[48] See Homes, 'Friend and Foe', pp. 248–251.

[49] Joseph Banks to Charles Blagden, 6 March 1784, RS CB/1/1/99; Joseph Banks, rough minutes of a meeting of the Board of Longitude, 6 March 1784, RS MM/7/41, both quoted in Homes, 'Friend and Foe', pp. 248, 318. Nothing of this appears in the confirmed minutes of the Board.

[50] McConnell, *A Survey of the Networks*, p. 162; see also Turner, 'The Government and the English Optical Glass Industry', although Turner does not mention the rewards offered by the Board of Longitude.

[51] BoL, confirmed minutes, 27 November 1779, CUL RGO 14/5, pp. 352–354.

with Banks, where successful 'navigation' meant maritime triumph in war as much as prosecuting trade or establishing routes, and a focus on national or imperial self-sufficiency in goods – whether wool, corn or telescopes.[52] The Board, which on this day consisted of the Earl of Sandwich, Maskelyne and the five professors, as well as Banks, unanimously agreed to a reward of £1000 offered by advertisement, on condition of fulfilling a strict series of requirements involving no less than three committees.[53]

Banks returned to the question of glass in 1788, on behalf of six eminent mathematical and optical instrument makers. Their petition objected to new excise duties affecting manufacturers of flint glass due to 'the different processes it must necessarily undergo to bring it to the desired state of perfection'.[54] Again, this interest in excise duties and government action aiding or constraining private enterprise is part and parcel of Banks's interests, which from 1787 included commentary on the Corn Laws. The proposed action, too, reflects Banks's typical mode of operating. While the Board's Secretary was to send on the instrument makers' memorial to the Lords of the Treasury, as a matter 'which appears to this Board of so much importance to Science, and to Astronomy in particular', he also referred them to Banks himself, who undertook 'to enquire into the Circumstances of the Manufactory and point out to their Lordships the Method by which the Makers of Flint Glass may be enabled to receive the greatest proportion of benefit, compatible with the least diminution of Revenue which this matter will bear'.[55] Banks stepped easily into his role of independent adviser to government, as undertaken for the Board of Ordnance, Excise Office or other departments.

Within the Board, the matter of instruments was usually down to Maskelyne but, whether on his own initiative, or in response to requests from within the Navy, Banks also initiated loans of instruments, one of the more significant things that the Board could be seen to do. He became closely involved in a business that was otherwise Maskelyne's by creating a joint storage arrangement for instruments belonging to the Board and the Royal Society. This he suggested on 13 July 1782, and quickly followed up.[56] Once the Board's collection was physically associated with the Royal Society's, Banks seems to have taken several opportunities to act unilaterally. As well as the loans to Bligh mentioned above, Banks reported on the same day that he had loaned H4 to William Roy, then undertaking the triangulation survey to link the Paris

[52] Gascoigne, *Science in the Service*, pp. 65–110.

[53] BoL, confirmed minutes, 27 November 1779, CUL RGO 14/5, pp. 354–355. It is not clear where or how this advertisement was distributed. A printed copy, bearing the original 27 November 1779 date, is at RS MM/7/14.

[54] BoL, confirmed minutes, 1 March 1788, CUL RGO 14/6, p. 2:123.

[55] BoL, confirmed minutes, 1 March 1788, CUL RGO 14/6, pp. 2:123–2:124.

[56] BoL, confirmed minutes, 13 July and 7 December 1782, CUL RGO 14/6, pp. 2:35, 2:39.

and Greenwich observatories.[57] He also reported in 1785 loans to William Herschel and to French ships departing on a voyage of exploration.[58] It was presumably also Banks who, in the last of these meetings, demanded on behalf of the Royal Society the return of instruments they had previously loaned to the Board. Coming as they did around the period of the Royal Society Dissensions, it seems possible that Banks was taking further opportunities, if not to exact revenge, then to assert strength. It was not until 1792/1793 that Banks appears as active in the minutes again, this time in cooperation with Maskelyne and Shepherd, and in defence of the former and the Board's actions regarding Thomas Mudge.[59] It was, however, to be differences over dealings with other watchmakers, as described in Chapter 6, that effectively ended Banks's relationship with the Board of Longitude until after Maskelyne's death.

Gaining a Hearing: Circles of Credit and Trust

Much of the work carried out under the auspices of the Board was done by individuals who had repeated interactions with the Commissioners. They included Maskelyne's network of observers and computers: men like Charles Mason, Israel Lyons, John Crosley and Michael Taylor, as well as Wales and Gilpin. There was also a set of instrument makers who were trusted to supply and maintain instruments or to judge and develop new ideas. Until his death in 1776, John Bird had been top of this list. In the later part of the century the group included George Adams, Jesse Ramsden, Larcum Kendall, Edward Nairne, Peter and John Dollond, Thomas Blunt and John Arnold. Ramsden was held to strict conditions in order to receive rewards for his dividing engine but the Board, which had previously ordered from him an instrument on Samuel Dunn's design, was keen that he should train other makers to divide instrument scales on the engine.[60] Later he was used to supply instruments to Dawes on the First Fleet and William Gooch on the Northwest Pacific voyage.[61] His was also a name that could focus the Commissioners' attention onto the work of others outside the elite group: certificates from Dollond and Ramsden, for example, convinced the Board

[57] BoL, confirmed minutes, 8 December 1787, CUL RGO 14/6, p. 2:117.

[58] BoL, confirmed minutes, 5 February, 5 March and 17 August 1785, CUL RGO 14/6, pp. 2:74, 2:80, 2:87.

[59] BoL, confirmed minutes, 1 December 1792 and 2 March 1793, CUL RGO 14/6, pp. 2:188, 2:192. When Mudge appealed to Parliament, Banks came to the defence of both Maskelyne and the Board, championing its jurisdiction in such matters. Howse, *Nevil Maskelyne*, pp. 173–174; correspondence and papers on the Mudge affair, RS MM/7/89–135.

[60] Maskelyne and the professors were asked to consult and report on what conditions Ramsden should be asked to meet: BoL, confirmed minutes, 27 May 1775, CUL RGO 14/5, p. 278; see McConnell, *Jesse Ramsden*, pp. 39–51.

[61] BoL, confirmed minutes, 3 February 1787 and 11 June 1791, CUL RGO 14/6, pp. 2:105, 2:170.

Figure 8.1 Ralph Walker's meridional compass, by George Adams, London, 1793 or later, NMM NAV0263. © National Maritime Museum, Greenwich, London

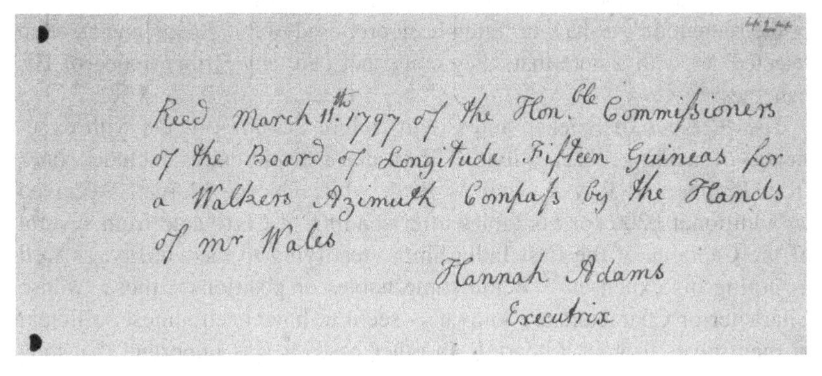

424

Recd March 11th 1797 of the Hon.ble Commissioners of the Board of Longitude Fifteen Guineas for a Walkers Azimuth Compass by the Hands of mr Wales

Hannah Adams
Executrix

Figure 8.2 Receipt from Hannah Adams for a Walker meridional compass supplied by George Adams to the Board of Longitude, CUL RGO 14/16, p. 424r. Reproduced by kind permission of the Syndics of Cambridge University Library

to trial James Weir's artificial horizon.[62] Adams's name gave credibility to Ralph Walker's azimuth compass (Figures 8.1 and 8.2), despite Maskelyne's dismissal of his magnetic theory and longitude scheme.[63]

[62] BoL, confirmed minutes, 7 December 1793, CUL RGO 14/6, p. 2:205. The trial and reward were discussed on 6 June 1795 (BoL, confirmed minutes, 6 June 1795, CUL RGO 14/6, pp. 2:247–2:248).

[63] BoL, confirmed minutes, 7 December 1793 and 6 December 1794, CUL RGO 14/6, pp. 2:206, 2:224–2:227. An example of Walker's compass is at NMM NAV0263.

Such testimonials were essential for those getting in touch with the Commissioners for the first time. Personal connections and the credibility of those who would write on your behalf remained as essential as they had been earlier in the century, although bureaucratised to the extent that, from 1772, the Board officially required supportive certificates. It was stated then:

> That all persons henceforward proposing Schemes of any Kind to this Board be told that no notice will be taken of them unless they are accompanied with a Certificate from person or persons of Science (whose Character or Characters are known) that such schemes are likely to answer the purposes intended.[64]

This resolution had been provoked by someone offering what appeared to be simply a sundial, although many with more plausible ideas were to be referred back to this filtering mechanism. The Commissioners usually agreed to see projectors who waited at the Admiralty to attend the Board personally, and all correspondence, whether addressed to the Board or individuals, seems to have been noted in the minutes, with an instant or more considered judgement recorded at the same or a later meeting. However, they were used to seeing off a range of proposals and individuals. Schemes for perpetual motion or squaring the circle (see Chapter 9), as well as magnetic and a range of astronomically inspired methods, were dismissed as 'not in the smallest degree deserving of their attention', as having 'often been proposed to this Board, and as often rejected' or with a note that 'they could not take any farther notice of His Schemes'.[65]

The 'Persons of Science and Credit' might include officers with experience of making observations.[66] Ramsden's testimonials included ones from Phipps and Roy as well as Bird, while George Margetts received an additional £200 for his tables after sending 'a Certificate from several of the Captains of the East India Ships' testifying to their utility, as well as noting his expenses.[67] While some names or positions – those 'whose Character or Characters are known' – seem to have been almost sufficient in themselves to gain a hearing, in other cases it was important that they offered the right kind of support. Sir Joseph Senhouse offered certificates from 'three Gentlemen of science', but this was not seen as satisfactory, and 'farther proofs' were demanded, probably because they did not include clear observational results.[68] The high bar insisted on in this case perhaps

[64] BoL, confirmed minutes, 13 June 1772, CUL RGO 14/5, p. 228.

[65] BoL, confirmed minutes, 1 December 1781, 7 March 1795 and 18 July 1786, CUL RGO 14/6, pp. 2:30, 2:238, 2:96.

[66] BoL, confirmed minutes, 12 July 1788, CUL RGO 14/6, p. 2:127.

[67] BoL, confirmed minutes, 4 March 1775, CUL RGO 14/5, p. 271; BoL, confirmed minutes, 1 March 1794, CUL RGO 14/6, p. 2:214.

[68] BoL, confirmed minutes, 3 December 1796, CUL RGO 14/6, p. 2:273. He sent a further testimonial from an East India Company captain, also deemed unsatisfactory: BoL, confirmed

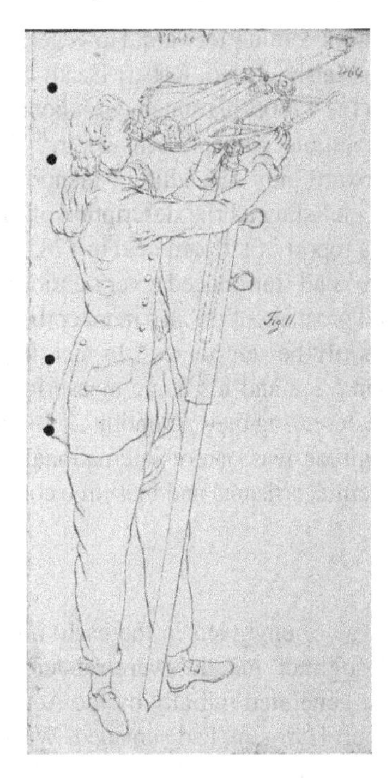

Figure 8.3 Christian Carl Lous, illustration of a new telescope to be used at sea, 1789, CUL RGO 14/30, p. 464r. Reproduced by kind permission of the Syndics of Cambridge University Library

reflected the fact that the proposal was for a marine chair to assist observations of Jupiter's satellites. Elsewhere, in a report on an observing instrument by Christian Carl Lous (Figure 8.3), professor at the Naval Academy in Copenhagen, Maskelyne emphasised the practical impossibility of this observation, noting: 'I can vouch from my own experience when I tried to observe Jupiters Satellites both in, and out of Irvin's [sic] Marine Chair by Order of this Board in my voyage to Barbados in 1763'.[69]

Ongoing but now waning public interest, alongside increasing awareness of mathematical and technical methods among some naval officers, ensured that

minutes, 4 March 1787, CUL RGO 14/6, p. 2:278. Senhouse's letter to Wales, as Secretary, and supporting material are at CUL RGO 14/30, pp. 471r–491v. See also Dunn, 'Scoping Longitude', pp. 145–147.

[69] BoL, confirmed minutes, 6 March 1790, CUL RGO 14/6, p. 2:152. Lous, who had produced a Danish translation of *The Principles of Mr. Harrison's Timekeeper* in 1768 (see Chapter 5), corresponded with Banks until 1795 about a possible reward; see Dawson (ed.), *The Banks Letters*, pp. 556–557, 879; Chambers (ed.), *Scientific Correspondence of Joseph Banks*, IV, pp. 187–188, 194–195.

there was still a range of hopeful projectors coming forward. However, the conversation by the end of the century was fairly closed. Robert Heath continued to attack the Board and Maskelyne into the 1780s, his statements about the lack of progress towards finding longitude solutions perhaps seen as fair.[70] Searches in *The Times* Digital Archive bring forward only a handful of mentions of the Board or Commissioners of Longitude, including a brief description of Mudge's petition in the House of Commons and a report of a Board held in 1785. The latter claimed that 'Mr. Viec [sic] of Truro' had 'introduced a very curious instrument' to which the Commissioners 'paid great attention' in a manner the reporter assumed was 'flattering' for it 'will shortly be sent to sea'. In fact, the Board minutes report the examination of John Vice and his angle-measuring instrument, and the judgement that 'it was not deserving their Attention'.[71] By the later eighteenth century the question of longitude was one of international interest, but it circulated largely in specific academic, artisanal and maritime contexts.[72]

Conclusion

When Hogarth's *A Rake's Progress* was re-engraved in the early nineteenth century, the iconic diagrams of the 'longitude lunatic' were rendered barely visible.[73] The bubble of public interest generated initially by the Act of 1714 and subsequently by projectors, above all Harrison, had subsided. While there was much interest in the voyages of exploration of the later eighteenth century, this was more focused on exotic exploits than the people and ideas behind the navigation and survey techniques. For all the rhetoric of public service, the Board's audiences were largely known networks of astronomers, artisans, mathematical practitioners and scientific servicemen. These connections were to be maintained in the nineteenth century but, after Maskelyne's death, Banks was determined to expand the Board's networks and, particularly, to render it 'more Effective' by developing closer links with the Admiralty and the Royal Society.[74] His changes also further increased the bureaucratisation of the Board, with more meetings, committees and salaried posts, cementing the change of direction evident after 1774.

The Board's networks and ways of working became in the late eighteenth century less ad hoc and more focused on its publications and the personal and

[70] 'Rotterdam' to the Editor of the Universal Register, *The Times*, 19 April 1787, p. 1.

[71] *The Times*, 30 April 1793, p. 2; *The Times*, 28 September 1785, p. 2; BoL, confirmed minutes, 17 August 1785, CUL RGO 14/6, p. 2:85.

[72] See the range of pan-European contexts for longitude work in Dunn and Higgitt, *Navigational Enterprises*.

[73] Barrett, *Looking for Longitude*, p. 173 and Figure 53.

[74] Joseph Banks to Charles Blagden, [13] March 1818, in Chambers (ed.), *Scientific Correspondence of Joseph Banks*, VI, pp. 273–274.

correspondence networks of Commissioners, especially those of Maskelyne. Known and trusted individuals took on activity for the Board or received its rewards, although new faces might demonstrate sufficient plausibility or credible support to gain trials or rewards. Between 1774 and 1810 the Board motivated or recognised some significant innovations in timekeeping and instrument making (especially in the work of Arnold, Earnshaw and Ramsden) and much effort in the improvement of astronomical tables. It continued to offer large rewards and to pay out small ones, but now put far more of its financial resources into paying for the labour and expenses of individuals undertaking assigned specialist work. These ranged from Commissioners and the Secretary to those who worked on the *Nautical Almanac*, added scientific manpower to voyages of exploration, or undertook trials, made instruments or prepared material for publication. While very few could make a living from the Board, at least for more than a few years, it was for several a significant element within the eclectic and piecemeal careers available to Georgian mathematical practitioners.[75]

[75] Higgitt, 'Mathematical Examiners'. See Bennett, 'The Rev. Mr. Nevil Maskelyne, F.R.S. and Myself', for an account of one such career that was never rewarded by the Board.

9 A Practical Institution Weighed Down by Impractical Proposals?

Sophie Waring

Partly because of the size and the publicity surrounding the rewards they oversaw, the Commissioners of Longitude received a range of projects that lay well beyond the narrowly defined fields of watchmaking and navigational tables. These visionary proposals seem to have become more common in the later decades of the Board of Longitude's activities. In 1828, the First Secretary of the Admiralty, John Wilson Croker, condemned the Board in Parliament as:

> wholly occupied in reading the wild ravings of madmen, who fancied that they had discovered perpetual motion and such like chimaera, stimulated with expectations of obtaining parliamentary rewards, held out for the encouragement of inventions, which every man of science knew to be perfectly ridiculous.[1]

As this chapter details, there was a significant association between longitude methods and devices that could move perpetually, not least since marine timekeepers, so it was often imagined, needed to function regularly and reliably. This becomes evident in the documents the Board preserved that detail such apparently unsound projects and enterprises. These later volumes in the Board's archive are an important window through which to view the development of popularisation and specialisation that permeated scientific knowledge in the lecture theatres, parlours and libraries of late Georgian society. This chapter describes this 'impracticable' activity, in order that it may be included in descriptive accounts of scientific activity and culture. Yet, in the ambition to make sense of this activity, an unchallenged call for the inclusion of these individuals and their work in accounts of scientific society and thinking in this period is potentially misleading, as this would overlook the exclusion they did endure from elite metropolitan science. Nonetheless, yet another reclassification of these individuals is to be avoided and instead this chapter will understand them in the context of an emergent scientific middle class. This is particularly significant for interpretation of those schemes communicated to the Commissioners that contained elements of sound theory and mathematics but lacked novelty.

[1] 'House of Commons, Friday, July 4', *The Times*, 5 July 1828, p. 5.

There was a transformation in reaction to proposals for self-moving machines in the institutions that supported Georgian scientific activity during the lifespan of the Board of Longitude. The Board is a prominent repository for perpetual motion schemes, as well as a wider range of impractical schemes, from an era that saw them consistently published and patented. While Harrison began work on H3, the Royal Society's demonstrator and curator, John Theophilus Desaguliers, discussed 'the common Herd of *Perpetual-Motion Men*, which every Age affords' in his published *Course of Experimental Philosophy*. He also felt obliged, as a person who had always 'declared against all Projects tending that way', to write up his 'Reasons why the thing seem'd impossible or impracticable' in the *Philosophical Transactions* to speak against the 'Wheel at Hesse-Cassel, made by Monsieur Orfireus', a notorious scheme for perpetual motion launched in the German lands by a physician and watchmaker Johann Bessler (who took the name Orffyreus), much publicised in the second decade of the eighteenth century.[2]

Over a century later, an 1851 Select Committee on Patents recorded the testimony of Matthew Davenport Hill, a barrister focused on patent law, who advocated the possibility of creating a searchable record of patents to assist inventors. Hill suggested keeping a record of 'unpatented inventions' and, when his interviewer queried how to control such a list, the topic of perpetual motion arose again. Hill suggested that a patent would not be issued 'for the abstract principle' but only for 'a particular machine', adding that, 'if such a machine were applicable to useful purpose, the inventor would make a valuable gift to society by throwing it open'. Hill was not proclaiming his faith in perpetual motion, yet it is interesting that it is still used as an example by the Select Committee as a thing to be prevented. In the testimony of a civil engineer, Richard Roberts declared that:

> Our patent list now contains a great number of very silly things, which no man, who had been long in a workshop, would ever think of patenting; and the reason is, that the patentee has money, though deficient in experience and mechanical talent; probably he thinks he cuts a figure by being in the patent list.[3]

Marsden and Smith have discussed the continuing fantasy of perpetual motion in the nineteenth century, particularly in the context of new engineering periodicals that featured schemes for machines 'developing power without end'.[4] With the ubiquity of these plans in patent applications and publications, it is unsurprising that schemes were regularly sent to the Board of Longitude. Responses to perpetual motion schemes function as an indicator of the changing

[2] Desaguliers, *A Course of Experimental Philosophy*, I, p. 183; Desaguliers, 'Remarks on Some Attempts'; Schaffer, 'The Show that Never Ends'.

[3] House of Lords Select Committee, *Report and Minutes*, pp. 285 (Hill), 186 (Roberts). On eighteenth-century patent systems, see Macleod, *Inventing the Industrial Revolution*; Hilaire-Pérez, 'Technical Invention'.

[4] Marsden and Smith, *Engineering Empires*, p. 65.

place in public life occupied by the Commissioners throughout the eighteenth century and into the nineteenth.

Fixing Longitude and Perpetual Motion

Perhaps the Board was the most fitting of all repositories for these schemes given the consistent coupling of perpetual motion with the discovery of longitude at sea. Stimulated by the passing of the 1714 Act, there was a flourishing of publications on perpetual motion and this continued as part of the larger development of public science throughout the century.[5] In a 1731 proposal for a new 'Astronomical Quadrant', Elias Pledger argued that this increase in publishing activity was caused by 'the extravagant Rewards that have been offered to any Person, that should solve the Grand Problem of finding the Longitude at Sea'. Comparing it to the Philosopher's Stone, he continued that the longitude problem had:

> set many Heads to work, and hath occasion'd a Multitude of trifling and fruitless Schemes and Projects … And the World hath so often been disappointed in its Expectations, when any Thing of this Kind hath been publish'd, that it hath occasion'd the Neglect of many good Inventions propos'd together with it, which might, in many other Cases, have been very serviceable.[6]

The challenge to filter this 'multitude' to find the 'good Inventions' became one of the responsibilities of the Commissioners of Longitude, as they started to meet to discuss the work of at first Harrison and then others who approached them for support.

The coupling of the quest for longitude with perpetual motion, as well as its association with the Philosopher's Stone and quadrature of the circle, was particularly prominent in the publications of eighteenth-century satirists. The phrase 'finding the longitude' became a way of expressing something that was either impossible or extremely difficult to accomplish.[7] The activities and writings of the Scriblerus Club are well documented. Devoted to satirising false learning, using characters that stepped unwisely and carelessly into a multitude of arts and sciences, Scriblerian texts ridiculed current trends in culture and scholarship. The only complete meetings of the club occurred in March and April 1714, just before the passing of Longitude Act in July. Taking place at the lodgings of the physician and natural philosopher John Arbuthnot, in St James's Palace, the meetings provided a space for intense collaboration that informed the future work of its members: Thomas Parnell, John Arbuthnot, John Gay, Jonathan Swift and Alexander Pope. The group was a significant

[5] Wigelsworth, 'Navigation and Newsprint'; Elliott, 'The Birth of Public'; Stewart, *The Rise of Public Science*, pp. 183–211.
[6] Pledger, *A Brief DESCRIPTION*, pp. i–ii.
[7] Howse, *Greenwich Time*, p. 61; Barrett, *Looking for Longitude*, pp. 105–112.

influence on later satirists, subverting contemporary politics and expressing pessimism about the future of society and human nature.[8] Longitude and perpetual motion were recurrent themes for the Scriblerian authors, linking them in political and scientific thinking while mocking the rhetoric employed by projectors to evoke the authority of natural philosophy and criticising exaggerated confidence in mathematics in both financial and scientific projecting.[9]

Jonathan Swift's Gulliver reflected on immortality as the only sure way to achieve such lofty ambitions: 'I should then see the Discovery of the *Longitude*, the *perpetual Motion*, the *Universal Medicine*, and many other great Inventions brought to the utmost Perfection'.[10] The projector Jeremy Thacker, who in his work *The Longitudes Examin'd* commented critically and expertly on longitude proposals following the 1714 Act, may well have been an invention of the Scriblerians, perhaps fabricated by Arbuthnot himself. Attacking the genuine 1715 horological proposal of James Clarke, which described a timekeeping instrument that poured mercury between various glass vessels, Thacker suggested that, '[i]f he would have this Instrument go, let him consult about it with the inventors of the perpetual motion'.[11] Thacker's own proposal was to place the sea-clock in a vacuum as a way to limit the effects that prevented clockwork from functioning at sea. The discussion of temperature change and the interruption of the pendulum's motion on board a ship was sound, demonstrating the employment of technical knowledge in satire.[12] The influence of this proposal was long in the institutional memory of the Board. As described in Chapter 6, the Commissioners recalled Thacker's scheme in 1809 when questioning Joseph Manton about his timekeeping proposal.[13]

The conjured character of Martin Scriblerus was the 'author' of several schemes devised by the group which resembled real schemes that aroused Scriblerian cynicism. Scriblerus wrote on the 'many precise and Geometrical Quadratures of the Circle … the Projects of Perpetuum Mobiles, Flying Engines … the Method of discovering the Longitude by Bomb-Vessels'.[14] The 1714 longitude proposal of the mathematical lecturers William Whiston and Humphry Ditton is clearly referenced here. As noted in Chapter 2, Whiston's proposals made him a prime target of Scriblerian satire, which blended fiction and contemporary projecting. Arbuthnot's fantastical *Humble Petition* suggested taxing sunbeams, since

[8] Rumbold, 'Scriblerus Club (act. 1714)'; Rogers, *Documenting Eighteenth-Century Satire*, pp. 45–62.
[9] Lynall, 'Scriblerian Projections'.
[10] Swift, *Travels into Several Remote Nations*, II, p. 136.
[11] Clarke, *The Mercurial Chronometer*; Thacker, *The Longitudes Examin'd*, pp. 6–7.
[12] Rogers, 'Longitude Forged'; Betts and King, 'Jeremy Thacker'.
[13] BoL, confirmed minutes, 2 March 1809, CUL RGO 14/7, p. 126.
[14] Kirby-Miller, *Memoirs of the Extraordinary Life*, pp. 166–167.

they could be captured to generate heat; the funds raised would pay 'the Commanders and Crew of the Bomb-Vessels, under the Direction of Mr. Whiston for finding out the Longitude, who by reason of the Remoteness of their Nations, may be reduc'd to streights for want of Firing'.[15] The work also foreshadowed the experiments of Swift's philosophers of Lagado on the flying island of Laputa, which in turn ridiculed the experimental demonstrations of the Royal Society.[16]

Satirists outside this elite group also observed and enforced the connection between the seemingly impossible search for longitude at sea and perpetual motion. In 1722, a pseudonymous pamphlet proposed draining the Irish Channel to reveal rare prizes such as old moons that had fallen into the sea, the rare fur of sea foxes and 'the particular Places of the Channel wherein (most likely) the Perpetual Motion, and the Longitude, are to be grappl'd up; for you will allow that 'tis matter of great Depth to find those out'.[17] The connection between longitude projects and Grub Street literature was again the subject of the 1771 comic poem 'The Ruffle', which linked longitude and perpetual motion:

> Look in the papers valu'd most,
> The London or the Gen'ral-Post,
> Or, in the Publick Advertiser,
> You'll find each day the world grow wiser …
> What cannot art of man find out?
> Astrologers no longer doubt
> The Longitude, as yet unknown,
> Will to the world be quickly shown.
> Each ign'rant Dunce will have a notion,
> How to construct Perpetual Motion.[18]

Such examples, from a very large literature, show that schemes for longitude and perpetual motion were often situated on the fringes of reputable mathematical knowledge. Innovation and insanity were regarded as being separated by only a thin veil, blown by the changing winds of speculation and projection that dominated industry, politics and society during the long eighteenth century.[19]

'Impracticable Schemes'

Decades after the final closure of the Board of Longitude, Astronomer Royal George Airy assembled its papers and proceedings. In 1840 he significantly 'found that the papers of the Board for Longitude were divided between the

[15] Arbuthnot, *To the Right Honourable the Mayor*, p. 2.
[16] Nicolson and Mohler, 'The Scientific Background', pp. 310–312; Lynall, *Swift and Science*.
[17] A. M. in Hydrostat, *Thoughts of a Project*, p. 30. [18] Lund, *The Mirrour*, pp. 39–40.
[19] Alff, *The Wreckage of Intentions*; Kareem, 'Forging Figures of Invention'.

Royal Society and the Admiralty' and got Admiralty approval to bring them all to Greenwich. Twenty years after this unification, Airy had them all 'stitched into books' and remarked that '[t]hey will probably form one of the most curious collections of the results of scientific enterprise, both normal and abnormal, which exists'.[20] In this reconstruction, Airy manufactured the principal resource for making memories of the Board's affairs. In the post-Harrison era, the Board was actively meeting to consider correspondence usually sent directly, rather than works published with a dedication to the Commissioners or the Board. Airy collected what he judged to be 'impracticable schemes' into two volumes, schemes for perpetual motion and quadrature of the circle into another, and 'irrational astronomy' into a fourth volume. In the second half of its existence, the Board was a clearing house for many navigational schemes, with letters forwarded from the Royal Observatory and the Admiralty and Navy Board, especially in improved optical instruments and precisely divided scales, magnetic compasses and a range of survey methods. Correspondents sought reward and sponsorship.[21]

Airy's category of 'impracticable schemes' features plans for a host of mathematical and navigational projects, ranging from annotated plans of variably practicable instruments for calculation and determination of position to complex schemes for computations and measurement. The Commissioners often judged such proposals unworkable and barely worth consideration, but the collection includes the work of claimants who also appear elsewhere in the Board's surviving archive. It can be difficult to discriminate between these allegedly impractical designs and those that were taken more seriously.

In February 1783, Walter Bedford described the Board as 'the Grand National Tribunal for such difficult and obstruse undertakings' in a letter accompanying his scheme for finding longitude based on the comparison of two revolving wheels. In a meeting of the Commissioners on 1 March 1783, Bedford's correspondence was passed to the Astronomer Royal, Nevil Maskelyne, who was 'desired to report if they contained anything worthy the attention of the Board'. A note on the back of Bedford's scheme signed by Maskelyne states: 'This paper does not appear to me to deserve the attention of the Board of Longitude.' A reply from the Secretary of the Board, Thomas Hurd, is not found in the Board's archive. Bedford wrote again a little over a year later with another scheme but this escaped the condemnation of 'impracticable' and is found in

[20] Airy, 'Report of the Astronomer Royal ... 1858', p. G5.

[21] Volumes of papers from the later eighteenth and early nineteenth century, CUL RGO 14/29 and RGO 14/30, contain documents on instruments for observations at sea, while petitions and correspondence from the same decades, in CUL RGO 14/11, RGO 14/12 and RGO 14/32 to RGO 14/41, offer an invaluable record of the activities of a vast number of practitioners, often otherwise little-known, engaged in cartography, mathematics, astronomy, navigation and magnetism.

'Correspondence regarding methods and instruments used to establish longitude and the use of chronometers at sea'. There is, again, no evidence of a reply and in this instance no mention in the minutes of the Board's meetings. Such curt rejection was characteristic of much of the Board's treatment of these petitioners.[22] Yet the Commissioners were not guilty of such negligence in all cases; Hurd made an effort to address a proposal sent by Arthur Hodge, of 12 Bells, Bride Lane, Fleet Street. When Hodge's proposal was laid before the Commissioners in June 1814, it was minuted that 'it was impossible to discover thereby the drift or intention of the writer'.[23] Hurd wrote to Hodge asking for clarity regarding the 'affinity' his plans might have 'towards the discovery of the Longitude at Sea which was the principal object on the original appointment of that Board by Parliament'.[24] Hodge peppered the Commissioners with letters setting out schemes for determining position using a tabular combination that somehow recorded the variation of the magnetic compass, the heights of tides and the direction of prevailing winds. Hurd wrote again to tell him they would be seen by the Board in December while also declining Hodge's request for an advance payment to aid his progress: 'on the subject of an advance of money in aid of experiments, the Board must be well convinced of the feasibility of your plan, before they give their sanction to such a measure'.[25] For all this engagement with Arthur Hodge, Hurd minuted in December 1814 that '[a] Variety of papers were laid before the Board with Letters signed Arthur Hodge but none of them were deemed worthy attention'.[26] Hodge was not dissuaded and continued with his work. He wrote again in 1820 to describe his method for measuring a circle, wrongly addressing his letter to Hurd, who was no longer Secretary. Hodge conveyed sheets and sheets of diagrams and calculations to the Board and, a month later in January 1821, wrote again to Hurd to complain that he was 'still without the pleasure of an answer' and included several more calculated tables. Hurd's replacement as Secretary, Thomas Young, did take Hodge's scheme to the next Board meeting on 1 February 1821. It was minuted that 'Mr A. Hodge is to be informed that his measurement of a circle is of no utility'.[27]

Responses to Bedford and Hodge by two of the Board's secretaries highlight the responsibility the Commissioners felt they had to consider and engage with the ideas of those who wrote to them. It is not possible to know how long the Commissioners spent considering these ideas, but knowledge that the Board

[22] Walter Bedford to BoL, 7 February 1783, CUL RGO 14/39, p. 17r; BoL, confirmed minutes, 1 March 1783, CUL RGO 14/6, p. 2:45; Maskelyne's note on the letter is at CUL RGO 14/39, p. 18v; Walter Bedford to BoL, 14 July 1784, CUL RGO 14/38, pp. 5r–5v.

[23] BoL, confirmed minutes, 2 June 1814, CUL RGO 14/7, p. 2:192.

[24] Thomas Hurd to Arthur Hodge, 8 June 1814, CUL RGO 14/39, p. 172r.

[25] Thomas Hurd to Arthur Hodge, 13 June 1814, CUL RGO 14/39, p. 182r.

[26] BoL, confirmed minutes, 1 December 1814, CUL RGO 14/7, p. 2:193.

[27] BoL, confirmed minutes, 1 February 1821, CUL RGO 14/7, p. 2:334.

of Longitude offered a system for any person to have contact with some of the most influential members of society drew attention from all sorts of projectors. Arthur Hodge's letters and calculated tables were considered by Joseph Banks, the Astronomer Royal and four professors of mathematics and astronomy from Oxford and Cambridge in the same meeting at which they debated the improved longitude tables of the Spanish astronomer and mathematician Joseph de Mendoza y Ríos.[28]

Even those advocating perpetual motion were given time and corresponded with appropriately. Hurd's annotations of a perpetual motion scheme sent by John Bell on behalf of his friend Philip Thompson Rutherford in 1817 note 'receipt acknowledged on 5th July' only three days after the letter was sent. Bell's letter was 'answered 10 December', a few days after the Board met on 4 December, when 'The Petition of Jno: Bell of Inkeltham was not deemed worthy of attention'.[29] Significantly, this meeting concluded with the passing of a resolution stating that 'the Board can give no opinion on unformed Instruments or without receiving satisfactory evidence of their having answered in practice'. Several communications were illustrated, but this resolution was an attempt to prevent the Commissioners having to consider the speculations of anyone in possession of pen and paper. Chapters 6 and 8 provide plentiful evidence that this policy, with respect to the evaluation of instruments proved in practice, was at least in principle applied by the Commissioners to marine timekeepers and navigation devices.

Mutiny by Bureaucracy

Not all those who wrote to the Board were seeking support for schemes. Alongside petitions for charity from the wives and families of those employed as calculators, there are several instances of men seeking their liberty during wartime.[30] A French prisoner of war named Blandin, held on the prison ship *Canada* at Chatham in 1812, promised the Commissioners a 'method as exact as it is simple' for securing longitude at sea, one 'capable of being comprehended by every Sailor & so simple that it requires frequently no Logarithmic

[28] BoL, confirmed minutes, 1 December 1814, CUL RGO 14/7, p. 2:193; correspondence relating to a reward to Joseph de Mendoza y Ríos, 1815, CUL RGO 14/1, pp. 216r–224r.

[29] BoL, confirmed minutes, 4 December 1817, CUL RGO 14/7, p. 2:242.

[30] Within the Board of Longitude papers there are also appeals for financial charity from wives, daughters and sometimes sons of past computers and comparers; for example, Margaret Mackay to BoL, 8 April 1822, CUL RGO 14/12, p. 326r, a polite and apologetic request for assistance after the death of her husband who had worked as a computer. See also Mary Mason to BoL, CUL RGO 14/12, p. 346r: Mary, wife of Charles Mason, formerly assistant at the Royal Observatory, petitioned the Board for renumeration for her husband's work on lunar-distance tables. She received three payments totalling £220 in 1791–1792; CUL RGO 14/18, pp. 139r–141r.

Calculations'.[31] Blandin described himself as an ex-professor of mathematics and hydrography, but insisted that he could not convey his method for 'determination of the Longitude at Sea without measuring distances' on paper as his freedom would not be guaranteed.[32] Hurd responded that:

> whatever papers may be entrusted to my care shall be faithfully laid before the Commissioners of the Board of Longitude who must be furnished with a compleat & satisfactory explanation of your proposed method for determining the Longitude at Sea otherwise your application will not be attended to – your particular situation as a prisoner of War will be a sufficient excuse for your non appearance before the Board, but should the Commissioners deem it necessary proper steps will of course be taken to procure you permission to come to town.[33]

The scheme was obediently sent by Blandin and referred to the Astronomer Royal, John Pond, during the meeting held in December 1812. No details of the scheme survive but it was found to be 'not worthy of further notice' alongside several other proposals.[34] Hurd directly quoted the minutes of the Board in his final letter to Blandin: 'The Commissioners having taken into consideration Mr Blandin's proposals for improving the method of finding the Longitude do not approve thereof.'[35]

Another petition was sent to the Board in 1813 from a Dutchman named Kohlmann, a prisoner of war on the *Brunswick*, a prison ship at Chatham. Like Blandin, he claimed that he had discovered a method of finding longitude and requested to be released from his imprisonment.[36] No reply from the Board survives in the archive. A year earlier, another French prisoner captured in April 1805, William Violaine, wrote from the *Buckingham*, a ship taken from the Spanish at Trafalgar a few months after Violaine's capture and converted for use as a prison ship. A draft reply from Hurd states that his proposal was considered at a meeting in June and, like Blandin's, was referred to John Pond for consideration in depth.[37] Violaine wrote an extensive paper on his scheme, a garbled and complex version of the method of lunar distances, including diagrams and many pages of calculated tables, from the *Buckingham*, but at the following meeting in December the Astronomer Royal condemned all schemes referred to him at the last meeting; they had been 'carefully examined' and found not worthy of the Board's attention.[38]

[31] M. Blandin to Thomas Hurd, 7 June 1812, CUL RGO 14/11, p. 41r.
[32] M. Blandin to Thomas Hurd, 7 June and 15 November 1812, CUL RGO 14/11, pp. 39r, 41v, 58r.
[33] Thomas Hurd to M. Blandin, 28 July 1812, CUL RGO 14/11, p. 55r.
[34] BoL, confirmed minutes, 3 December 1812 and 4 March 1813, CUL RGO 14/7, pp. 2:167, 2: 172.
[35] Thomas Hurd to M. Blandin, 25 June 1813, CUL RGO 14/11, p. 66r.
[36] D. Kohlmann to BoL, 24 January 1813, CUL RGO 14/11, pp. 282r–282v.
[37] Thomas Hurd, draft letter to Monsieur Violaine, CUL RGO 14/33, p. 468r; BoL, confirmed minutes, 3 June 1812, CUL RGO 14/7, pp. 2:162–2:163.
[38] BoL, confirmed minutes, 2 December 1813, CUL RGO 14/7, p. 2:183; Violaine's extensive work is at CUL RGO 14/33, pp. 469r–501v.

Alongside these attempts to secure liberty is a proposal from James Davis, who in 1805 attempted to use the Board to secure passage back to London from where he was stationed in the East Indies. On board HM Sloop *Victor*, Davis forwarded a letter to the commander-in-chief of the East India Station, Admiral Edward Pellew, Viscount Exmouth, with the forged signature of Hurd's predecessor as Board Secretary, Harry Parker. The forgery states that 'the plan of your "perpetual motion" has been found to answer every purpose' and seemed likely to prove 'essentially serviceable to Navigation, and Commerce' as a means of discovering longitude.[39] The letter conveniently summoned Davis back to London to receive his 'Reward'. Pellew did nothing with the letter for two years, but in 1807 asked his civil secretary, the watercolourist, administrator and naval captain Edward Hawke Locker, to obtain the opinion of Davis's immediate commanding officer. Captain George Bell responded: 'My opinion of him, is that his ignorance and stupidity are such, as to make it absurd ever to imagine him Capable of discovering Longitude or anything of the kind.'[40] James Davis was summoned to have his perpetual motion machine examined by Commander Pellew on board the *Culloden* and was found 'to be a poor ignorant creature apparently wanting in Common Capacity'. Davis failed to convince Pellew with 'absurd papers, scrawled with rude drawings, and some miserable fragments in Wood which he pretended to be parts of his projected machine for perpetual motion'.[41] Hawke Locker forwarded the correspondence generated by Davis's ploy to Parker in London for the Board's records. The delay of two years by Pellew in dealing with the letter was most likely caused by his own disagreement with the First Lord of the Admiralty, Lord Melville, which was not resolved until January 1807, when a newly elected government reunited the East India Station under Pellew's command.

These attempts to gain liberty or abscond are at first rather perplexing. In the knowledge that their scheme or proposal would eventually be found out, perhaps Davis and the prisoners of war reckoned they could abscond upon arrival in the capital or perhaps they genuinely believed they had knowledge of value to the Board. While the motives of individual correspondents remain hidden, conclusions can be still be drawn from this correspondence. Appeals to the specialist knowledge needed for finding longitude at sea or realising perpetual motion show that this knowledge was little understood by many in the Navy, yet the longitude problem and the activity of the Board were sufficiently well known for a manipulation of this kind to be attempted. While writing to Davis's captain could

[39] Harry Parker to James Davis, 21 September 1805, CUL RGO 14/54, p. 79r. The name is spelt both Davis and Davies.

[40] Edward Hawke Locker to Captain George Bell, 8 February 1807, and Bell to Locker, 10 February 1807, CUL RGO 14/54, pp. 81r–82r.

[41] Edward Hawke Locker to Secretary of the BoL, 11 February 1807, CUL RGO 14/54, pp. 83v–84r.

be regarded as due diligence, Pellew was not able to dismiss Davis's letter at the first mention of perpetual motion. Furthermore, Davis presumably felt that his scheme might work, and that judgement of his perpetual motion machine could only be performed by the Commissioners of Longitude, which gives insight into his estimation of scientific knowledge among his commanding officers. Davis was attempting to obtain a discharge from the Navy by manipulating the new bureaucratic systems that were being established to manage scientific knowledge in the service of the Navy. The 'consultative committee' had emerged as a new tool of administration favoured by many government departments at the turn of the century, and they reinforced the increasing connection between expert knowledge and social or political authority over a subject.

James Davis and these prisoners of war were not the only petitioners to attempt to gain something by manipulating the increasingly formalised and bureaucratic nature of the Board of Longitude. The Board itself used the defined limits of its remit to fend off the increasing number of impracticable schemes it received. Rejection was consistently achieved by suggesting that the schemes fell outside the Board's remit rather than the realms of plausibility. The quadrature of the circle and perpetual motion were not judged impossible but rather seen as lying beyond the formal responsibilities of the Board. Responses that evoked bureaucratic routine rather than scientific discussion to dismiss irregular or irrational schemes and projects were sent by all the Board's secretaries.

Hurd's responses as Secretary (1810–1818) are often retained in the Board's archives, and a general attitude towards perpetual motion schemes emerges as he converses with correspondents. In response to a scheme sent by Henry Browse, a gentleman in Paignton, Devon, in 1814, Hurd stated that his papers had been 'taken into consideration' but that 'the discovery of a Perpetual Motion does not properly come under the cognizance of the Commissioners'.[42] This sentiment was repeated in Hurd's reply to another perpetual motion machine from 1812, which stated: 'I have to inform you that as the discovery of a perpetual motion has nothing to do with the discovery of the Longitude the Commissioners did not think it necessary to take your Letters into consideration'.[43] Hurd stated fully the position of the Board in July 1813 when advising a Bloomsbury petitioner, Charles Jacomb, to edit his scheme for the next meeting of the Board, 'unless you can at the same time show how this discovery can be made useful towards finding the Long[itu]de at Sea it is not likely the Board will take it into consideration; such however has been the answer invariably given to others who like yourself have come forward with discoveries of a similar nature'.[44]

[42] Thomas Hurd to Henry Browse, 9 March 1814, CUL RGO 14/54, p. 53r.
[43] Thomas Hurd to Henry Holden, 10 March 1812, CUL RGO 14/54, p. 100r.
[44] Thomas Hurd to Charles Jacomb, 9 July 1813, CUL RGO 14/54, p. 110r.

The Board minutes suggest that Jacomb never submitted an improvement on his scheme but four other perpetual motion machines were presented to the Board in December 1813. Summarising all the proposals into one action, Hurd recorded again that he would 'inform them that perpetual motion did not come under the consideration of the Commissioners'.[45]

Hurd was not the first to be irritated by such schemes. As noted in Chapters 2 and 3, a century earlier, longitude proposals had been sent directly to Isaac Newton, who had taken the time to demonstrate where projectors were erroneous in their thinking.[46] In the absence of a specific remit to protect him from public obligation, Newton informed those proposing improvements relating to clockwork that, without observations of the Moon or satellites of Jupiter, a clock could only keep longitude at sea, not find it from scratch or recover it, if lost. Newton took the time to correct the philosophy of his correspondents.[47] As Secretary to the now formalised Board of Longitude, Hurd, and subsequently Young, would simply dismiss schemes as outside their jurisdiction. While engagement and written responses seem to be the lifelong habit of recipients of perpetual motion schemes for finding longitude, the manner of their rejection changed as bureaucratic and administrative reform instilled a social hierarchy in scientific lives and circles.

The Cold Indifference of Dr Young

The public nature of the Board, its broad remit of 'improving navigation' and the attention drawn by its transformation by the 1818 Longitude Act, made it one of the most prominent public institutions for science in Regency Britain. An activity that was traditionally the reserve of the socially elite now had a very visible, publicly financed, façade. Through intensified printed news about invention and investment in the markets, practices of useful knowledge and technical improvement seemed somehow to offer remarkable profits to their aspiring and ambitious authors.[48] The number of individuals contacting the Board of Longitude rose. Letters, projects, pamphlets and schemes had flowed towards the Commissioners since 1714, with a larger proportion preserved following the formation of a Board of Longitude with a Secretary from the 1760s, yet the establishment of a 'new Board' in 1818 drew increased attention and there was a surge in correspondence

[45] BoL, confirmed minutes, 2 December 1813, CUL RGO 14/7, p. 2:185 The four schemes sent to the Board between June and December were from: Henry Briggs, Private 17th Light Dragoons (CUL RGO 14/54, pp. 37r–39v); John Dalling, joiner of Castle Douglas (CUL RGO 14/54, pp. 70r–73r); William Martin, Northumberland Light Infantry (CUL RGO 14/54, pp. 135r–141r); and James Stoat of Ashburton (CUL RGO 14/54, pp. 183r–185v).

[46] Hall and Tilling, *The Correspondence of Isaac Newton*, VII, p. 172. The original is at CUL MS Add 3972 fol. 37r, dated about 1721.

[47] Stewart, *The Rise of Public Science*, p. 197.

[48] Topham, 'Publishing Popular Science'.

following the reorganisation. Thomas Young, made eminent by his achievements in optics and public lectures on natural philosophy as well as administrative labours at the Royal Society and on public commissions, was appointed Secretary of the Board as well as Superintendent of the Nautical Almanac in 1818. Young's previous work as secretary on other consultative committees and his loyalty to Banks secured him one of the only paid scientific posts in London and is further discussed in Chapter 11. He replaced Thomas Hurd, who was however appointed Superintendent of Timekeepers and continued to serve as Hydrographer to the Navy until his death in 1823 at the age of seventy-six. Hurd wrote to Banks, thanking him for sending papers for the Commissioners but breaking the news of his dismissal, complaining at the injustice and lack of clarity at the situation. He suggested that the presence of so many army offices in the reconstituted Board was an 'indignity ... to the Naval Service'.[49]

The Board's central position in science during the late Georgian era has been underestimated. After 1818 it had an annual budget of £1000 and was one of the main sources of government money, through the naval budget, for scientific work in this period. Along with providing knowledge, instruments and training to personnel on scientific voyages, the Commissioners rewarded the work of the Prussian astronomer Friedrich Bessel in 1819 and granted funds to Captain Edward Sabine for his gravimetric pendulum work, while Joseph de Mendoza y Ríos was given £1200 for improved longitude tables in 1814–1815 and a reward of £500 was given to the Woolwich Royal Military Academy teacher and technical expert Peter Barlow in 1824 for examination of the Navy's compasses.[50] Prominent members of the Board, including Thomas Young, the Second Secretary to the Admiralty John Barrow, and the engineer, author and politician Davies Gilbert all choreographed and assisted in major scientific projects in this period, such as the establishment of the Cape Observatory (described in Chapter 10), the provision of instruments, instructions and training for a series of scientific and exploratory voyages, in addition to the production and maintenance of the *Nautical Almanac* and its network of computers.

Aware of the obligations associated with being publicly funded in a period of deep financial retrenchment following the end of war with France, Young kept and organised all the Board of Longitude's incoming correspondence, even refusing to send letters back when asked for them. Edward Naylor wrote to Young in 1820 complaining that after his scheme for reducing variation in a compass had

[49] Thomas Hurd, draft letter to Banks, 5 November 1818, CUL RGO 14/55, pp. 86r–v.

[50] Correspondence relating to a reward to Joseph de Mendoza y Ríos, 1815, CUL RGO 14/1, pp. 216r–224r; correspondence regarding the award of £500 to Peter Barlow for the examination of compasses, CUL RGO 14/1, pp. 231r–236r. Upon the closure of the Board, Sabine was rewarded £1000 for pendulum work done in the three years to 1826. He wrote to the Board to ask for 'some reward for his services' and a warrant was signed for £1000 in February 1826; BoL, confirmed minutes, 4 February, CUL RGO 14/8, p. 39.

been judged as not 'of any utility' his original papers had not been returned to him. Pursuing the matter, Naylor quoted Young's letter back to him: 'it was not the practice of the Board to return Books or Papers presented to them' and, with reference to the Longitude Act, 'nothing in the clauses of that Act which sanctions the detention of tables, proposals or inventions presented for the consideration of the Board' and demanded his property.[51] Young replied stating that Naylor might ask for the papers as a particular 'favour' and the absence from the archive of a 'published memoir' mentioned by Naylor suggests that Young may have returned some of his papers, while retaining all correspondence, an outline of his proposal on magnetic variation in compasses and several tables of calculations.[52]

Young was constantly on the defensive, attempting to show himself as transparent and fair in his dealing with applicants. Retention of incoming correspondence and proposals was a newly significant feature of civil life in a period where calls for transparency and accusations of corruption were widespread and defined the political landscape after the Napoleonic Wars. The composition of the Board of Longitude archive reflects this change in practice, as retained material comes to dominate the later composition of the collection. Additionally, the volume of incoming correspondence would have prevented Young returning all letters. Generic responses could be sent back and original letters filed in case they were required later to demonstrate the fairness or objectivity of Young's conduct and the Board's decisions. This sense of public obligation extended deeper into the meetings of the Board as Young presented the incoming correspondence to the newly organised Commissioners. Young does not appear to have selected the incoming correspondence presented to the Commissioners for its utility or potential, instead bringing a selection of letters that represented the spectrum, from the potentially fruitful to those that could be immediately rejected as nonsense or entirely impracticable. A proposal discussed in a meeting of the Board could have one of four outcomes in the post-1818 era: first, polite rejection or, secondly, a request for more information or evidence regarding the scheme or instrument described in the letter. These two categories occasionally overlapped, as sometimes the reply requesting more information about a project were clearly written with the hopes of silencing the author. Third, a letter could be referred to one of the Commissioners for closer examination. Fourth, the rarest outcome, was funding being promised or given to the correspondent in exchange for either anticipated work or services already rendered. The fourth response did not occur unless the applicant, and often their work, was already known to the Commissioners. In those cases the archive is often lacking formal written correspondence regarding the scheme,

[51] Edward Naylor to Thomas Young, 8 February 1820, CUL RGO 14/12, p. 376r; Naylor to Young, 15 February 1820, CUL RGO 14/12, p. 378r.
[52] Edward Naylor to BoL, 29 February 1820, CUL RGO 14/12, p. 380r.

Figure 9.1 The discretionary power of the kind exercised by the Board of Longitude was satirised by George Cruikshank's print, *Waiting Room at the Admiralty – (*No Misnomer)*, 1820, which carried a quote from Shakespeare's *Othello*, 'Tis the curse of service that preferment goes by favour and affection'. Yale Center for British Art, Paul Mellon Collection

as contact with the Board was made through polite and private conversation or direct correspondence with one of the Commissioners. The minutes of the Board meeting simply note what had been discreetly decided beforehand.

Without the familiarity of personal connection, a very different experience of dealings with the Board is evident. Captain Peter Heywood, who had conducted surveys of the coast of Ceylon after being court martialled as one of the mutinous crew of the *Bounty*, criticised at length his experience of attending a meeting of the Board (Figure 9.1): 'The cold room he was shown into, the time he waited there, the Omission … of the common civil offer of a Chair to a Stranger [and] the cold indifference of Doctor Young'.[53] This discreet power exercised by the Board over payment made and work commissioned extended directly from the informal networks that underpinned Georgian society and hierarchy. Payments to Sabine, Barlow and Mendoza y Ríos among others contrast with the somewhat peremptory nature of the Board's dealings

[53] Peter Heywood to Thomas Young, n.d., CUL RGO 14/11, pp. 240r–241r.

with others who wrote to it. This discretionary authority was also used by the Commissioners to discredit or diminish the character of individuals. The Polish mathematician Józef Maria Hoene-Wroński travelled to London and applied to the Board for support in 1820–1821 with a new mathematical system to solve longitude. Contained in a letter to Young, Banks forwarded an anonymous note sent to him, declaring Wroński to be 'a very artful dangerous character not without knowledge, particularly in mathematics, but making it his trade to dupe the ignorant by talking unintelligibly'.[54] Wroński's interaction with the Board quickly descended into hostility, particularly after instruments owned by him were delayed by customs and sent to Young's private residence on Welbeck Street. Wroński petitioned King George IV, claiming the Commissioners had rejected his proposal in order to secure the glory of discovering longitude for the British. In 1820 he published a pamphlet and booklet criticising the Board.[55]

While successfully frustrating Wroński, who moved to Paris to attempt a perpetual motion machine after his dealings with the Board in London, the Commissioners were not always able to silence their correspondents. Despite polite but firmly stated rejections from the Board's Secretary, several correspondents would pepper the Commissioners with letters for years. Corresponding with the Board on various topics on and off for over twenty years, the Reverend William Mitchel wrote to Secretary Harry Parker between 1786 and 1788 about one particular scheme to discover longitude using the satellites of Jupiter and an improved quadrant.[56] The Spanish astronomical writer Joseph Emanuel Pellizer wrote to the Board for over twenty years regarding his lunar theory. In 1803 Pellizer justified communicating with the Board yet again, claiming that his proposal had been forwarded by the Admiralty and that he was therefore not contravening the Board's order, ten years earlier, never to contact them further.[57]

The variety of ways the Commissioners engaged with incoming correspondence are all visible in the first meeting of the Board after its restructuring, in February 1819. At that meeting held at the Admiralty, the Commissioners discussed twenty-seven proposals. Eleven were considered eligible for consideration

[54] Anonymous note, n.d., CUL RGO 14/12, p. 502r.

[55] Józef Maria Hoëné-Wroński, petition to King George IV, 23 May 1820, CUL RGO 14/12, pp. 536r–538v; Hoëné-Wroński to 'The Enlightened Men of the British Empire' (printed petition), January 1821, CUL RGO 14/12, pp. 565r–565v. See also Hoene-Wroński, *Address of M Hoene Wronski*.

[56] Papers of Rev. William Mitchel, 1787, CUL RGO 14/36, pp. 5r–22v; Mitchel on his improved quadrant, CUL 14/38, pp. 117r–133a(v); correspondence with Mitchel concerning remuneration for work on an improved quadrant, CUL RGO 14/12, pp. 364r–373v.

[57] Joseph Emanuel Pellizer on the correction of time and a new system of the universe, 1793–1804, CUL RGO 14/53, pp. 125r–146v; Pellizer, letters on a method of finding longitude, 1803–1804, CUL RGO 14/12, pp. 399r–407r; Pellizer to BoL, 12 September 1789, CUL RGO 14/35, p. 364r.

and full response by the Commissioners. Young was requested to write to two correspondents requesting more evidence of the functioning of their schemes. Only one correspondent was awarded funds: Friedrich Bessel was given £150 for the 'great labour and diligence employed … in computing and reducing the Observations of Dr Bradley'. Three communications appeared to be entirely sound in their proposals but were devoid of 'novelty' or offered 'no improvements on the methods in common use'. Young was instructed to report the Board's judgements to the authors of the schemes. Two letters had been previously referred to John Pond to examine more closely. He reported that both should be rejected while a further three were referred to him for more detailed consideration. The other sixteen communications were dismissed, falling into the most common outcome, polite rejection. These schemes were deemed as not requiring 'further consideration', condemned as 'wholly unadmissable' or 'unworthy of attention'.[58]

Two of the letters rejected were for projects relating to quadrature of the circle, 'not a subject in which the Board of Longitude is at all concerned' and a perpetual motion machine, schemes about which 'the Board is not disposed to listen'.[59] Perpetual motion was singled out by the Commissioners so that correspondents could be explicitly told that the Board did not consider such proposals. This is in contrast to other schemes regarded as equally illogical or unlikely, which were given specific responses by Young, most often based on their scientific merit. At the February meeting, Young was instructed to inform Henry Liston, that the 'Board is of the opinion that … he had not succeeded in rendering it probable that his marine table would be free from the inconvenience of vibration'.[60] The rejection of Liston's device to facilitate observations of Jupiter's satellites was a specific response to his proposal and dealt with its scientific and theoretical failings. In contrast, responses from Young to the authors of perpetual motion schemes took their justification from the legislation that circumscribed the remit of the Board, rather than Young composing a standard letter containing references to scientific work that challenged the possibility of perpetual motion.

Receiving and organising this correspondence, reporting the content at Board meetings and then sending out replies were significant tasks for Young and evidence of his organisation practice in order to control the volumes of correspondence is found in the archive. Young developed a habit of writing in red the author and content of each letter at the top of the first page of correspondence; this is clearly private as harsh terms, such as 'nothing new', are common (Figure 9.2).[61] In his responses, Young often put his private address

[58] BoL, confirmed minutes, 4 February 1819, CUL RGO 14/7, pp. 2:272–2:274.

[59] BoL, confirmed minutes, 4 February 1819, CUL RGO 14/7, p. 2:275.

[60] BoL, confirmed minutes, 4 February 1819, CUL RGO.14/7, p. 2:273.

[61] W. Waldron to Thomas Young, 7 April 1819, CUL RGO 14/37, p. 267r. See also: 'J. Tyrell Baylee – again', Baylee to Thomas Young, 2 August 1824, CUL RGO 14/37, p. 291r; 'Answered in the negative', John Gresley to BoL, 9 February 821, CUL RGO 14/11, p. 216r.

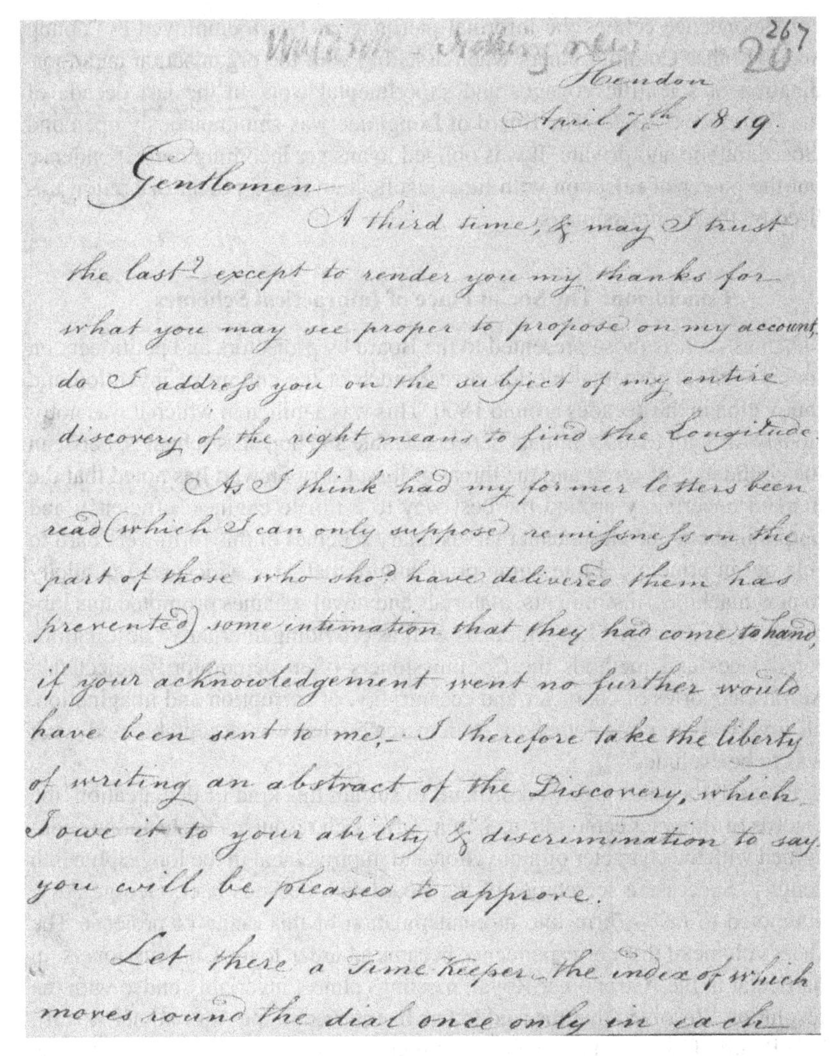
267

Hendon
April 7th 1819.

Gentlemen

A third time, & may I trust the last? except to render you my thanks for what you may see proper to propose on my account, do I address you on the subject of my entire discovery of the right means to find the Longitude.

As I think had my former letters been read (which I can only suppose remissness on the part of those who shod have delivered them has prevented) some intimation that they had come to hand, if your acknowledgement went no further would have been sent to me, — I therefore take the liberty of writing an abstract of the Discovery, which I owe it to your ability & discrimination to say you will be pleased to approve.

Let there a Time Keeper the index of which moves round the dial once only in each

Figure 9.2 An example of Thomas Young's cursory additions in red to incoming correspondence, dismissing a correspondent's idea as 'Nothing new'; William Waldron to the Board of Longitude, 7 April 1819, CUL RGO 14/37, p. 267r. Reproduced by kind permission of the Syndics of Cambridge University Library

at the top, as he worked from his home in Welbeck Street and, from 1825, Park Square. As a result, replies would be sent direct to Young's household and his private rooms would be transformed into the public offices of the Board. This blurring of the boundary between private and public spaces in Young's correspondence echoes the informal patronage network employed by Young and the other Commissioners when assisting with the organisation and coordination of scientific voyages and experimental work in the last decade of the Board's existence. The Board of Longitude was simultaneously open and closed, public and private. It was obliged to answer incoming correspondence but the power of rejection with little justification was a potent discretion utilised by the Commissioners.

Conclusion: The Social Place of Impractical Schemes

Schemes such as those presented to the Board by projectors and petitioners on topics such as perpetual motion reveal much of the culture of invention and innovation in the decades around 1800. This was a milieu in which it was notoriously difficult to discriminate between viable and hopeless schemes, between the confidence of profit and the threat of loss. Larry Stewart has noted that the marked uncertainty around the best way to estimate engines' efficiency and output in these decades meant that visionary schemes of this form were hard to rule out in principle. Flourishing print culture that gave widespread publicity to new machines, instruments, materials and novel schemes prompted this lobbying of the Board of Longitude, as well as providing alternative destinations for devices and methods the Commissioners often peremptorily rejected.[62] Moral categories of character and eccentricity, of corruption and imagination, all involved attempts somehow to demarcate what was acceptable and what was to be excluded.[63]

Precisely because it proved difficult to sustain this kind of demarcation, the records of 'impracticable schemes' may offer rich resources for historians concerned with the character of innovation and improvement in the long eighteenth century. Successive secretaries of the Board and, decades later, George Airy, attempted to take a firm line in condemnation of this range of projects. The sheer volume of this correspondence became a burden to the Commissioners, in particular to the Astronomer Royal; meetings almost invariably ended with the resolution 'Resolved that the said letters be referred to the Astronomer Royal'. The Board of Longitude was an exercise in bureaucracy, testing the limitations of the consultative committee, a new political mechanism developed in the eighteenth century. When the social experiment was ended in 1828, Thomas Young commented in a letter to a friend that he mourned the loss of the Board but reckoned nonetheless that it did 'encourage' the 'ravings of madmen'.[64]

[62] Dircks, *Perpetuum Mobile*; Stewart, *The Rise of Public Science*, p. 286.
[63] Carroll, *Science and Eccentricity*, p. 19; Morrell, 'Individualism and the Structure'; Secord, 'Corresponding Interests'.
[64] Thomas Young to Hudson Gurney, 11 January 1829, in Wood, *Thomas Young*, p. 312.

10 What Is an Observatory? From the Metropolis to the Cape

Simon Schaffer

In autumn 1849, the Astronomer Royal, George Airy, completed the first published history of the early career of the Royal Observatory at the Cape of Good Hope. 'Most readers think historical statements on the origin of institutions interesting,' Airy declared, 'and I confess that I myself feel a desire to give due credit to the first proposers of this Observatory.'[1] The characteristically scrupulous Airy was well-placed to track relations between the Observatory and the Board of Longitude, through his requisition of surviving papers of the Board, on which he had served in its final years. As outlined in our introductory chapter, Airy got these documents from the Admiralty in late 1840. During 1846 he examined what they revealed of the Cape Observatory's business and over the next three years compiled a documentary account, collated with Admiralty proceedings, as well as information from some surviving Board members.[2] The Cape establishment was in significant respects a direct initiative and responsibility of the Board of Longitude. The early history suggests how the Board and its official interlocutors understood the character of a properly working astronomical observatory.

Proposed by leading Board members from late 1819, the Cape Observatory, initiated in 1821 under the management of the Cambridge mathematician Fearon Fallows, was the southern hemisphere's first state astronomical establishment and its systematic work at last launched in 1829, a year after the Board's disappearance. The Cape's position at the hub of colonial enterprise and principal node of long-range maritime networks seemed crucial to metropolitan interests. Eminent astronomer and Resident Commissioner of the Board, John Herschel, observing the southern skies from the Cape in the 1830s, judged it was 'the zero point of African and South American longitudes'.[3] Yet instruments were delayed and often unreliable and staff problems and budgetary pressures were endemic; Fallows was frequently ill and died in July 1831. Having seen the late Cape Astronomer's 'most valuable' observations, Airy

[1] Airy, 'Results', pp. 1–18; George Airy to Francis Beaufort, 19 July 1849, CUL RGO 15/31, p. 39.
[2] George Airy, Journal, 3 December 1840, 6 February 1846 and 9 June 1846, CUL RGO 6/24, pp. 51r, 107r, 111r; George Airy to Francis Beaufort, 9 June 1846, CUL RGO 15/31, p. 23.
[3] John Herschel to George Gipps, 26 December 1837, RS HS.19.72.

hoped that, 'in the same manner as those of European observers', efficient publication would soon follow.[4] It did not. The Cape Observatory was treated as a kind of model but was also a caution about the challenges of long-range management of precision sciences from the imperial metropolis.

An Observatory for the Cape

The first quarter of the nineteenth century witnessed expanded European provision for publicly funded astronomical establishments. These projects helped define the public observatory in that period, with a robust transit instrument in pride of place, a mural circle for planetary and stellar angular heights, and a reliable pendulum clock. An authoritative contemporary guidebook simply defined an astronomical observatory under its entry on transit instruments. Scheduled production of reduced meridian tables, giving times and positions of satellites, planets and stars carefully corrected for a range of observational and physical factors, was supposed to be the observatory's principal concern. Astronomers such as Airy and Herschel publicised these institutions as exemplary sites for fixing positions and times. 'Every astronomical observatory which publishes its observations,' Herschel declared, 'becomes a nucleus for the formation around it of a school of exact practice … a centre from which emanate a continual demand for and suggestion of refinements and delicacies.'[5]

In the German-speaking lands, where this development was most marked, the quarter-century from 1800 saw new observatories at Königsberg, Jena, Altona, Dorpat and Hamburg, along with major overhauls at Göttingen and Munich. In Britain, new public observatories were founded in Edinburgh and Glasgow, while Cambridge University set up its own. The East India Company, already administering an observatory at Madras, sponsored new establishments at St Helena and Bombay (Mumbai). The Company's pugnacious Bombay astronomer unsuccessfully demanded transit equipment from the Board of Longitude in imitation of Greenwich. At just the same period as the Cape initiative in 1819–1821, the Board of Longitude's leader, Joseph Banks, used his interests with the Colonial Secretary, Henry Bathurst, to aid his colleague, the New South Wales governor Thomas Brisbane, to commission an observatory at Parramatta.[6] Not all these observatories survived intact. Airy took a blunt view: '[I]n several instances of similar institutions, no difficulty has been experienced in the first construction of the Observatory, purchase of instruments, etc. The difficulty which has soon begun to show itself is in the maintaining of it.'[7]

[4] Airy, 'Report … 1858', pp. 129–130.

[5] Dewhirst, 'Meridian Astronomy', p. 150; Herschel, *Essays*, p. 640.

[6] Chapman, 'Astronomical Revolution', pp. 40–42; Schaffer, 'Bombay Case', p. 166; Saunders, 'Brisbane's Legacy', p. 181.

[7] George Airy to John Davies, 31 January 1840, in Clarke, *Reflections*, p. 135.

Routines of meridian astronomy and transit catalogues, sidereal times to rate marine chronometers and the provision of baselines for geodetic surveys preoccupied these observatories, their workforces and their high-class if fragile hardware. Though often troubled, these labours were underwritten by ideologies of empire and state power.[8] Reflecting on her work at the Cape Observatory, the nineteenth-century astronomy writer Agnes Clerke held that 'British empire on the seas led directly to British empire over the skies, the one gaining completeness as the inevitable consequence of the other'. The early career of the Cape establishment became a proving-ground for timekeepers and positional instruments, and for the Board's surveillance of artisans and projects that typified observatory sciences in colonial enterprise. Clerke declared that, 'in proportion as England's colonial empire became consolidated, the need of a supplementary establishment to that at Greenwich was rendered more and more imperative'.[9]

Occupied by British forces between 1795 and 1802, permanently annexed from the Dutch in 1805, the Cape was seen by metropolitan authorities as a site of commercial exploitation and indispensable for their links with South Asia.[10] Principal Board members had immediate concerns with the Cape Colony's population, its economy and scientific surveys run from there. Banks had been at Cape Town in 1771, witnessed East India shipping in Table Bay, then compiled an account of the colony's organisation and natural history. After British seizure of the Cape, he proposed settlement of impoverished British migrants and in 1800 insisted on a thorough coastal survey.[11] Admiralty Secretary and Longitude Commissioner John Barrow had even more extensive experience of the colony's affairs. He had been secretary to the Cape's governor from 1797, married a Cape Dutch heiress, and lobbied for the colony's political and military importance. Equipped with a Ramsden sextant, he constructed a map of the territory, which in 1801 and 1804 he published in a two-volume survey.[12] Banks at once noted positional errors in this publication, suggesting Barrow send back to London his original measures 'for at present the public want confidence in the Longitudes of the different parts of the Colony'.[13]

Once appointed Admiralty administrator, nicknamed 'Mr Chronometer', Barrow continued his interests in the Cape. Encouraged by the autocratic Cape Governor Lord Charles Somerset, during summer 1819 Whitehall funded a Cape Emigration Scheme in attempted response to British mass poverty and

[8] Aubin, Bigg and Sibum, 'Introduction'; McAleer, 'Stargazers', pp. 391–395.

[9] Clerke, 'Southern Observatory', pp. 381, 384; Ruskin, *Herschel's Cape Voyage*, pp. 37–38, 47.

[10] McAleer, *Britain's Maritime Empire*, pp. 57–73.

[11] Banks, 'Account of the Cape of Good Hope'; Gascoigne, *Science in the Service*, pp. 126, 177–178.

[12] Barrow, *Account of Travels*.

[13] Penn, 'Mapping the Cape', pp. 116, 121; Liebenberg, 'Unveiling South Africa'; Joseph Banks to Sylvester Douglas, 1 February 1801, in Chambers (ed.), *The Letters of Joseph Banks*, p. 226.

Figure 10.1 The hut first used as the observatory at the Cape, from a watercolour by Thomas Bowler. © South African Astronomical Observatory

insurrection. Wooden huts shipped to the Cape for these migrants also provided provisional housing for the Cape Observatory (Figure 10.1).[14] In a long essay published in a Tory journal in late 1819, Barrow defended such emigration as the sole answer to metropolitan unemployment and hunger, stressed the Cape's potential as 'the great entrepot of the eastern and western hemispheres' and demanded a systematic survey of the colony. There were thus urgent colonial interests among senior Board members and its Admiralty representatives in backing precision scientific labours at the Cape and in the connections between that territory and the metropole. An astronomical observatory there might somehow help secure property and rational order.[15]

The Josephites and the Observatory

Plans for the Cape Observatory were launched at the same moment as Barrow's Cape article during December 1819. Issues of governance and economy raised by the Admiralty Secretary applied both to the Board of Longitude and to the Royal Observatory, Greenwich. Financial management of Greenwich was shifted from the Board of Ordnance to the Admiralty, 'as the Royal Observatory was originally instituted with a view chiefly to the promotion of the knowledge of ascertaining the Longitude at Sea'. The change was backed in 1816 by Banks as head of the Observatory's Board of Visitors, and, though initially vetoed by the Treasury, approved by 1818.[16] Chapter 11 details the

[14] Benjamin Disraeli to John Murray, 22 September 1825, in Smiles, *A Publisher*, II, p. 189; Patrick Scully to Thomas Young, 8 August 1822, CUL RGO 14/48, p. 190v.

[15] [Barrow], 'Cape of Good Hope', pp. 212, 244; Dubow, *A Commonwealth of Knowledge*, pp. 20–21, 25–26; Bulstrode, 'Eye of the Needle', pp. 14–15.

[16] Minute of the Council of the Royal Society, 7 November 1816, CUL RGO 6/1, p. 41; George Harrison to John Wilson Croker, 23 December 1816 and 27 June 1818, CUL RGO 6/1, pp. 44, 49.

major overhaul at the same time of the Board of Longitude's organisation and budget. In manoeuvres typical of his patronage system, Banks was now able to nominate three Royal Society Fellows to the Board, including his ally the Cornish engineer and Tory MP Davies Gilbert (known as Davies Giddy until 1817). Gilbert actively backed the transfer of Greenwich to Admiralty control, as well as the 1818 Longitude Act. That Act also provided for three Admiralty nominees as salaried Resident Commissioners: these were initially the Board's Secretary, Thomas Young, the instrument-designer Henry Kater and the chemist William Wollaston. When Young was also made Superintendent of the Nautical Almanac in late 1818, he was replaced as Resident Commissioner by the military engineer William Mudge, director of the Ordnance Survey. Through the Board's new Committee for Examining Instruments and Proposals, these Resident Commissioners played a significant formal role in the Cape scheme's early management.[17] The Admiralty Secretary and Board member John Wilson Croker spelt out connections between the Cape Observatory, Resident Commissioners and public patronage. Unlike 'branches of science' directly connected with 'trade and manufactures', he reckoned that 'the study of the higher mathematics cannot produce such immediate results'. So the Admiralty had decided to fund 'an Observatory at the Cape of Good Hope and the salaries attached to the situations of the Resident Commissioners of the Board of Longitude, which besides the direct benefit they produce to the public, have the ulterior advantage of offering a few objects of scientific and pecuniary distinction to which the Mathematical and Astronomical student may aspire'.[18]

The Cape project was thus linked to imperial and utilitarian schemes in which the Board became more involved. As described in Chapter 11, the 1818 Act additionally set out a reward for voyages to establish a Northwest Passage to Asian entrepots and northern whaling stations, naming the reformed Board to manage these rewards. An Order in Council of March 1819 defined a scale for the rewards depending on the west longitude reached by these expeditions.[19] In the event, the Board of Longitude acted officially just once, in the wake of the Arctic voyage under William Parry that sailed in May 1819. The Board was thus concerned both with northern and southern routes to Asia. Formal approval of the Cape Observatory and decisions about Parry's Arctic rewards coincided.[20]

[17] BoL, confirmed minutes, 5 November 1818, RGO 14/7, p. 2:255; 'Minutes of Committee of the Board of Longitude for Considering Instruments and Proposals Consisting of Resident Commissioners and Open to the Board', 1819, CUL RGO 14/10, p. 20r.

[18] John Wilson Croker to William Brande for the Council of the Royal Society, 14 March 1823, CUL RGO 6/22, p. 124.

[19] Order in Council, 19 March 1819, CUL RGO 14/1, pp. 113–114.

[20] 'Correspondence Relating to the Reward of 5000£ to the Officers of H M Ships Hecla and Griper', CUL RGO 14/1, pp. 126–142; BoL, confirmed minutes, 27 November 1820, CUL RGO 14/7, pp. 2:321–2:330; Levere, *Science*, pp. 67–74.

Banks and Gilbert launched the Cape scheme at the end of 1819. Gilbert claimed the plan 'originated entirely with me', remarking that 'it will be the first permanent observatory that has ever existed in the Southern Hemisphere'. Fallows later addressed Gilbert as 'the Father of the Observatory (pardon me for thus addressing you)'.[21] With the labours of Gilbert, Barrow and Young, the lobby survived Banks's death in June 1820 and navigated complications presented by the new Astronomical Society of London. This society, a rally of metropolitan astronomers with strong commercial and financial interests, denounced what they called 'the Josephites', Banks's cohort and its alleged hostility to exact sciences. The Society's promoters recalled the conflicts of the 1780s, described in Chapter 8, which had pitted Maskelyne and his colleagues against Banks's regime and affected significant aspects of the Board's conduct.[22] One of the Astronomical Society's more truculent founders, Charles Babbage, was prudently advised in early 1820 somehow to link his new organisation to the Board of Longitude. The designated Cape Astronomer Fearon Fallows was a very early member of the Astronomical Society and soon joined its campaign to get the Board to improve the *Nautical Almanac*.[23] Another of the Society's founders and its second president, Henry Colebrooke, held land at the Cape, wrote on the colony's legal and political conditions, proposed a botanic garden there and, while visiting the Cape, actively advised Fallows on the new observatory site.[24]

The Admiralty did nominate two of the Astronomical Society's leading lights to the Board as Resident Commissioners: the military surveyor Thomas Colby succeeded Mudge both at the Ordnance Survey and as Resident Commissioner, while Herschel took Wollaston's place.[25] Nevertheless, Babbage claimed he himself was refused Board membership by Banks simply because he helped start the Astronomical Society.[26] Banks ensured that the well-connected Edward Seymour, Duke of Somerset, turned down that Society's presidency. Banks's close ally Gilbert declined too. In his carefully phrased letter to Babbage refusing the Astronomical Society's presidency in February 1820, Gilbert also revealed the Board of Longitude would soon recommend

[21] Joseph Banks to Charles Blagden, 20 December 1819, RS CB/1/1/169; Davies Gilbert to Mary Anne Gilbert, 25 December 1819, in Todd, *Beyond the Blaze*, p. 212; Fearon Fallows to Davies Gilbert, c. October 1829, CUL UA Obsy G.6.iii.

[22] Edward Ffrench Bromhead to Charles Babbage, 25 March 1820, BL Add MS 37182, p. 242; Francis Baily to Charles Babbage, 11 March 1820, BL Add MS 37182, pp. 237–238; Ashworth, 'Calculating Eye', p. 414.

[23] Edward Ffrench Bromhead to Charles Babbage, January 1820, BL Add MS 37182, p. 201; Fearon Fallows to John Herschel, 19 February 1820, RS HS.7.156; Turner, '1820–1830', p. 14.

[24] Fearon Fallows to John Herschel, 30 May 1822, RS HS.7.157; Fearon Fallows to Francis Baily, 15 June 1822, RGO 15/29, p. 2; Colebrooke, *Life*, pp. 330–344.

[25] BoL, confirmed minutes, 1 February and 7 June 1821, CUL RGO 14/7, pp. 331, 363.

[26] Babbage, *Passages*, p. 474.

to government the establishment of the Cape Observatory. Following the previous month's accession of George IV, this would be 'a commencement favourable to science of a new reign'. The Cape project was thus represented by the Josephites as prestigious state support for astronomical sciences just when their attitude to those sciences was publicly in question.[27]

A Southern Greenwich

The comparative ease with which negotiations proceeded during 1820 with the Admiralty and the Colonial Office depended on close links binding Board members to government. Chapter 8 showed how these ties of personal trust gradually transmuted to more regulated official conduct. As Banks and Gilbert had planned, on 3 February 1820 the Board with its Admiralty officials and Resident Commissioners endorsed the Cape plan in general terms as 'highly conducive to the improvement of Astronomy' and 'the glory of this country'.[28] Through July, messages about budget and Admiralty approval passed between Young at the Board, Barrow at the Admiralty and Bathurst's Colonial Office. Barrow helped Young get the plan 'into official shape'. At the end of that month, faithfully copying these messages, Bathurst ordered Charles Somerset, the Cape governor preoccupied with land assignment and security for new British settlers under the Cape Emigration Scheme, to allot land for the Observatory and 'afford every assistance in your power'.[29] The plan received royal approval in October 1820. Seven months later Fallows embarked for the Cape. These were comparatively speedy and deceptively uncontroversial manoeuvres. They reveal something of the interests served by the scheme.

Young's memorial, sent to Barrow on 22 July 1820, claimed that 'nothing could more essentially promote the glory of the British nation' than a southern observatory to provide 'the astronomers of Europe' with 'a series of comparative observations made under circumstances the most favourable for correcting the unavoidable imperfections depending on the instruments employed, and on the materials surrounding them, by a countervailing tendency to equal and opposite errors'. The identical expression was endorsed by the Admiralty, approved by the Crown and minuted by the Board.[30] On this showing, the Cape Observatory might be understood as a remote testing station for the Royal

[27] Turner, '1820–1830', pp. 8–11; Davies Gilbert to Charles Babbage, 1 February 1820, BL Add MS 37182, pp. 205–206.

[28] BoL, confirmed minutes, 3 February 1820, CUL RGO 14/7, pp. 297–298.

[29] John Barrow to Thomas Young, 14 July and 24 July 1820, CUL RGO 14/48, pp. 35, 38; John Barrow to Henry Goulburn, 24 July 1820, and Lord Bathurst to Lord Charles Somerset, 31 July 1820, in Theal, *Records of the Cape Colony*, XIII, pp. 201, 223.

[30] Thomas Young to John Barrow, 22 July 1820, in Theal, *Records of the Cape Colony*, XIII, p. 202; Warner, *Royal Observatory*, pp. 6–7; BoL, confirmed minutes, 2 November 1820, CUL RGO 14/7, p. 2:314.

Observatory at Greenwich. In 1812–1816 Edward Troughton had provided the Astronomer Royal John Pond at Greenwich with a six-foot-diameter mural circle and ten-foot transit instrument taken as exemplary by many contemporary observatory managers. Troughton transit instruments were built for Edinburgh from 1814 and Parramatta from 1822. Encouraged by the Board of Visitors in 1819–1820, Pond also commissioned from Troughton a device to check more accurately the mural circle alignment by overhead observations, a twenty-five-foot zenith sector not completed or installed at the Royal Observatory for more than another decade.[31]

There was much debate at the Astronomical Society about the design and cost of such apparatus. Fallows praised this new Society, 'that illustrious body', for 'furnishing practical astronomers with hints for their future attention and drawing the notice of various observatories to the examination of the same class of phenomena. By a comparison of results obtained in different parts of the Globe,' wrote the Cape Astronomer, 'a great mass of valuable knowledge must be brought to light.'[32] Instruments' remote deployment, especially in the southern hemisphere, could act as trials to be exploited in the metropolis by rival interest groups. Such tests were important features of the Board of Longitude's programme. In June 1821, as explained in Chapter 6, the Board's Committee for Examining Instruments and Proposals, then made up of Pond, Kater, Herschel and Wollaston, otherwise concerned with Cape affairs and a range of novel instrument schemes, also agreed to launch chronometer trials at Greenwich under Pond's supervision.[33] It was proposed that the Cape Observatory, like Greenwich, be funded through the Navy Board under the Admiralty, with the Board of Longitude giving direction and advice. Emphasis on the value of instrument tests, even on the scale implied by the Cape initiative, matched these plans. Closer alliance between the Admiralty and Greenwich offered through the Board would provide oversight of a range of navigational, computational and astronomical equipment.

The project to construct an observatory at the Cape that would act as check and comparison for the Greenwich and other northern establishments was a remarkably ambitious enterprise for the Board. During the 1820s, when naval estimates show annual Board budgets up to £5000, the separate Cape project alone would cost the Admiralty an estimated £2500 for instruments and £11,000 for construction, along with an annual wage bill of almost £1000.[34] The plan was first discussed in any detail in February 1820 by the Board's Committee for Examining Instruments and Proposals. Present with Young

[31] McConnell, *Instrument Makers*, pp. 16–19; Howse, *Greenwich Observatory*, pp. 26–29, 38–40, 66–67; Bennett, *Divided Circle*, pp. 169–171.
[32] Fearon Fallows to Francis Baily, 15 June 1822, CUL RGO 15/29, p. 1.
[33] Ishibashi, 'Government Chronometers', p. 55.
[34] Warner, *Royal Observatory*, pp. 103, 130.

as Secretary were Resident Commissioners Kater and Wollaston, Barrow for the Admiralty, and 'added on the present occasion', Banks, Gilbert and Astronomer Royal Pond, a rally of almost all active members of the Board. It is telling that they assumed the Cape plan would be approved and funded, and swiftly agreed that 'the principal instruments be ordered to be put in hand for the observatory', emphasising significantly that these should be 'on the same scale as those at Greenwich and as much as possible on the same construction'. In early April, the Board ordered its committee to get estimates from metropolitan makers for the principal fixed instruments supposed to match those at the Royal Observatory. But the committee was well aware of the time these commissions would take. Because of challenges to siting the observatory at the Cape, through sandy soil and 'other local difficulties', it was therefore vital swiftly to hire an astronomer to be sent out with 'portable instruments'.[35]

There was thus a pair of interlinked Cape initiatives managed by the committee and the Board during 1820–1821: commissions of sophisticated instruments and recruitment of a reliable astronomer. Both relied on patronage alliances between Board members and their confederates. The appointment of Fallows was somewhat the easier. The Anglican minister and Fellow of St John's College, Cambridge, had previously passed the 1813 mathematics tripos two places behind the senior wrangler Herschel, a member of the same college. Fallows joined the Astronomical Society early in 1820. Varsity connections with metropolitan elites helped his cause: Fallows learnt after just over a fortnight of the Board of Longitude's approval of the Cape plan, told Herschel he planned to apply and asked for more information.[36] In March 1820 Herschel, Babbage, Colby and Troughton nominated him to a Fellowship of the Royal Society, while the same month his college head got Herschel to lobby another well-placed Johnian, Lord Palmerston, then in charge of the military budget as Secretary at War.[37] It seemed evident to the gentlemen of science that a Cambridge clergyman and mathematician should run the new observatory. It is a sign of the Board of Longitude's links with these well-placed protagonists that, though Young had as yet recorded no such decision, on 11 April 1820 Fallows nevertheless heard in Cambridge that the Board had backed him 'for the situation of chief observer at the Cape'.[38] Fallows was soon being paid the generous annual salary of £600, a stipend agreed by the

[35] Minutes of the Committee for Examining Instruments and Proposals, 17 February 1820, CUL RGO 14/10, p. 29v; BoL, confirmed minutes, 6 April 1820, CUL RGO 14/7, pp. 2:302–2:303.

[36] Airy, 'Results', p. 8; Warner, *Royal Observatory*, pp. 24–28; Fearon Fallows to John Herschel, 19 February 1820, RS HS.7.156.

[37] James Wood to John Herschel, 20 March 1820, RS HS.18.291; John Herschel to Lord Palmerston, 31 March 1820, RS HS.17.343.

[38] Fearon Fallows to Humphrey Hervey, 12 April 1820, CUL UA Obsy G.6.ii; Warner, *Royal Observatory*, p. 23.

Board and the Admiralty between Young and Barrow, at just the same rate as that of Pond, Astronomer Royal at Greenwich.[39]

The Board, the Observatory and the Instrument Trade

The notional equivalence set out at the Board between the Cape and Greenwich somewhat guided Fallows's immediate programme. He spent the year from April 1820 touring British observatories, working with Pond at the Royal Observatory and at the grand amateur Sir James South's observatory in Southwark, both of which boasted new Troughton transit instruments.[40] The Board planned to match the Royal Observatory's extant instruments with commissions from Troughton and his fellow makers. Fallows visited his friend Troughton at his Fleet Street workshop, while Gilbert met the instrument maker at the launch of the Cape project in early February 1820 to ask 'what part [Troughton] would take in making the instruments for the new establishment'.[41] In response to Gilbert's request and the agreement a week later of the Board's Committee for Examining Instruments and Proposals to proceed with instrument commissions, Troughton agreed to make a twenty-five-foot zenith sector like that underway for Pond. During April there were exchanges about plans and costs with the Board's Secretary, Thomas Young. The optician George Dollond at St Paul's Churchyard, recently elected member of both the Royal and the Astronomical Societies, quickly agreed to make a five-inch lens for Troughton's Cape zenith sector for £100. Troughton told Young that the zenith sector could be assembled within eighteen months for £650. But Troughton eventually declined the commissions for the capital equipment, transit instrument and mural circle.[42]

Dollond therefore also took on the work on the entire transit instrument, just as he soon would for Cambridge's observatory. Dollond consulted Troughton on the otherwise inaccessible structure of the Greenwich transit and told Young that a similar device for the Cape would cost £500.[43] The Cape's six-foot mural circle, also to be based on the Greenwich apparatus, was instead commissioned from Troughton's former assistant Thomas Jones, now based at Charing Cross, who would continue Troughton's schemes for mural circles after the master's death. Jones was a close contact of the Board, which now

39 John Barrow to Thomas Young, 14 July 1820, CUL RGO 14/48, p. 35.

40 Babbage, *Exposition*, p. 156; Warner, *Royal Observatory*, p. 30.

41 Edward Troughton to Thomas Young, 27 November 1820, CUL RGO 14/48, p. 200; Edward Troughton to Thomas Young, 12 April 1820, RGO 14/48, p. 193.

42 Minutes of the Committee for Examining Instruments and Proposals, 17 February 1820, CUL RGO 14/10, p. 29; Edward Troughton to Thomas Young, 12 and 19 April 1820, CUL RGO 14/48, pp. 193, 195.

43 George Dollond to Thomas Young, 19 April 1820, and Edward Troughton to Thomas Young, 12 April 1820, CUL RGO 14/48, pp. 83, 193.

granted him use of the dividing engine originally developed by Jesse Ramsden. On 10 April 1820 Jones told Young it would take two years to build the Cape mural circle, at a cost of 750 guineas (£787 10*s*). The Board began to authorise payments on account from the Admiralty to Jones.[44] During April, Young and Kater also discussed telescopes for the Cape Observatory with Dollond, who learnt the Board wanted two forty-six-inch instruments with improved eyepieces and micrometers, giving a price of 300 guineas (£315) for the pair to be completed within two years. There was some talk of a reflecting telescope to be added to the schedule. Dollond offered to construct a seven-foot Newtonian instrument 'from my own manufactory'.[45]

The day after these messages from Troughton and Dollond, an instrument budget of around £2500 and a two-year timetable were both agreed by the Board's Committee for Examining Instruments and Proposals, with only Pond, Wollaston and the Secretary, Young, in the room at the Admiralty. These were significant interventions in the instrument market. Though ultimately Cape costs would be met from the naval estimates, the sums involved could instructively be compared with the total of around £7500 the Board spent on instrumentation throughout its century-long career (see Appendix 2). At this meeting Pond offered to reduce the cost slightly by substituting for Dollond's Newtonian his own smaller reflector, originally made by James Short. The estimated instrument budget was cut to £2300, but in the event neither reflector was sent to the Cape.[46]

Over the next year, while Fallows habituated himself to demands of observatory management, the Board's members discussed the Cape Observatory's conduct. The Admiralty meanwhile recruited its favoured naval engineer, John Rennie, to design the Observatory's buildings, the sole astronomical or indeed architectural commission this expert on canals, lighthouses and docks ever executed. Rennie died in autumn 1821, his epitaph composed by the Admiralty's John Wilson Croker and his observatory plans carried out posthumously at the Cape over the next eight years. While the Board of Longitude and its committee both formally approved Rennie's design for a somewhat neoclassical building, its façade rather typical of earlier nineteenth-century observatories, with separate rooms for the transit instrument, mural circle and zenith sector, they intervened little in its details.[47] At stake instead for the Board were plans for observatory projects, provisions for assistant staff and

[44] BoL, confirmed minutes, 1 June 1820, CUL RGO 14/7, pp. 2:309–2:310; Thomas Jones to Thomas Young, 10 April 1820, CUL RGO 14/48, p. 180.

[45] George Dollond to Thomas Young, 11 and 19 April 1820, CUL RGO 14/48, pp. 81, 83.

[46] Minutes of the Committee for Examining Instruments and Proposals, 20 April 1820, CUL RGO 14/48, p. 211; copies at CUL RGO 14/7, p. 308, and RGO 14/10, p. 32; Warner, *Royal Observatory*, pp. 130–131, 147. The instrument cost of £2300 is given in CUL RGO 14/48, p. 45.

[47] BoL, confirmed minutes, 1 February 1821, CUL RGO 14/7, p. 2:332; Minutes of the Committee for Examining Instruments and Proposals, February 1821, CUL RGO 14/10, p. 34; Warner, *Royal Observatory*, pp. 92–100.

the instruments Fallows would take while awaiting delivery of hardware from metropolitan workshops.

In late January 1821, a somewhat impatient Fallows lobbied Barrow for provision of transport to the Cape. The astronomer had 'chalked out' some plans for work and would take advice from the Board of Longitude on the instruments to bring before the major commissions were ready. Specifications were agreed by Gilbert, Pond, Young and the Resident Commissioners at the Board's committee in early February 1821 then sent to Barrow. The committee made some superficial comments on the observatory's site, to avoid possible disturbance by the Cape's dusty atmosphere and to permit observation of a bright zenith star. More significantly, Fallows was ordered to compile a southern star catalogue, as well as maintaining records to be sent to Young every six months for publication. These were to be managed just as Bradley and Maskelyne previously had from the Royal Observatory, 'as much as possible of the same kind and in the same manner as the Greenwich Observations have been made: to employ the same stars, where it can be done conveniently, and to draw up the register in the same form'. Consistent with the model of the Cape as a southern Greenwich, the aim was to produce 'two corresponding series capable of comparison in all their parts'.[48]

To establish such matching series, the Board members realised, it would be necessary to secure the delivery of hardware just commissioned. Until then Fallows would make do with a portable twenty-inch Dollond transit instrument (Figure 10.2) and an equatorial telescope adapted from a thirty-inch altazimuth circle by Ramsden. Troughton told Fallows this circle was stored at Greenwich but certainly did not belong there, and that with its loan 'you may do a great deal of preliminary business previously to your own shop being furnished with better tools'. The Astronomical Society's official view was that Fallows's improvised instruments were 'of a very humble description' and that the Board had underestimated 'the length of time which must necessarily elapse between the design and completion of a first-rate Observatory, in a foreign station'.[49] These were the devices Fallows used to complete the southern star catalogue demanded by the Board's instructions in early 1821, sent by Fallows to the Board in June 1823 and published by the Royal Society in February 1824.[50] Fallows also got a Robert Molyneux astronomical clock, apparently purchased from his erstwhile

[48] Warner, *Royal Observatory*, pp. 33–34, 44; Minutes of the Committee for Examining Instruments and Proposals, 3 February 1821, CUL RGO 14/10, p. 34; 'Instructions for the Astronomer at the Cape Observatory', CUL RGO 14/3, p. 151.

[49] Edward Troughton to Fearon Fallows, 4 December 1820, CUL UA Obsy G.6.i; [Obituary of Fearon Fallows], *Monthly Notices of the Royal Astronomical Society*, 2 (February 1832), p. 64. For portable transits by Troughton and Dollond, see Bennett, *Divided Circle*, pp. 171–174; Warner, *Royal Observatory*, p. 47.

[50] 'Instructions for the Astronomer at the Cape Observatory', CUL RGO 14/3, p. 151; Fallows, 'A Catalogue', pp. 458–459, 462, copy in CUL RGO 14/48, pp. 119–125; BoL, confirmed minutes, 6 November 1823, RGO 14/8, p. 9.

Figure 10.2 George Dollond, repeating transit, about 1820, used at the Cape Observatory before the completion of the new building. © South African Astronomical Observatory

host James South, while the Board's committee recommended the Admiralty provide him with a pocket chronometer, a device perhaps to be identified with the gilt chronometer by Charles John Cope later listed in the observatory's inventory. Both London clockmakers, Cope and Molyneux, were former apprentices of Thomas Earnshaw, whose fraught relations with the Board in 1803–1808, as well as with Gilbert's parliamentary committee, are described in Chapter 6. The Board was regularly informed that their timepieces were somewhat unreliable for Fallows's attempts at regular timekeeping.[51]

The Board, the Observatory and Assistant Staff

As significant as the variable performance of this provisional equipment was that of the Cape Observatory's assistant staff. It was agreed by metropolitan authorities that for meridian astronomy an observatory should employ at

[51] Minutes of the Committee for Examining Instruments and Proposals, 3 February 1821, CUL RGO 14/10, p. 33; Warner, *Royal Observatory*, pp. 149–150.

least two assistants, so that clock, transit instrument and mural circle could be read simultaneously. Only Greenwich, it seemed, in fact commanded such resources.[52] Were the Cape to emulate this regime it would need comparable personnel. No trace survives of any detailed discussion of Cape Observatory staffing by the Board or its committee during 1820. Troughton told Young he was surprised by the Board's reticence: '[I]n many conversations which I have had with members of the Board of Longitude respecting the Cape of Good Hope Observatory … I have not once heard an assistant mentioned.' The sole record is a brief statement by Barrow and Young in July, reiterated in October, that annual salaries of £250 for an assistant and £100 for a labourer be added to the budget.[53]

These wages, like those of Fallows, set the Cape scheme at comparable levels to the Royal Observatory. Under Pond's regime the number of Greenwich assistants eventually rose from one to six. Thomas Taylor, a naval master hired as assistant by Maskelyne in 1807, stayed in post under Pond, who also brought with him a second assistant, while in late 1817 Pond started lobbying through the Board of Longitude to hire two more. Eminent Board members including Gilbert, Herschel and Young agreed that 'improvements in the general construction and accuracy of modern Instruments' demanded continuous observation and rapid reduction. By August 1822, Pond had secured posts for his first assistant's son, Thomas Glanville Taylor, and for William Richardson, former employee of South and recommended by Troughton. Troughton told Fallows that with these two extra Greenwich staff 'the Mill is continually to be kept grinding'.[54] From 1822, at the same time as the Board's oversight of the Cape establishment, its principal members used their positions in the Royal Society and Admiralty to debate increases in the Greenwich workforce and their wages, especially the social status and technical competence to be demanded from observatory staff. In March 1824, it was briefly contemplated that Taylor be replaced as first assistant by a more socially and technically superior candidate.[55] Pond's judicious response in April 1826 to

[52] Dewhirst, 'Meridian Astronomy', p. 153.

[53] Edward Troughton to Thomas Young, 6 May 1820, CUL RGO 14/48, p. 198; John Barrow to Thomas Young, 14 July 1820, CUL RGO 14/48, p. 35; BoL, confirmed minutes, 2 November 1820, CUL RGO 14/7, p. 2:315.

[54] BoL, confirmed minutes, 4 December 1817, CUL RGO 14/7, p. 2:238; Minutes of the Council of the Royal Society, 4 July 1822, CUL RGO 6/22, p. 120; [Obituary of Thomas Glanville Taylor], *Monthly Notices of the Royal Astronomical Society*, 9 (1849), p. 62; James South, [Presentation of the Society's Gold Medal to William Richardson], *Monthly Notices of the Royal Astronomical Society* 1 (1830), p. 165; Chapman, 'Airy's Greenwich Staff', pp. 7–8; Edward Troughton to Fearon Fallows, 13 September 1822, CUL UA Obsy G.6.i.

[55] Humphry Davy to John Wilson Croker, 13 March 1823, in Fulford and Ruston (eds.), *Collected Letters*, III, pp. 395–396; John Wilson Croker to William Brande for the Council of the Royal Society, 14 March 1823, CUL RGO 6/22, p. 125; Minutes of the Council of the Royal Society, 18 March 1824, CUL RGO 6/22, p. 130.

Wollaston, that the observatory rather needed 'obedient drudges' to engage in 'the mechanical act of observing' and the 'dull process of calculation', became notorious through its publication in Babbage's incendiary essay on the decline of English science.[56] During the 1820s, the senior Greenwich assistant Taylor was paid at least £300 a year, while the Admiralty allowed Pond's ordinary assistants £100 per annum, rising each year by £10. To reinforce this hierarchy, Airy later raised senior assistants' salaries and abolished ordinary assistants' annual increments. At Cambridge University the observatory was only granted a total of £150 annually to meet the stipends of two assistants together. It is instructive to note that Young's yearly salary as Secretary to the Board of Longitude was just £100.[57]

Metropolitan instrument makers assumed such observatory assistants would principally maintain and repair astronomical hardware. Instrument users in the colonies often agreed. Soon after Fallows's arrival at the Cape in August 1821, he met the energetic East India Company surveyor George Everest, who reckoned that 'in Countries where instrument-makers are not to be procured, such as … Cape of Good Hope, Australia, India and others, artists with an adequate supply of tools should be attached to the establishment'. In a report on the Indian survey he directed, Everest observed that, if 'any accident arise in one of the mural circles at Greenwich, that instrument would only be powerless until the coach from Charing Cross had brought up … Mr Jones or Mr Dollond'.[58] Once at the Cape, Fallows learnt why proximity mattered, especially 'the very great difficulties I had to encounter from the simple circumstances of not having a Troughton or a Dollond &c at hand'.[59]

During 1820–1821, both Troughton and Dollond lobbied the Board of Longitude, proposing candidates for Cape support staff. Just as he backed Richardson for Greenwich, so for the Cape post Troughton proposed his young kinsman and employee James Fayrer, who made sextant frames and clocks and could reportedly clean and fix astronomical instruments. In November 1820, Fallows met Fayrer at Troughton's shop, while Young recommended him to Barrow as Cape assistant. When Fayrer provided a list of equipment for the new observatory, Fallows vetoed what he judged an expensive request for a

[56] John Pond to William Wollaston, 5 April 1826, RS MS 371, p. 49, and Minutes of the Council of the Royal Society, 6 April 1826, CUL RGO 6/22, p. 175, cited in Babbage, *Reflections*, p. 126.

[57] Minutes of the Council of the Royal Society, 12 June 1823, RGO 6/22, p. 127; Meadows, *Greenwich Observatory*, p. 8; Hutchins, *British University Observatories*, pp. 66–69; BoL, confirmed minutes, 5 November 1818, CUL RGO 14/7, p. 2:255.

[58] George Everest, 'Memoir regarding the Survey Establishment in India', June 1826, BL IOR L/MIL/5/402, p. 369; George Everest, Surveyor-General's Report, 3 August 1839, in Phillimore, *Historical Records*, IV, p. 124. For Everest at the Cape in 1821, see Fearon Fallows to John Barrow, 5 September 1821, CUL RGO 14/48, p. 99v; Warner, *Royal Observatory*, pp. 60–61.

[59] Fearon Fallows to Francis Baily, 9 December 1830, CUL RGO 15/29, p. 7; McAleer, 'Stargazers', p. 402.

dividing engine, then left the rest to the Board's discretion.[60] In spring 1821, Dollond also proposed one of his men, Thomas Smith, working on the Cape transit instrument, as assistant. Other Dollond employees came forward as possible labourers.[61] The Board considered and rejected these applications and decided to hire Fayrer as assistant, but to recruit a labourer from among Cape residents.[62]

Fallows and Fayrer left Britain in May and reached the Cape in August 1821. Fallows took the Board's instructions and rudimentary equipment. Fayrer had his own four-foot Troughton achromatic telescope, which Fallows borrowed so as to determine, without success, longitude at sea by eclipses of Jupiter's moons.[63] When they arrived, Fayrer at once had the difficult task of shepherding the fragile instruments ashore. It immediately became apparent that Cape Town officials were almost entirely unaware of provisions made in London for the observatory scheme.[64] From then, members of the Board of Longitude, especially Barrow, Young, Pond and the Resident Commissioners, notably Herschel, became involved in long-range decisions about personnel, hardware and practice at the Cape Observatory. This was somewhat untried and often backstage business, requiring sustained judgement over errant or frustrated long-range initiatives in meridian and survey astronomies. Management also involved more familiar tasks, such as artful dealings with Admiralty paymasters and instrument makers. Leading Board members dealt with an observatory under slow construction, its staff and structures under constant negotiation, and shifting and somewhat overoptimistic visions for the Observatory's work.

A characteristic example of long-range surveillance over the Cape Observatory run from the metropolis and its frustrations during the 1820s was the management of Fallows's assistant staff. Scrutiny of observatory labour proved fraught in Greenwich and challenging at the Cape. In either case it was only possible in a formal sense from Whitehall. There were endemic troubles of instrument maintenance, which relied on embodied skills hard to oversee. Thus when the telescope's micrometer crosswires, key to good transit measures, were destroyed by a moth, Fayrer vainly tried to replace them with a spider's web matching the former scale divisions. This was a technique that Fayrer's kinsman and former employer Edward Troughton had introduced into

[60] Edward Troughton to Thomas Young, 5 May and 27 November 1820, CUL RGO 14/48, pp. 198, 200; Fearon Fallows to Thomas Young, 24 March 1821, CUL RGO 14/48, p. 93.

[61] George Dollond to Thomas Young, 1 March 1821 and Fearon Fallows to Thomas Young, 26 February 1821, CUL RGO 14/48, pp. 85, 95.

[62] Minutes of the Committee for Examining Instruments and Proposals, February 1821, CUL RGO 14/10, p. 34; Warner, *Royal Observatory*, pp. 34, 78.

[63] Fearon Fallows to John Barrow, 5 September 1821, CUL RGO 14/48, p. 104; Fallows, 'Communication of a Curious Appearance'.

[64] Fearon Fallows to John Herschel, 30 May 1822, RS HS 7.157.

widespread astronomical use.[65] More crucial to such labour was the stress the observatory regime placed on the correction and reduction of observations. Pond's 1826 comment on 'the dull process of calculation' highlighted this principle, while Herschel agreed. When he was at the Cape, he declared that 'the reduction and printing of the observations is a matter quite as important as the making them, it being fully understood at present that unreduced observations are scarcely worth transmitting home'. Fallows very similarly told Young that 'we consider the time spent in the Observatory as mere play, compared to that of computing afterwards'.[66]

It soon became apparent that Fayrer was incapable of such reliable computation. In late 1821, Fallows therefore decided to use the £100 salary assigned to the labourer's post to hire as computer a recently arrived Catholic priest, Patrick Scully, whose government grant to preach at the Cape had just been cancelled by the governor, Somerset. The Board of Longitude formally approved the decision to hire Scully in April 1822.[67] The following November, the Committee for Examining Instruments and Proposals, with Herschel and Pond present, endorsed Fallows's further propositions that Scully be promoted to first assistant in Fayrer's place. Two days later the Board itself approved Fallows's scheme to change the assistants' stipends, paying Scully £200 a year, and that Fayrer correspondingly take on the labourer's role at £150.[68] There was talk of tensions involving Fayrer's and Fallows's wives that affected the good order of the establishment. It was agreed, eventually, that 'Mr Fayrer will be the workman only, having no part in observing'. Fayrer made up his income by advertising his services in Cape Town as clockmaker and instrument repairer, with Fallows providing him with true solar time. It was said he did little work as labourer and took to drink.[69] Barrow sternly insisted that, though Scully could indeed be made first assistant, the Board was forbidden to change salaries fixed by an Order in Council, thus part of state regulation of the Admiralty's budget.[70] Hierarchy mattered both at the Admiralty and the

[65] Fearon Fallows to John Barrow, 7 January 1822, CUL RGO 14/48, p. 107; [Nicholson], 'Spider's Webs'.

[66] Minutes of the Council of the Royal Society, 6 April 1826, RGO 6/22, pp. 176–178; John Herschel to George Gipps, 26 December 1837, RS HS.19.72; Fearon Fallows to Thomas Young, 6 March 1823, CUL RGO 14/48, p. 114.

[67] BoL, confirmed minutes, 4 April 1822, CUL RGO 14/7, p. 384; John Barrow to Fearon Fallows, 8 April 1822, CUL RGO 15/27, p. 22; Warner, *Royal Observatory*, pp. 78–79.

[68] Fearon Fallows to Thomas Young, 13 June 1822, CUL RGO 14/48, p. 108; Minutes of the Committee for Examining instruments and Proposals, 5 November 1822, CUL RGO 14/10, p. 40; BoL, confirmed minutes, 7 November 1822, CUL RGO 14/7, p. 391.

[69] Fearon Fallows to Thomas Young, 6 March 1823, CUL RGO 14/48, p. 114; *Cape Town Gazette*, 16 November 1822, p. 1; Charles Piazzi Smyth to George Airy, 29 November 1840, CUL RGO 15/31, p. 90.

[70] John Barrow to Thomas Young, 20 November 1822, CUL RGO 14/48, pp. 54, 56; BoL, confirmed minutes, 6 February 1823, CUL RGO 14/7, p. 396.

Cape, while the mix of bureaucracy and morals became characteristic of the Observatory's fate.

It is telling that the affairs of the Cape Observatory staff were not minuted by the Board of Longitude in the last five years of the Board's life. Rather, employment decisions taken at the Observatory were handled in correspondence between the Board's Secretary, Thomas Young, his colleagues such as Pond at Greenwich, and Barrow's office at the Admiralty, as matters of more discretion. Fallows initially praised Scully's behaviour: possessed of 'a tractable, mild and amiable disposition, never giving me a cross word', the astronomer testified, the new first assistant was highly competent at transit observations and improving computational skill. Just as Herschel would describe the astronomical observatory as 'a school of exact practice', so Fallows told Barrow in June 1823 that he sought to make 'the temporary observatory a kind of school for the instruction of my assistant'.[71] The following year, however, confidence collapsed. In July 1824, Scully was sacked and left the colony after what was judged 'immoral conduct' with a female servant: iniquity undermined competence. In a message to Fallows, Troughton described the scandal in typically earthy terms: 'I hear you had turned away your late assistant for using your servant maid's Transit Instrument.' None of this appeared formally before the Board of Longitude.[72]

After failures in the cases of Fayrer and Scully, employment control shifted to authorities at the Admiralty. As noted above, during their debates about appropriate qualifications for Greenwich staff, it was recommended in May 1824 by Pond and the Board of Visitors that Taylor retire from his post as senior assistant at the Royal Observatory and be replaced with a person 'of the superior class'. With Croker's influential backing, an army captain, William Ronald, at once petitioned for the position.[73] Meanwhile, following Scully's dismissal in July 1824, the anxious Fallows wrote directly to Taylor asking whether his son might be willing to take the vacant Cape job.[74] When Fallows's official request reached London, the Admiralty approved the employment of a new Cape assistant, while in November the well-informed Ronald at once offered himself for both Cape and Greenwich posts, throwing the familiar issue of identity between the two observatories into unusually clear light. Pond quickly interviewed Ronald, reckoning his absence of a Cambridge education, thus lack of scholarly superiority, told against his employment at the Royal Observatory. The Astronomer Royal also noted Ronald's lack of experience

[71] Fearon Fallows to Thomas Young, 18 June 1823, and to John Barrow, 20 June 1823, CUL RGO 14/48, pp. 115, 117.

[72] Fearon Fallows to Thomas Young, 17 July 1824, CUL RGO 14/49, p. 154; Edward Troughton to Fearon Fallows, 20 December 1825, CUL UA Obsy G.6.i.

[73] Minutes of the Council of the Royal Society, 18 March and 6 May 1824, CUL RGO 6/22, pp. 130–132.

[74] Fearon Fallows to Thomas Taylor, 17 July 1824, in Warner, *Royal Observatory*, p. 145.

with 'management of large Astronomical Instruments'. But Pond approved his 'talent and activity' and commended him for the Cape post.[75]

Ronald was thus not hired at Greenwich. In the event, Taylor stayed at the Royal Observatory for another decade until the end of Pond's regime. Pond later wrote that, though 'he has been a faithful servant', nevertheless 'his sight is imperfect; he has grown petulant and has latterly taken to drinking'. Airy, who sacked him immediately, concurred that Taylor was a drunkard and took bribes from clockmakers. Meanwhile, Ronald was officially hired for the Cape on 1 December 1824 and trained in observatory duties on the mural circle at Greenwich during 1825.[76] Troughton told Fallows that Ronald would 'attend at the observatory for the purpose of learning the business of an astronomer, who if he was able and liked it would be sent to the Cape as your assistant'. Ronald eventually sailed for the Cape in August 1826, in company with twenty cases of equipment, including the Dollond transit instrument and Jones's mural circle. 'With so able an assistant as Captain Ronald is apprehended to be,' Barrow instructed the Cape Astronomer, the Admiralty 'expect you will in conjunction with him take every opportunity of making observations for the advancement of Astronomy in particular and also for the benefit of science in general.'[77]

The Cape Observatory and the Fate of the Board of Longitude

As the Admiralty took a more direct and commanding role in instructions for the Cape Observatory, including employment of staff, the Board of Longitude's final deliberations attended rather to the equipment. The mural circle on which Ronald was trained at Greenwich was in fact the instrument first commissioned for the Cape from Jones in April 1820 but then installed at Greenwich in 1824 and retained by Pond at the Royal Observatory.[78] Fallows had passionately lobbied the Board, through Young, for an altazimuth circle to observe satellites and determine latitudes. He knew Dollond had made such a device for the Board's Resident Commissioner Kater. 'Besides,' he wrote to Young, 'who knows how long we may wait for the Mural Circle?'[79] During 1825, while substantial further sums were approved by the Board of Longitude to be paid

[75] John Barrow to Fearon Fallows, 5 October 1824, CUL RGO 15/27, p. 36; William Ronald to William Thomas Brande, 6 November 1824, CUL RGO 6/22, p. 135; John Pond to John Barrow, 14 November 1824, CUL RGO 6/22, pp. 135–136; Warner, *Royal Observatory*, p. 146.

[76] John Pond, 'Statement respecting the Assistants at the Royal Observatory' [1835], CUL RGO 6/72, p. 233; George Airy to Lord Auckland, 15 June 1835, CUL RGO 6/1, p. 158; John Pond, 'A Memorial relative to the Employment of New Assistants at the Royal Observatory', 20 April 1825, CUL RGO 6/1, p. 52.

[77] Edward Troughton to Fearon Fallows, 20 December 1825, CUL Obsy G6.1, p. 6; John Barrow to Fearon Fallows, 5 July 1826, CUL RGO 15/27, p. 109; Warner, *Royal Observatory*, pp. 143–144.

[78] Warner, *Royal Observatory*, pp. 137.

[79] Fearon Fallows to Thomas Young, 6 March 1823, CUL RGO 14/48, p. 113.

to Jones for this instrument, the maker told the Board that he wanted to incorporate many changes he'd made on the mural circle at Greenwich into the Cape device, 'to produce greater accuracy and certainty in making the observations'.[80] At the same time, Dollond completed the Cape transit instrument, based on Troughton's Greenwich original.[81] Yet again, juxtaposition between Greenwich and Cape equipment and practice seemed an ideal.

The Board's attention during this period was also directed both to Fallows's complaints about the performance of his Molyneux clock and his demands for a reflector with which to track the southern nebulae. The clockmaker William Hardy, who made his reputation with the fine clock he delivered to Greenwich in 1811, and produced a similar device for Cambridge Observatory, was eventually commissioned to produce such a clock for the Cape. The Board got Admiralty approval to purchase one for 100 guineas (£105) and it was then rapidly despatched to the Cape.[82] The question of a reflecting telescope was less straightforward. As already indicated, in the initial Cape scheme it had been suggested that Dollond construct a seven-foot Newtonian telescope for the observatory, and Pond had offered a six-foot reflector in his possession, but neither of these proposals was put into effect. In early 1822, the Committee for Examining Instruments and Proposals, with Barrow, Pond, Colby, Herschel and Kater all in attendance, asked Kater to inquire about a reflector owned by the ailing Henry Englefield, cometary astronomer and antiquarian, and at the same time to find out about the fate of the fourteen-foot reflector cast by William Herschel himself in April 1810 and installed in Glasgow at a cost of around £400 in the new observatory opened that year.[83] By 1822 it was evident this observatory would shut, and from that point the Board continued to lobby for its Herschel instrument. A generous price of around £300 was agreed but it was not until late 1826 that the reflector reached the capital, it was not shipped to the Cape until considerably later and it was never used by Fallows there.[84]

[80] BoL, confirmed minutes, 6 February 1823, CUL RGO 14/8, p. 28; Jones to Young, May 1825, CUL RGO 14/48, p. 184 (annotated 'approved by the Board of Longitude, 2 June 1825'); BoL, rough minutes, 2 June 1825, CUL RGO 14/3, p. 116 ('Jones' instrument').

[81] Warner, *Royal Observatory*, pp. 142–143.

[82] William Hardy to Thomas Young, 6 September 1822, CUL RGO 14/48, p. 173; BoL, confirmed minutes, 7 November 1822, CUL RGO 14/7, p. 391; John Barrow to Thomas Young, 20 November 1822, CUL RGO 14/48, p. 54; Fearon Fallows to Thomas Young, 10 October 1823, CUL RGO 14/48, p. 131.

[83] Minutes of the Committee for Examining Instruments and Proposals, 3 January 1822, CUL RGO 14/10, p. 37r; Warner, 'The William Herschel 14-foot Telescope'; Clarke, *Reflections*, pp. 126–132.

[84] Minutes of the Committee for Examining Instruments and Proposals, 5 November 1822, CUL RGO 14/10, p. 40r; BoL, confirmed minutes, 6 April 1826, CUL RGO 14/8, p. 42; Andrew Templeton to Thomas Young 14 January 1822, and to John Herschel, 10 March 1826, CUL RGO 14/48, pp. 203r, 204r; John Barrow to Thomas Young, 7 November 1826, CUL RGO 14/48, p. 70; Clarke, *Reflections*, pp. 132–134.

By summer 1826, many astronomical instruments destined for the Cape were ready for shipment, including Jones's mural circle, Dollond's transit and associated clocks and other apparatus. Fallows warned the Board that the instruments should not be delivered until the observatory buildings were completed, but this prudent advice was ignored. In November 1826, twenty cases of equipment arrived at the Cape, watched over by the newly arrived astronomical assistant William Ronald.[85] This despatch might have represented the belated launch of the Cape Observatory's organised observational regime but it was also one of the final actions of the Board and accompanied the reinforcement of firmer Admiralty control. As detailed in the next chapter, during summer 1828 Croker and Barrow, with Davies Gilbert's stated support, decided to bring affairs until then managed by the Board decisively under the Admiralty's institutional remit. The Board was replaced with a Resident Scientific Committee consisting of Thomas Young and, so it was originally intended, the Board's former Resident Commissioners Kater and Herschel. Cape Observatory business and reputation played an important role in these debates about the Board's fate. As already noted, Croker's expressed view was that provision for Resident Commissioners and for the Cape Observatory were both aspects of state support for otherwise ill-served students of mathematical and astronomical sciences. This made the Cape Observatory's condition a matter of urgent concern. In 1825, after conversations with Young at the Board, Humphry Davy, as then president of the Royal Society, had already told the eminent observatory manager Heinrich Schumacher that, while 'our observatory at the Cape is beginning to give us some results', 'I do not think the situation favourable for astronomy'.[86] Fallows himself was well aware of London challenges to the Cape Observatory's viability. He told Gilbert, from 1827 Davy's successor as president of the Royal Society, that the English must be wondering at delays in completing the observatory, and wrote in October 1828 that 'the late Board of Longitude conceived I was taking too much trouble about the Piers', the massive stone bases on which the telescopes were to be erected. 'With all due respect to so learned a Body I think not so,' exclaimed the astronomer, protesting that 'I should have cut my right hand off' rather than compromise on these pillars of stability.[87]

London press debates in late 1828 about the fate of this 'learned Body', the Board of Longitude, were entangled with real threats to the survival of the

[85] Warner, *Royal Observatory*, pp. 143–150; John Barrow to Fearon Fallows, 5 July 1826, CUL RGO 15/27, p. 109; George Dollond to Thomas Young, 21 September 1826, CUL RGO 15/27, p. 69; McAleer, 'Stargazers', pp. 396–397.

[86] Humphry Davy to Heinrich Schumacher, 27 April 1825, in Fulford and Ruston (eds.), *Collected Letters*, III, p. 537.

[87] Fearon Fallows to Davies Gilbert, October 1829 and Fearon Fallows, draft letter, 14 October 1828, CUL MS Obsy G/iii; Warner, *Royal Observatory*, pp. 189–190.

Cape Observatory. On 4 October, the under-secretary at the Colonial Office Horace Twiss wrote to Davies Gilbert at the Royal Society asking which of the Cape Observatory or the astronomical establishment founded by Brisbane in New South Wales should survive, since their latitudes were so similar. The Royal Astronomical Society's lobbyists insisted that the Cape should not be shut, because of 'its very advantageous situation in the same meridian with the principal observatories of Europe'. Once again, co-ordination between the Cape and Greenwich seemed decisive. South took to the pages of *The Times* against 'the shame of the astronomical character of the British government' and was answered at once that the government 'has established an Observatory at the Cape of Good Hope and furnished it with the best instruments that British ingenuity can construct', while *The Times* itself reported on the establishment of the Resident Scientific Committee as a 'new Board of Longitude'.[88] Fallows at the Cape seemed already aware that his observatory was under direct attack, contacting friends to lobby for employment back in Britain if he lost his post (and telling them to burn his letters if he survived as Cape Astronomer): 'I little suspected that in a distant Colony I might have to regret the want of faith of my countrymen at home. However I am ready to hear the worst.'[89]

In late November 1828, South prepared a long polemic about the condition of public astronomy, with the Cape Observatory and the Board of Longitude as his principal concerns. He praised the ambitions of his friend Fallows: 'but the Cape Astronomical Establishment has cost the country, as may be seen by various parliamentary documents, upwards of 20,000l. sterling. *Cui Bono*? I would ask.' South's sarcastic answer to his own question was blunt: 'To afford Englishmen the enviable happiness of feeling, that owing to the scientific *energy* of the British Government, for any observations which might have hitherto been transmitted to Europe by the Cape Astronomer, the Cape Astronomer might as well have remained at home.' In the *Morning Chronicle* he spelt out the costs the Board had lavished on the Cape establishment, the frustrations of its astronomical enterprise, and what he saw as the laxity of the government's support for the science both in Europe and in the southern hemisphere. The Cabinet itself was in any case debating a complete overhaul of state-supported astronomical staff. The Chancellor of the Exchequer, Henry Goulburn, discussed with the Admiralty the replacement of the Astronomer

[88] Horace Twiss to Davies Gilbert, 4 October 1828, RS MS DM 4/65; William Stratford for John Pond to Davies Gilbert, 11 November 1828, RS MS DM 4/67; James South, 'The Appearance of Encke's Comet', *The Times*, 7 November 1828, p. 2; 'Encke's Comet', *Morning Chronicle*, 17 November 1828, p. 3; 'New Board of Longitude', *The Times*, 24 November 1828, p. 3.

[89] Fearon Fallows, draft letter, 14 October 1828, CUL UA Obsy G/iii; Warner, *Royal Observatory*, pp. 189–190.

Royal by Herschel: 'I fear that without some arrangement for getting rid of the stagnant Pond our observatory will always disgrace us.'[90]

The critical condition of the Cape Observatory continued for at least the next two years. In October 1830 Ronald fell ill and left the Cape forever; in July 1831 Fallows died of scarlet fever contracted by the observatory staff, and in the meantime John Barrow at the Admiralty decided to resuscitate the proposal to close one of the southern observatories. The Royal Society at once set up a Southern Hemisphere Observatories Committee, with the young George Airy, then Cambridge Observatory manager, as member alongside the Astronomer Royal, Pond, and former Board of Longitude Commissioners Kater and Herschel.[91] The documents first assembled in 1828, and then amplified by this committee, provided Airy with the initial material he used for his historical account of the foundation of the Cape Observatory. While he was composing this survey of the Cape Observatory's establishment in the later 1840s, he told the then Admiralty Secretary that the existence of 'the Cape Observatory in a most efficient state (as respects both its instrumental and its present personal establishment)' would give it pride of place in the southern latitudes of the British astronomical imperium.[92] As the complex choreography of administration and patronage, both institutional and personal, outlined in this chapter amply demonstrates, at this period the character of an efficient observatory was always under intense negotiation and never unambiguous. These ambiguities and complexities played an important role in the Board of Longitude's fate.

[90] James South, 'To the editor of the Morning Chronicle', *Morning Chronicle*, 27 November 1828, p. 3, and South, 'Astronomy versus the Government', *Morning Chronicle*, 12 December 1828, p. 3; Henry Goulburn to the Duke of Wellington, 25 August 1829, University of Southampton Library, MS61/WP1/1040/2.

[91] John Barrow to Peter Mark Roget, 8 December 1830, RS MS DM 4/105; 'Minute of the Southern Hemisphere Observatories Committee', 23 December 1830, RS MS DM 4/108.

[92] George Airy to George Ward, 30 April 1847, CUL RGO 6/140, pp. 99–100.

11 The Death and Rebirth of the Board of Longitude

Sophie Waring

Regency Britain witnessed a major series of projects to overhaul and reorganise significant public institutions in finance, the military and public welfare. It also saw powerful and important restatements of the virtues and values of entrenched and allegedly traditional national systems in church and state. Fights about reform and its consequences often appealed to, and as frequently affected, the principal institutions of the sciences, including the Royal Observatory and the Royal Society, along with the emergence of newly specialist scientific organisations, clubs, societies and institutions. The Astronomical Society of London, founded in 1820, and the Political Economy Club, established the following year, exemplified this institutional and disciplinary innovation. Models of expert advice and of scientific authority were critically at issue in this period: it was not self-evident where the state should find its advisers and counsellors, or what appropriate forms of validation and experience might warrant reliance on rival groups of practitioners and experts.

These crises were of considerable significance for the conduct and ultimate fate of the Board of Longitude, one of the few long-term institutions of state that could claim to offer advice and judgement on matters of national importance in navigation and technical knowledge. The government decision, prompted from within the Admiralty, that the Board should be discontinued in 1828, might easily be interpreted as another example of reformist overhaul of public budgets and patronage. It has been common simply to read that decision as part of a general pattern of financial retrenchment in the period after the Napoleonic Wars. But this reading ignores what protagonists gave as reasons for the 1828 measure and the striking fact that the range of activities sponsored by the Board, including the administration of the *Nautical Almanac* and the regulation of naval chronometers, did not cease in 1828. Nor, indeed, did the continuing need for scientific advisers attached to the naval administration. The ambiguities of the Board's ultimate destiny in the 1820s, its status as a target for reformers' criticisms and, simultaneously, a possible avenue through which at least some state patronage could be directed, must be located firmly within the milieux of political crises and fiscal management of the reformist moment.

Furthermore, the remit of the Board was systematically connected with the Admiralty's administrative structure and post-war ambitions. As a result, the Board often became involved in decisions affecting ambitions for technique, science and economy, in the development of newfangled institutions within colonial settings and the co-ordination of such institutions around the globe. As the previous chapter demonstrated in the case of the Cape Observatory, and as is evident in several of the episodes discussed in this chapter, imperial ambitions involving long-range networks linking Britain with Asia and the Americas, dependent on secure navigation and maritime commerce, helped guide the Board's initiatives, and often provided resources for its work.

The Board of Longitude as a scientific institution, and the curious events surrounding its demise, have been neglected in accounts of the reform of science in the early decades of the nineteenth century. As well as the outfitting of the new Cape Observatory, the Regency period saw the Board's intensified participation in scientific and navigational activities including pendulum and optical experiments and the trigonometric determinations of longitude between different locations. Despite this apparent flourishing of activity, challenges were made to the Board's utility, its domination by the Royal Society and the regime of Joseph Banks, and the quality and content of editions of the *Nautical Almanac* produced after Nevil Maskelyne's death in 1811.

Central to the workings of the Board was a complex relationship between the principles of public administration and the vital significance of personal connections in webs of patronage and obligation. This was by no means unique to the Board and has been well observed by historian David Philip Miller as dominating a set of institutions within what has been called the 'Banksian learned empire'.[1] In the period after Maskelyne's death, Banks moved quickly to use the Board of Longitude to strengthen alliances between the Royal Society and the Admiralty. Continuing in the traditions of the eighteenth-century Commissioners, the Board still functioned through informal social connections. Personal acquaintances remained the most effective way to engage the Commissioners. Banks was well placed for this task. In the opening decades of the nineteenth century, the Board initially weathered the storms of the age of reform. Disparate ambitions for the reorganisation of the social and financial administration of the sciences in this period are comparable to broader Benthamite campaigns for reform in poor laws, public health and statistics in the 1830s.[2] Amid calls for transparency and accountability in commissions from the public purse, and the reconfiguration of the Board by an Act of Parliament in 1818, Banks pulled the

[1] Miller, 'The Royal Society of London', p. 10. Gascoigne, *Science in the Service*, pp. 1–33.

[2] Bentham, *The Rationale of Reward*, p. 214, stated that, '[t]hough discoveries in science may be the result of genius or accident, and though the most important discoveries may have been made by individuals without public assistance, the progress of such discoveries may at all times be materially accelerated by a proper application of public encouragement'.

Board firmly into his domain. After his death in 1820, the networks established by Banks continued, maintained by his admirer, Thomas Young.

Banksian Ambitions

The 1818 Longitude Act was the most significant transformation of the Board of Longitude since the Act that had established longitude rewards a little over a century before, fundamentally shifting the position of the Board within political, naval and scientific circles. Peace with France after 1815 provided an occasion for official reflection on the utility of the Board as it currently stood. At this earlier nineteenth-century conjuncture, the Commissioners were the personification of the various networks and connections linking individual men of science and politics. Scrutiny and criticism associated with the aims to reform and overhaul the institutions of the Regency state ensured that the Board was subject to investigations alongside other bureaucratic organisations. These challenges developed as part of an enterprise to reorganise the state apparatus, as the British government dismantled the great war machine that had allowed the Admiralty to dominate Whitehall during the fight against Napoleon.[3]

In the period after Maskelyne's death in 1811, the activity of the Commissioners was preoccupied with the continuing production of the *Nautical Almanac*. There was also a small collection of instruments under their care and an unrelenting amount of incoming correspondence containing proposals for improving navigation. The majority of this correspondence, regarded as an unfortunate consequence of an offer of financial reward, was dismissed as being of little interest. The reorganisation of the Board and its Commissioners in 1818 drew attention to its obligations regarding this correspondence and, indeed, the volume of letters retained in the Board's archives increased in the last decade of its public life. The record shows this was because of a general rise in administrative activity in the post-war period, along with reference to the Board of matters until then more often considered by Admiralty institutions.

The reorganisation of the Board was undertaken in the context of a re-examination of its administrative structure and mechanism. Admiralty Secretary John Wilson Croker steered the bill through Parliament, seconded by the Tory MP and gentleman of science Davies Gilbert, who stressed the importance of perfecting the scientific method of longitude. He argued that ongoing improvement in navigation by astronomy rather than mechanical devices justified the continued existence and expansion of the Board of Longitude. The reputation of the *Nautical Almanac*, and therefore the nation, had become 'a by-word among the literati of Europe' and needed attention and repair. Since the death in

[3] Brewer, *The Sinews of Power*, pp. 181–204; O'Brien, 'The Triumph and Denouement'; Rodger, 'From the "Military Revolution"'.

1809 of Malachy Hitchins, who had overseen much of the work of the *Nautical Almanac* computers, and the effective discharge from employment of the pre-eminent computer and comparer Mary Edwards and her family, he added, 'the publication had fallen into other hands, and was not so well conducted'.[4]

More significantly, the Board's position, utility and loyalty were being reimagined in the context of the hegemony exercised by Joseph Banks over London's scientific society. The Board had been dominated by Maskelyne and the university professoriate, but in the aftermath of Maskelyne's death Banks began once more to attend meetings (which he had not done since March 1806), recognising an opportunity to refashion the Board, along with its budgets and authority, as another pillar of his 'learned empire'.[5] Together with Admiralty secretaries Croker and Barrow, Banks used a reorganisation of the Board simultaneously to consolidate and expand scientific activity with the Admiralty.[6] Writing to his friend, the chemist Charles Blagden, Banks called the Board the 'most Compelet Twaddle I have Ever attended' and expressed his ambition that the new Board would prove 'active and usefull'.[7]

Banks's strategic ambitions for reshaping the Board of Longitude were connected to his interests in the fate of the Royal Observatory. The first move had been made when John Pond, a Banksian ally, succeeded Nevil Maskelyne as Astronomer Royal, while Banks continued to push for the enlargement of the observatory to take advantage of newly installed instruments and to maintain the high standards of the *Nautical Almanac*. More money would be required. When the Board of Ordnance refused to increase its funding, the Board of Visitors suggested that the observatory's finances should become the responsibility of the Admiralty. In 1815, Davies Gilbert advocated this move in the House of Commons and suggested an increase in funding to train the mathematicians required to facilitate the observatory work done in the service of naval, and therefore national, ambition. This move allowed Banks to bring the Royal Society and the Admiralty closer together, further strengthened by the designation of a place for one of the Secretaries of the Admiralty on the Council of the Royal Society. In March of the following year, Croker, Banks and Lord Melville, First Lord of the Admiralty, attended a quarterly meeting of the Board of Longitude.[8] Their attendance betrays an interest in the activity of the Board and a renewed interest in its potential to form another link between the Royal Society and the Admiralty, just as the observatory had done.[9]

4 Longitude Discovery Bill, 6 March 1818, *Hansard*, 37, cc. 876–880.
5 Gascoigne, *Science in the Service*, p. 23.
6 Gascoigne, 'The Royal Society', pp. 180–181; Miller, 'The Royal Society', pp. 130, 141; Jackson, *Scientific Advice*, pp. 43–48.
7 Joseph Banks to Charles Blagden, [13] March 1818, in Chambers (ed.), *Scientific Correspondence*, VI, pp. 273–274.
8 BoL, confirmed minutes, 7 March 1816, CUL RGO 14/7, p. 2:209.
9 Higgitt and Dunn, 'The Bureau and the Board', pp. 203–204.

With the success of developments at Greenwich, Banks's attention shifted to the Board and its Commissioners. The Board had been neglected by the Admiralty, with no representative sent from 1813 to 1815, but was still under its auspices. Banks recognised that reform could build another bridge between the Admiralty and the Royal Society, further securing his dominion of scientific work and society. A new Longitude Act in 1818 brought six new Commissioners to the Board to work alongside Banks, who was already entitled to attend in his capacity as president of the Royal Society: three Fellows of the Royal Society and three salaried Resident Commissioners, who were to increase the Board's breadth of knowledge and proficiency. These six Commissioners would be joined by several salaried positions, notably the previously established position of Secretary, to facilitate the administration of the Board, take minutes at the quarterly meetings and handle finances, as well as produce an annual estimate of expenditure to be placed on the ordinary estimate of the Navy. The Secretary was to be assisted by two clerks already employed by the Admiralty. In addition, two new positions were created to remove responsibilities from the seemingly over-burdened John Pond. A Superintendent of the Nautical Almanac took over the management of the system of calculators and computers established by Maskelyne to produce the tables for the *Nautical Almanac*. A Regulator (or Superintendent) of Timekeepers assumed responsibility for observing and checking the rates of chronometers kept at Greenwich and co-ordinating their assignment and issue to officers at ports across the country.[10]

Contemporary accounts of Pond praised his skill in astronomical observation but he had been simultaneously publicly berated for errors and inconsistencies in the *Nautical Almanac* and *Greenwich Observations*. In the reading of the Longitude Discovery Bill on 6 March 1818, Croker reported to the House on the recent standing of the *Nautical Almanac* 'after his [Maskelyne's] death, the reputation of that book greatly declined, and it had latterly fallen to a very low state'. Croker continued: '[T]he latter publications, however, were very incorrect, and he was sorry to be obliged to say, that the volume for the present year did not contain less than eighteen grave errors, and the publication for the next year not less than forty.'[11] The *Greenwich Observations* for 1821, published under Pond's supervision, were heavily criticised by the astronomer Stephen Lee, assistant secretary at the Royal Society since 1810. The attack was so vitriolic that a committee was formed to investigate Lee's claims. Reporting to the Council in November 1825, the committee found Pond not 'culpable' but judged him to have been negligent for allowing errors to enter the publication. The council suggested that Pond that appoint 'superior assistants' rather than his preferred 'drudges'. Pond ignored

[10] Discovery of Longitude at Sea, etc. Act 1818 (58 Geo 3 c 20), copy at CUL RGO 14/1, pp. 79r–82v.

[11] Longitude Discovery Bill, 6 March 1818, *Hansard*, 37, cc. 876–880.

the request. Lee refused to apologise for the attack, which was condemned as 'highly inproper and indecorous', and resigned his position as assistant secretary.[12] Evidence of Pond's frustration at the hindrance of his work caused by his administrative duties can be found in the archives of the Board: in early 1825, he wrote that he could not attend a committee meeting because the weather was just right for testing a new mural circle by Thomas Jones, whose scale divisions Pond deemed to be 'perfect'.[13] His neglect of the *Nautical Almanac*, along with his own admission that he was not vigilant enough as an editor, resulted in the superintendence of the publication being handed to Thomas Young in 1818 at the same time as he became Secretary to the Board of Longitude.

The Board's new appointees were all men of science and politics intimate with Banks, from either the Royal Society or parliamentary committees, most often both. The three Fellows of the Royal Society were all from Banks's intimate circle. Two were noted politicians. Charles Abbot, Lord Colchester, the half-brother of Jeremy and Samuel Bentham and recently retired Speaker of the House of Commons, was a popular political figure. His addition and that of Davies Gilbert added political clout to the Board of Longitude. Any perceived absence of scientific knowledge in such Fellows of the Royal Society joining the Board of Longitude was somewhat diminished by the presence of Colonel William Mudge, who as indicated in the previous chapter became a Resident Commissioner to replace Thomas Young on the latter's appointment as Superintendent of the Nautical Almanac. During time serving in the Royal Artillery at the Tower of London, Mudge had entertained himself with the construction of clockwork and was tutored in mathematics by Charles Hutton. In 1791, Mudge was appointed to the Ordnance Survey. In 1798, he was promoted to its directorship, and in 1809 he was appointed as lieutenant-governor of the Royal Military Academy at Woolwich, where he encouraged the practical training of cadets in surveying and topographical drawing.

The positions of salaried Resident Commissioners were also filled by those close to Banks, all Fellows of the Royal Society as well. Chemist William Hyde Wollaston offered knowledge missing from the current Commissioners. He had been secretary of the Royal Society since 1804 and was prominent on both the parliamentary and Royal Society committees concerned with the standardisation of weights and measures. The second Resident Commissioner, Captain Henry Kater, had also been involved with these committees, developing a gravimetric pendulum and encouraging refinement of geodetic measurement in the pursuit of standards for weights and measures. During his military career in the Royal Engineers, Kater assisted William Lambton while stationed

[12] Council of the Royal Society, minutes, 3 March, 5 May, 16 June, 17 and 24 November, 15 December 1825, RS CMO/10, pp. 102, 114–115, 185–199, 206–208.

[13] John Pond to BoL, 29 January 1825, CUL RGO 14/48, p. 29v.

in India, surveying a region of Madras. With peace in 1815, he had left military service on half pay and with Banks's sponsorship his focus moved to his scientific interests and the Royal Society, to which he was elected the same year.

Thomas Young was the third Resident Commissioner, while also taking on the role of Secretary of the Board and responsibility for the *Nautical Almanac*. Previous secretaries of the Board John Ibbetson (1765–1782) and Sir Harry Parker (1782–1795) were both clerks under the Secretaries of the Admiralty, emphasising the link to the Board of Longitude's parent department of state. The incumbent Secretary at the time of Banks's coup was Thomas Hurd, who had succeeded Alexander Dalrymple as Hydrographer of the Navy in 1808. Hurd had enjoyed an illustrious career from 1768, amassing substantial personal wealth in prize money during wars against France and America. Positioning Young so centrally within the activities of the Board was a reward for his loyalty to Banks as well as for his demonstrated ability as a mathematician, which would allow him to edit and interpret the tables required for the *Nautical Almanac*. Young's efficiency as an administrator and adviser had been demonstrated by his previous work consulting on proposed changes to warship construction and his secretarial work for the Royal Society and parliamentary committees for weights and measures.

The presence of Resident Commissioners chosen by Banks and three Fellows of the Royal Society allowed Banks to outweigh the influence of the university professors that had dominated the Board for much of its recent past. After Banks's death in 1820, Young continued to maintain the Board as a key component of the establishment linking the sciences with state apparatus. In this new position, Young joined a small group of individuals who sat between politics and science in the turbulent era of the 1820s, when the complex network that joined various committees stretching across the Royal Society, Board of Longitude and Admiralty was almost exclusively governed by Gilbert, Croker, Barrow and Young.

With this drastic change in personnel came a new remit. Prior to the 1818 Act, the Board's activity was mostly absorbed in judging applications for rewards for general improvements in navigation. This tradition of engagement survived the overhaul of 1818 and was expanded. New and old Commissioners alike were asked to consider and suggest potential subjects or problems to Young that would be motivated by the offer of financial recompense. Davies Gilbert suggested that a reward of £10,000 be offered for new solar and lunar tables and another £10,000 for improved chronometers, a suggestion more akin to the rewards available under the Act of 1774 than the more permissive wording of the Act of 1818 regarding possible rewards.[14] Henry Kater offered a list of desirable improvements to various instruments. He made suggestions for further perfecting the chronometer, and rewards for any effort to facilitate 'the improvement of Tables for calculating the place of the moon,

[14] Davies Gilbert to Thomas Young, 17 December 1818, CUL RGO 14/1, pp. 98v–99r.

improvements in dividing circular instruments and rendering the sextant free from central error, making the reflecting circle lighter and a machine or marine chair to enable the observer to take Lunar Distances more accurately'.[15]

Alongside the continuing desire to improve and maintain the instrumental hardware propping up British imperial ambitions in peacetime, John Barrow's interest in improving surveying and cartographic work also came to occupy the attention of the Commissioners. Legislation had been passed in 1745 for two rewards for a passable northern route between the Atlantic and the Pacific and for the approach to the North Pole.[16] Two expeditions were already being organised as the 1818 Act brought the rewards within the remit of the Commissioners of Longitude. The move was an effective demonstration of the new position and significance of the Board of Longitude. Navy-sponsored expeditions, for which the Royal Society provided advice and instruments, instructions and personnel, would now hand their logbooks to Thomas Young upon their return.[17]

Instruments and Voyages, Experiments and Personnel

This plethora of activity and the appointment as Commissioners of several prominent men of science, society and politics, drew attention to the newly restructured Board of Longitude. Its archive expands correspondingly in this era. Young was meticulous in keeping and filing incoming correspondence, which may have been lost or neglected by previous secretaries. The reform took place at a time of increasing scrutiny of public resources and spending. Young was well aware of the responsibility of holding a public budget in a period shaped by deep financial retrenchment, resulting from the expense of warfare on a previously unprecedented scale with France and the United States. He became responsible for receipts and reimbursements, and all the Board's finances from this point onwards passed through an account at Coutts held in Young's name. The drafts and vouchers of the Board are, without exception, signed off by him.[18] The assignment of public money for its new activities perhaps led to some expectation that the activity and governance of the Board of Longitude would now be somehow more open, providing an opportunity for men of science to encourage the state to utilise or sponsor their work financially or fund the provision of instruments. In reality, the Board in this final incarnation was tied firmly to the Admiralty. After years of rarely sending a representative to attend meetings, the 1818 Act imposed a clear hierarchy dictating that a quorum had to include a Navy representative; this was almost exclusively one of the Admiralty secretaries. The Commissioners,

[15] Henry Kater to Thomas Young, 1818–1819, CUL RGO 14/1, p. 105r.
[16] Discovery of North-West Passage Act 1744 (18 Geo 2 c 17).
[17] Higgitt and Dunn, 'The Bureau and the Board', p. 206.
[18] Stubs from Thomas Young's cheques, 1819–1823, CUL RGO 14/18, pp. 487r–580r.

Figure 11.1 'H. M. Ships Hecla & Griper in Winter Harbour', from William Edward Parry, *Journal of a Voyage for the Discovery of a North-West Passage from the Atlantic to the Pacific; Performed in the Years 1819–20* (London, 1821). Nasjonalbiblioteket/National Library of Norway

who had enjoyed relative autonomy before 1818, could no longer act without Admiralty permission. For activity to be sponsored by them, the utility to Navy and nation needed to be made clearly apparent.

The 1818 Act laid out clear parameters for claiming the Board's new Arctic rewards. For navigating the Northwest Passage, it promised rewards on a sliding scale from £20,000 for reaching the Pacific to £5000 for reaching 110 degrees west or 89 degrees north and £1000 for reaching 83 degrees north. There was only one award, £5000 for reaching 113 degrees west, given to Lieutenant William Edward Parry and the crews of the *Hecla* and *Griper* (Figure 11.1). In an extraordinary meeting of the Board in late November 1820, Lieutenants Parry, Matthew Liddon and Henry Parkins Hoppner and Captain Edward Sabine were interviewed by Young to validate and compare their claim for a reward of £5000 (Figure 11.2).[19] This proved to be the only application for the Arctic rewards.

[19] BoL, confirmed minutes, 27 November 1820, CUL RGO 14/7, pp. 2:321–2:330.

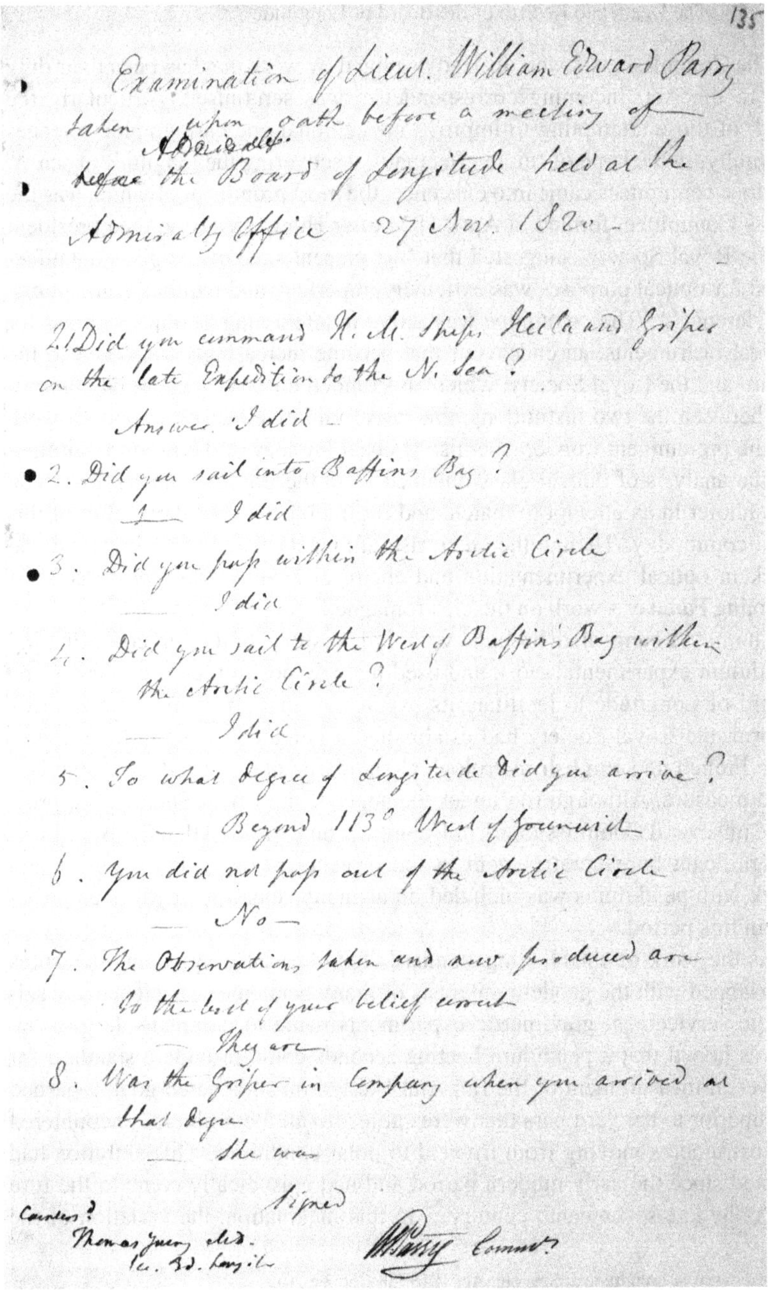

135

Examination of Lieut. William Edward Parry taken upon oath before a meeting of the Board of Longitude held at the Admiralty Office 27 Nov. 1820.

1. Did you command H. M. Ships Hecla and Griper on the late Expedition to the N. Sea?

Answer. I did —

2. Did you sail into Baffins Bay?

— I did

3. Did you pass within the Arctic Circle

— I did

4. Did you sail to the West of Baffin's Bay within the Arctic Circle?

— I did

5. To what degree of Longitude did you arrive?

— Beyond 113° West of Greenwich

6. You did not pass out of the Arctic Circle

— No —

7. The Observations taken and now produced are to the best of your belief correct

— They are —

8. Was the Griper in Company when you arrived at that degree

— She was

Signed

W E Parry Commr.

Countersd?
Thomas Young Sec.
Bd. Longitude

Figure 11.2 Questions put to William Parry by the Board of Longitude, 27 November 1820, CUL RGO 14/1, p. 135r. Reproduced by kind permission of the Syndics of Cambridge University Library

The Commissioners were instead occupied by work that had been their duty before the Act. Incoming correspondence was scrutinised, particularly the work of those attempting to improve navigational and cartographic sciences generally, in the hope of 'more effectually discovering the longitude at sea'.[20] Various committees came into existence, the most prominent of which was the Glass Committee, formed in April 1824 after Humphry Davy, then president of the Royal Society, suggested that 'the present state of the glass manufactured for optical purposes was extremely imperfect, and required some public interference'.[21] The committee was active in attempting to improve glass for optical instruments, an endeavour that became increasingly important to the Board and the Royal Society, which also funded the work. As a joint committee between the two institutions, this novel initiative backed important work by the pre-eminent London chemist Michael Faraday at the Royal Institution on the analysis of optical glass obtained from the Bavarian maker Josef von Fraunhofer in an attempt to match, and if possible surpass, the quality of this rival commodity. Though the enterprise ultimately failed, it involved critical work in optical experimentation and chemical analysis, and was crucial in enabling Faraday's work on the electromagnetic properties of light.[22]

Thomas Young also brought with him responsibility for co-ordinating pendulum experimental work and used his new position as Secretary of the Board of Longitude to facilitate its expansion. In 1816, before the Board's reform, the Royal Society had established a Pendulum Committee to compare French and English metrological standards and to produce a standard yard measure. Although run under the aegis of the Royal Society, the presence of several Commissioners of Longitude on it ensured that the Board was a significant bureaucratic agent in the project, meaning that experimental work with pendulums was included on as many scientific voyages as possible in this period.

As the work of establishing standards for imperial weights and measures overlapped with the geodetic interests of many gentlemen of science and scientific servicemen, gravimetric experiments came to dominate the project. It was hoped that a pendulum beating seconds could provide a standard for universal measurement of the Imperial Yard, a much-coveted goal, regarded as superior to the yard bars that were subject to all the problems encountered by instruments moving from tropical to polar conditions. This ambition had existed since the early modern period and had episodically come to the fore since the late seventeenth century.[23] In this incarnation, the variation of the

[20] Discovery of Longitude at Sea, etc. Act 1818 (58 Geo 3 c 20).
[21] BoL, confirmed minutes, 1 April 1824, CUL RGO 14/8, p. 20.
[22] Accounts of the Glass Committee, 1826–1829, CUL RGO 14/16, pp. 474r–497r; James, 'Michael Faraday's Work'; Jackson. *Spectrum of Belief*, pp. 145–162.
[23] Dew, '*Vers la ligne*', pp. 58–60; Dew, 'Scientific Travel', pp. 1–3.

length of a seconds pendulum corresponding to variation in the Earth's gravity was well understood. Any investigation into the possibility of the pendulum as a length standard necessitated investigation of global gravity variation. In this capacity, the Board of Longitude began to co-ordinate an experimental research programme in the physical sciences at least as large as the magnetic crusades later in the nineteenth century. Under the somewhat controversial management of Edward Sabine and the ingenious enterprise of Henry Kater's apparatus design, pendulum projects were conducted in many maritime and colonial settings, including bases in the south and north Atlantic, on the coasts of South America, in South Asia and the Indian Ocean, in Australasia and in the polar regions. The Board functioned as one of several public institutions involved in regulating, funding and co-ordinating these measures. The aim was to use a pendulum of standard and allegedly invariable length, and to determine how its period changed at different sites on the planet. The Board once again became involved in the thorny issue of standardising instrumental design and function. Furthermore, instruments were provisioned from a complex entanglement of public and private property, with personnel recruited through the informal social networks that linked the Royal Society, the Admiralty and the Board of Longitude. Financial backing for the project came exclusively from public money in a time of extraordinary austerity. The sums provided by the Treasury, the number of people involved, both in performing experimental work and in administration, and the number of committees, reports and, eventually, legislation produced by the Royal Society and Parliament make pendulum work very visible in the archives of both institutions and individuals. At least £5000 was spent on the project by the Board alone during the 1820s (see Appendix 2).[24]

The scale and budgets for the pendulum enterprise also made it visible to the public. Outside the sphere of scientific publication, voyage narratives, parliamentary debates and reports, summaries and opinion pieces in reviews, magazines and pamphlets all discussed the progression of pendulum work. Pendulum research was also debated in the more esoteric realms of private correspondence and committee minutes. Particularly visible is the sudden change in numbers of relevant publications in the *Philosophical Transactions*. Between 1800 and 1817, one paper was published on the problem of determining the figure of the Earth, while twenty-three appeared in the post-war period, between 1817 and 1832. This excitement was followed by another fall in output. No papers on pendulum work appeared after 1832 for the rest of that decade, although work continued to be published privately and in the *Memoirs* and *Monthly Notices of the Royal Astronomical Society*. This sudden increase was the direct result of state concern with standardisation. In response to the

[24] Howse, 'Britain's Board of Longitude'.

requirements of increasing trade and fiscal demands, metrological regulation was energetically pursued.[25]

The work of determining the distance between Dover and Calais in the early 1820s was undertaken though an alliance between the French and British longitude boards and the Board of Ordnance. A collaborative project, its outcome proved somewhat unsatisfactory in terms of the precision of the triangulation and the co-ordination of measures between British and French systems.[26] However, from the perspective of the Board, such a project had several potentially positive outcomes: 'the general improvement of geography, astronomy and navigation', the 'important step towards making the maps of Europe more perfect' and the promotion of 'harmony and liberal intercourse' between the two nations were all cited in the proposal drafted by Young on behalf of the Board of Longitude. The proposal also suggested that the project would cost the 'trifling' amount of less than £500, if it received the 'sanction of the British Government', as 'most of the instruments necessary' were already owned by the Royal Society and 'there can be no doubt that the English Men of Science would volunteer their services on the occasion'.[27] This activity and the experimental pendulum work performed at this time are examples of the entanglement of public and private interest, ambition, resources and personnel that was characteristic of partially state-funded scientific work in this period. Instruments belonging to the Royal Society, the Board of Longitude and individual officers and gentlemen moved around the globe as part of a complex network of private and public funds and interests.

As set out in Chapter 10, the work of establishing and maintaining an Observatory at the Cape of Good Hope was the administrative responsibility of the Board of Longitude during the 1820s. The founding of the Cape Observatory provides some insight into the inefficiencies and abuses of poorly organised bureaucracy. At the start of the Cape project, Barrow had to remind Young to write him a letter 'as Secretary of the Board of Longitude, recommending the Establishment', as well as estimates of appropriate salaries for an astronomer, an assistant and a labourer. This letter was required so that Barrow could transmit the proposal to Lord Bathurst, who in 1820 was Secretary of State for War and Colonies, for his 'sanction'.[28] By the time Young wrote this and several other letters, the decision to create a Cape Observatory had already been taken. The material preserved in the Board's archive is inevitably misleading, therefore, since it testifies to the production of an official and institutional memory rather than to the complexities of personal negotiation characteristic

[25] Miller, 'The Royal Society', p. 179; Ashworth, *Customs and Excise*, p. 281.

[26] Higgitt and Dunn, 'The Bureau and the Board', pp. 209–211.

[27] Thomas Young, draft proposal for the determination of the distance between Dover and Calais, 10 July 1821, CUL RGO 14/49, pp. 200r–200v.

[28] Thomas Young to John Barrow, 14 July 1820, CUL RGO 14/48, p. 35r.

of Regency administrative cultures. The complete absence of debate or negotiation in the Board's minutes suggests that all arrangements for establishing the new observatory had been negotiated and mapped out well before Barrow or Young committed to regular and official action, such as appealing to the Secretary for the Colonies, the Treasury or the King's Council.[29] The Cape Observatory was established using informal mechanisms of sociability concurrently employed by Young to control and cultivate pendulum investigation and to secure provisions for voyages attempting the Northwest Passage and performing strategic cartographic work for the Admiralty.

The Board had been successfully positioned as the Admiralty's 'scientific branch' alongside the Hydrographic Office and Naval College. Barrow's ambition was for the Board to function as a scientific council for the Admiralty and he increased his budget to facilitate this. Scientific spending becomes more visible as it starts to appear in the naval estimates. The Board's budget expanded rapidly from 1818 to 1821, alongside expenditure to establish the Cape Observatory. Board accounts suggest annual average expenditure for 1797–1828 to be around £2300. In reality the Board spent more between 1818 and 1828 than in the two decades from 1797. The annual average in the 1820s was just over £4990, which was more than double that of the previous six decades, which saw spending of around £2324 per decade.[30] After 1820, the Board's annual budget is recorded in the naval estimates as being anything up to £5000, with additional sums for one-off rewards and the Cape Observatory.

Despite this flourishing of activity in the 1820s, the Board of Longitude in the wake of the 1818 Act must not be understood as a proto-professional body of experts but rather as a collection of scientific and naval men with vested interests and personal opinions. It was a Board less than the sum of its parts, riddled with unspoken conflict; particularly with regard to Thomas Young's superintendence of the *Nautical Almanac* and the Board's right to make demands of the Admiralty. Upon joining the Board, the younger Cambridge-trained mathematicians John Herschel and George Airy had both been optimistic about its influence on the Admiralty, particularly with regard to the *Nautical Almanac*. But Herschel's correspondence with Young revealed the reality of the situation, as Young reminded him repeatedly that balance of power lay with those acting for the Admiralty rather than with those representing the scientific arts. Young's letters and the Board meetings revealed to Herschel with whom the power actually lay: Commissioners of Longitude could not use the Board as a way to negotiate significant funds, instruments, changes to the

[29] Copy of a petition to the king concerning the establishment of the Cape Observatory, 20 October 1820, CUL RGO 14/4, pp. 44r–46v; BoL, confirmed minutes, 3 February 1820, CUL RGO 14/7, p. 297.

[30] Howse, 'Britain's Board of Longitude'; Howse, 'The Board of Longitude Accounts'.

Figure 11.3 John Wilson Croker as active and enraged First Secretary of the Admiralty, while other Admiralty officials sleep. George Cruikshank, *The Merchants Memorial to Alley Croker*, September 1814 (detail). Loyola University Chicago Archives and Special Collections

Nautical Almanac or further autonomy.[31] Many decades later, Airy reminisced about the Board's conduct: 'usually the political secretary, Mr Croker, presided. It was always understood that the Board could not meet except under the presidency of one of the Lords of Admiralty, and thus the Board of Longitude became, in fact, a Committee of the Admiralty' (Figure 11.3).[32]

Significantly the 1818 Longitude Act had moved the care of the *Nautical Almanac* from Greenwich and the Astronomer Royal into the hands of an already select and somewhat privileged individual, Thomas Young. This move saw the superintendence move from a public and more transparent space into a private home on Welbeck Street, in London's medical district. Young's editorial choices for the *Nautical Almanac* were heavily criticised from both outside

[31] Thomas Young to John Herschel, 7 November 1825, RS HS/18/334; Herschel to Young, 8 November 1825, RS HS/18/335; Young to Herschel, 15 November 1825, RS HS/18/336; Young to Herschel, n.d. [May 1828 or earlier], RS HS/18/357.

[32] Airy, 'Remarks on the History', p. 31.

and within the Board of Longitude, as he focused on preserving content that prioritised its use by the navigator rather than the astronomer, as explained in Chapter 7. Individuals such as James South and Francis Baily became increasingly irate as the Astronomical Society was consistently unable to intervene or influence the content and organisation of the publication, which they regarded as a significant opportunity to demonstrate the utility of the specialised knowledge and methodology cultivated by certain members of the Society, particularly South and Baily as well as Charles Babbage, John Herschel and the veteran astronomer and antiquarian Henry Thomas Colebrooke.[33]

1828: The Admiralty in Command

The supposed end of the Board of Longitude in 1828 was apparently a rather unremarkable occurrence at a poorly attended parliamentary session. On the third reading of a bill written predominantly by Croker, Davies Gilbert rose to 'bear testimony to the uniformly creditable conduct of the Board of Longitude', yet he did not plan to oppose the bill, 'as it had been brought into the house on the recommendation of the finance committee'. The Commons report in *The Times*, after summarising Gilbert's speech, states that 'the measure was adopted in compliance with a due regard for economy'. This is where many historical investigations have ended; the Board is thus seen as a victim of post-war retrenchment. One of the other Commons speeches about the bill was from the political economist and Fellow of the Royal Society Henry Warburton, who hoped that the 'Board of Admiralty would, in future, have recourse to the Royal Society for the purpose of receiving their assistance and beneficial co-operation in nautical and scientific discoveries'.[34] Perhaps unbeknown to Warburton when he made this speech, the Admiralty had already prepared for life after the Board of Longitude by attempting to consolidate three of its Commissioners into an internal consultative committee. John Barrow offered positions on the Resident Committee of Scientific Advice to Thomas Young, John Herschel and Henry Kater. Only Young accepted the new role and afterwards arranged for Edward Sabine and Michael Faraday to join him. The Admiralty reduced the Board of Longitude to an internal Resident Committee to create a new model of interaction between the Admiralty Board and the select men of science it cared to endorse.

The changes brought about by this reorganisation were specific and somewhat undramatic. The power-grab by Barrow and Croker was done to ensure a change in the public visibility of the regime of scientific advice by its sequestration more firmly within the Admiralty and its consequent removal from

[33] Ashworth, 'The Calculating Eye', p. 415.
[34] 'House of Commons, Friday, July 4', *The Times*, 5 July 1828, p. 5.

public controversy. First, by shutting down the public aspect of the Board of Longitude, letters regarding impracticable schemes for improvements in navigation and instruments would in principle cease to arrive, or could be redirected to flow to the Admiralty itself, and indeed to the Astronomer Royal. Second, those who had criticised the Board and its actions, as well as Young's superintendence of the *Nautical Almanac*, would not be able to see their target so clearly, as the Resident Committee of Scientific Advice was internal to the Navy. Without a public face, the new committee could not be so easily attacked for wasting public money. The Admiralty adapted its relations with men of science in order to rebalance power closer to itself and the Royal Society, continuing with Young as its central consultant and Superintendent of the Nautical Almanac. The day the bill was read in July 1828, Croker had the last words, and used them to emphasise this coup, stating that the bill was 'not introduced with a view merely to economy … It was long since ascertained, that the Board of Longitude was of no use whatever. It met only four times a-year, and was then wholly occupied in reading the wild ravings of madmen'.[35]

Croker's words demonstrate the way in which the Admiralty had come to regard the Board of Longitude: a failed model of interaction between science and state interests. Unable to liaise efficiently with certain individuals and attracting attention from both those who were convinced they had designed perpetual motion machines or squared the circle and from others with a more reformist agenda, Croker and Barrow thought to try something new, an internal consultative committee. As in 1818, when Croker had moved to redress the balance of views on the Board, there was a repetition of the principle of reduction to a three-man committee, in an echo of the 1818 Resident Commissioners. By removing the Board's public face, Croker hoped to gain more from interactions with scientific men than the filtering of what he saw as useless correspondence, while simultaneously removing the public target for criticism of Admiralty interventions in the scientific world. The men of science who made up the Resident Committee would only be at liberty to advise exclusively on what the Admiralty Board wanted them to, but would be more protected from public scrutiny. Furthermore, Croker's ambition in attacking the Board in the public forum of the House of Commons was to send a message to those who had criticised it. The Admiralty was under no obligation to fund science and would no longer continue to do so in a way that had proved, in its opinion, so ineffectual and inefficient. Public commentary demonstrates that the establishment of a new model for state–science interaction was to the advantage of the Admiralty's power over scientific men rather than to its budget. A letter to *The Times* entitled 'New Board of Longitude' commented

[35] Ibid. Compare Jackson, *Scientific Advice*, pp. 51–53.

satirically on the benefit to the Admiralty that was created by the passing of the Repeal Bill 1828:

> But mark how this job has been turned into a snug little bit of patronage for the Admiralty … So that the important business of that highly scientific and responsible society (the late Board of Longitude) is in future to be intrusted to a fleeting and uncertain body of men – the chance-medley of the Council of the Royal Society. To advise with the Admiralty! We shall watch the progress of this new and curious machine; unless it be (as we anticipated) blown to atoms by some explosion in the House of Commons.[36]

Croker and Barrow would now only interact with those men of science they had specifically chosen. Without the space for appeal the Board provided, power over patronage was no longer negotiable and lay firmly with the Admiralty again. Further to this, there was no money saved by the closure of the Board but, rather, a slight increase in cost for the running of the Resident Committee of Scientific Advice. Croker had suggested the Board as a money-saving opportunity to the Finance Committee with a different agenda in mind entirely; concerned with the retention of power for the Admiralty, the Board had become a dangerous forum with reformist sympathy now present with Herschel and Airy. When the Navy estimates were examined in close detail by the House of Commons, Sir George Clerk was forced to defend the lack of savings and to justify the fact that the transformation of the Board into a private advisory committee had not saved the public any money at all; 'there was no diminution, but some increase in expense'.[37]

A New Board of Longitude

Henry Kater and John Herschel both declined positions on the Resident Committee that replaced the Board of Longitude. Herschel was outraged at the treatment of the Commissioners, while Kater was equally aggravated. Herschel wrote to Kater after the Board's abolition asking if he had information on the fate of its various activities.[38] Kater replied the following day, agreeing that the current situation was not good for state recognition of science and expressed the hope that the Royal Society would not take on the Board's work without being well paid by government.[39]

The motivation behind the invitation to Herschel to join the committee is worthy of consideration. Traditionally understood to be an advocate of reformist ambitions for science in the 1820s, Herschel was yet invited to join the

[36] A Correspondent, 'New Board of Longitude', *The Times*, 24 November 1828, p. 3.
[37] 'House of Commons, Friday, Feb. 27', *The Times*, 28 February 1829, p. 2.
[38] John Herschel to Henry Kater, 17 July 1829, RS HS/21/19.
[39] Henry Kater to John Herschel, 18 July 1829, RS HS/11/10.

Resident Committee and had been a Commissioner of Longitude beforehand. Croker and Barrow were perhaps interested in the utilisation of both scientific knowledge and scientific celebrity. If Herschel, already one of the Resident Commissioners of Longitude, could be brought into the new Resident Committee, it would benefit from his endorsement.

Upon Herschel and Kater's refusal to be brought into the Admiralty fold, Young was asked to recommend replacements. In the absence of Kater's surveying and pendulum knowledge, he suggested the recruitment of Sabine, now settled in London after a decade of voyaging. Young also suggested Michael Faraday, currently enjoying fame among men of science for his research at the Royal Institution in chemistry and the phenomena of electromagnetism, although, for the Admiralty, his work for the Glass Committee was of more use and interest.[40] This choice of a chemist over an astronomer is a significant moment, representing a shift in what the Admiralty judged as relevant specialist knowledge. Young's ability to edit and maintain the *Nautical Almanac* apparently offered sufficient astronomical knowledge; astronomy's application to the determination of longitude was shut down with the closure of the original Board of Longitude.

For Young and the Admiralty secretaries, it was perhaps not appropriate for disinterested scientific inquiry to be funded from the public purse. It was certainly of importance that Faraday, from the Royal Institution, had already begun his long association with the joint Board of Longitude–Royal Society glass project: optical improvements promised much for the development of navigational instrumentation. Significant, too, was the resulting absence of a specialist astronomer from the committee's membership. This absence or the implication that Young would suffice, particularly given his explicit endorsement of the status of the generalist within the realms of the sciences, which contrasted so sharply with increasing public emphasis on the virtues of highly specialist knowledge, ensured that the Resident Committee was criticised at least as much as the Board of Longitude.[41]

Letters to Young, Faraday and Sabine invited them to join the Resident Committee appointed to advise the Admiralty Board 'on all questions of discoveries, inventions, calculations and other scientific subjects'.[42] Their position was made explicit. The King's Council authorised 'the continuance of the Resident Committee to advise with their Lordships on all scientific subjects in the same manner and with the same duties and salaries as the Resident Committee of the late Board of Longitude'.[43] The activities of the Board of Longitude did not,

[40] James, 'Michael Faraday's Work'. [41] Jackson, *Spectrum of Belief*, pp. 158–161.

[42] John Barrow to Thomas Young, Edward Sabine and Michael Faraday, 7 January 1829, CUL RGO 14/1, pp. 240r–240v; copy at TNA ADM 1/3469.

[43] John Barrow to Thomas Young, 30 December 1828, CUL RGO 14/1, p. 242r.

therefore, cease with its supposed demise. The *Nautical Almanac* continued to be computed and compared under the superintendence of Thomas Young. Another essential duty was the Board's maintenance of the 'system of sending out chronometers'. In practice, this had passed to the staff of the Royal Observatory in 1821, although the Commissioners retained some interest in its correct operation. Mentioned during the third reading of the Longitude Repeal Act in Parliament, Croker stated that 'the Board of Admiralty had much more connexion with chronometers than the Board of Longitude'.[44] Indeed, the extensive correspondence about the care and issue of naval chronometers are retained in the papers of John Pond and are, for this period, exclusively between Croker, Barrow, Beaufort, Pond and his assistant Thomas Taylor.[45]

With the transition away from public scrutiny of the superintendence of the *Nautical Almanac* and the advisory role of the Resident Commissioners, the absorption of the Admiralty's scientific needs within its own various departments was complete. In this transition the Resident Committee became more formally a part of a public institution, yet simultaneously less publicly responsible, as it gained a large amount of discretionary power over maritime astronomical and navigational sciences. This small committee was, as the Board of Longitude had been before it, an important example of how the Admiralty as an organ of the military-fiscal state could aid in the construction of community in metropolitan scientific circles. Young noted the need for the new advisers to at least have the appearance of unity, just as the Board of Longitude had done before it under his guidance, in order to ensure that its advice was not ignored. 'It would,' he cautioned John Herschel, 'have a bad effect for any one member of the committee to address the Admiralty separately, as it would imply a want of unanimity that would greatly tend to weaken our hand.'[46]

Conclusion

Historians have tended to regard the Board of Longitude as a victim of financial retrenchment and reformist ambition. This oversimplified reading masks a more complex narrative in which the interests of a multitude of individuals were at play both in the 1818 reconstruction of the Board and in its dismantling a decade later. The variety of activities the Commissioners of Longitude choreographed and facilitated demonstrates the significant truth that the Board

[44] 'House of Commons, Friday, July 4', *The Times*, 5 July 1828, p. 5.

[45] For example, see John Barrow to John Pond, 17 December 1824, CUL RGO 5/229, p. 230; John Jones Dyer to Thomas Taylor, 4 January 1825, CUL RGO 5/229, p. 233; John Croker to John Pond, 22 January 1825, CUL RGO 5/229, p. 237; John Barrow to John Pond, 27 January 1825, CUL RGO 5/229, p. 238. See also Ishibashi, 'A Place for Managing Government Chronometers'.

[46] Thomas Young to John Herschel, n.d. [May 1828 or earlier], RS HS/18/357.

was a useful and important institution well after the end of its interactions with John Harrison. The Board should be central to discussions of state–science interaction in the Regency era and the fate of science in reformist Britain.

Within a decade of the reforms brought about by the Act of 1818, it had become apparent that a model for state–science interaction that had functioned with relative efficiency in the eighteenth century was no longer adequate in the turbulent context of the 1820s. The Board of Longitude was reformed once more and its duties continued by the Resident Committee that replaced it. This new consultative committee occupied a space deep within the Admiralty. It was a move motivated by the Admiralty's desire to remove the Board from public scrutiny and prevent further breaches by reformists such as Herschel and Airy, who became Commissioners in 1821 and 1826, respectively. Absorption also offered protection for Thomas Young from the increasing criticism surrounding the content and errata of the *Nautical Almanac* and provided him with a discreet space to continue his work as Superintendent. The urgent issues of patronage and judgement, of technical expertise and public authority, which had characterised the eleven decades since the 1714 Longitude Act, continued until and well beyond the official end of the Board of Longitude.

Epilogue

The fate of the Board of Longitude raises intriguing questions about the function and reputation in Georgian Britain of this kind of state organisation, devoted to a range of tasks of utility and administration, of patronage and of judgement, at a period of major political, economic and social change. This book charts many striking features of the Board's century-long career and in its closing chapters devotes attention to the contingencies of its disappearance in 1828. The *Nautical Almanac* is still published, although it and its use are much transformed. Fundamental issues of navigation and of position-finding evidently remain of national, political and military significance.[1] Equivalent institutions to the Board elsewhere, such as the French Bureau des Longitudes, survived much more dramatic regime changes.[2] Had it not been, perhaps, for the peculiar circumstances of Georgian scientific culture and the discretionary character of British government patronage, it is possible to imagine that the Board, reformed as it was in 1818, might have survived for many years, to be overhauled once again to suit new political and administrative circumstances. In the event, however, its various functions were shifted to a number of different institutional settings. The Resident Committee of Scientific Advice, established on the suppression of the Board in 1828, did not itself last long. That committee, and Thomas Young's superintendence of the *Nautical Almanac*, continued to be attacked by the Astronomical Society reformers, who took their complaints directly to the Prime Minister and the Chancellor of the Exchequer, recommending the foundation of a new and improved Board of Longitude. The insurgents' recommendation failed. It was, rather, Young's death in May 1829 and the lack of energetic leadership to replace his combination of networks and expertise that prompted further change.[3]

[1] On the *Nautical Almanac*, see Sadler, 'Bicentenary of the Nautical Almanac'; Seidelmann and Hohenkerk (eds), *History of Celestial Navigation*. For recent developments in navigation, see Ceruzzi, *GPS*; Czaplewski and Goward, 'Global Navigation Satellite Systems'. For the broader background, see Roland, 'Science, Technology, and War'.

[2] On the Bureau des Longitudes, see Schiavon, 'Bureau des Longitudes'; Schiavon and Rollet (eds), *Le Bureau des longitudes*; Schiavon and Rollet (eds), *Pour une histoire*.

[3] Dreyer, 'Decade 1830–1840', pp. 58–63; Miller, 'Royal Society', pp. 317–326. On the work and criticism of the Residential Committee and Young's death, see Waring, 'Thomas Young', pp. 183–217.

The superintendence of the *Nautical Almanac* devolved back to John Pond, as Astronomer Royal, but he was no more suited to this task than under earlier circumstances. He was forced to ask for help from individuals associated with the Astronomical Society. Ultimately there was little option for the Admiralty but to turn to that society for advice, implementing its suggestions and appointing its secretary, Lieutenant W. S. Stratford, as Superintendent. Now welcomed as advisers into the extended sphere of the British government, five representatives of the Society were also to be included on the Royal Observatory's Board of Visitors, alongside five fellows of the Royal Society and two astronomy professors. This change was a formal result of the requirement for a new royal warrant on the accession of William IV in 1830, but gave opportunity to the Admiralty and astronomical reformers alike to push for changes to the Observatory's output and, in 1835, to demand the resignation of Pond and help back the installation of George Airy as Astronomer Royal.[4] The reforming astronomers, and Astronomical Society (from 1831, after receiving its charter, the Royal Astronomical Society), had found routes to influence the Admiralty, and some parity of status with the Royal Society. The Admiralty benefitted from both connections but, as had been attempted with the abortive Advisory Committee, also continued to expand its in-house expertise.

In the 1820s, the Admiralty's expenditure on astronomy and navigation (including the expenses of the Royal Observatories at Greenwich and the Cape) had been grouped under the heading of the Board of Longitude.[5] In the following decade, after the Board's abolition, this instead appeared under the heading 'Scientific Branch'. Founded in 1831, the Scientific Branch budget was placed under the care of Francis Beaufort, who had been Hydrographer to the Navy since 1829.[6] Higher naval education was also brought under this heading, as were sums for rewards, experiments and extra pay. Altogether, nearly £20,163 was voted for by Parliament for the Scientific Branch for the year 1831–1832, a sum that steadily increased thereafter.[7] In 1857–1858 it was £63,091, divided among (in decreasing order of cost) the Hydrographic Office, Royal Observatory, Nautical Almanac, the Establishment for Scientific Education at the Royal Naval College at Portsmouth, Miscellaneous (including Rewards and Experiments and the Meteorological Office), Royal Observatory Cape of Good Hope, Chronometers, Compass Department and the Libraries and Museums of Haslar and Plymouth Hospitals. In terms of seniority, if salary

[4] Laurie, 'Board of Visitors', p. 334.

[5] See e.g. House of Commons, 'Estimates of Ordinary of the Navy', p. 6.

[6] Reidy, *Tides of History*, p. 140.

[7] House of Commons, 'Navy Estimate no. II', pp. 9–10. For this year, the Royal Naval College and School for Naval Architecture were included under this head but were shortly to be closed. Naval education likewise underwent much change in the 1830s; see Dickinson, *Educating the Navy*.

is a fair marker, the most significant scientific employees were the Astronomer Royal at Greenwich (£1000 p.a.), the Hydrographer (£800 p.a.), the Engineer of the Harbour Branch of the Hydrographic Office (£800 p.a.), Her Majesty's Astronomer at the Cape (£600 p.a.), the Superintendent of the Nautical Almanac (£500 p.a.) and the professor at the Royal Naval College (£500 p.a.). Each oversaw an increasing number of scientific and other staff.[8]

The production of the *Nautical Almanac* became a fully bureaucratised affair, rather than a product of the personal correspondence network of the Astronomer Royal. Under Stratford the Nautical Almanac Office was set up during the winter of 1831–1832. Until 1842 it was located at Somerset House, where the Navy Board and Royal Society were based, and 3 Verulam Buildings, Gray's Inn, London, briefly in 1832 and then again between 1842 and 1917.[9] It had a permanent staff of computers, rather than the pieceworkers of previous years. This prompted a move to London from Truro for long-term *Nautical Almanac* computer William Dunkin, at the request of Davies Gilbert.[10] While Dunkin somewhat regretted the move, it provided his son, Edwin, with the opportunity to develop a forty-six-year career at the expanding Royal Observatory.[11] Yet the growing professionalisation of scientific labour within the Admiralty spelt the end for another kind of worker. Eliza Edwards, whose parents, John and Mary, had both been *Nautical Almanac* computers, had assisted their work and continued it after her mother's death in 1815 until the creation of a London office forced her to stop in 1832.[12] Increasingly, routes to patronage or reward for those without institutional connections or formal scientific training were being closed down, a theme worth emphasising in any evaluation of the Board of Longitude's fate.

In its final years the Board had been 'a valuable and flexible resource', if a contested one, for astronomers and other practitioners of the sciences.[13] It had offered employment, paid for publications, loaned instruments and supported research into optical glass. However, after 1828, there were other and increasing opportunities offered to ambitious inquirers and scientific travellers. Funding, loans and support were offered, at least to some degree, by the Royal Society and specialist formations such as the Royal Astronomical Society and Royal Geographical Society.[14] From the 1830s, the British Association for the Advancement of Science provided another public forum and lobby, distributing

8 House of Commons, 'Navy Estimates for the Year 1857–58', pp. 21–25.
9 For an account of the Nautical Almanac Office in the nineteenth century, see Perkins and Dick, 'British and American Nautical Almanacs', pp. 157–178.
10 HM Nautical Almanac Office, 'The Nautical Almanac'; Dunkin, *A Far Off*, p. 45.
11 Chapman, 'Dunkin, Edwin'; Daston, 'Calculation', pp. 13–16.
12 Croarken, 'Nevil Maskelyne', p. 150.
13 Miller, 'Royal Society', pp. 314–315.
14 Wess, 'Role of Instruments', p. 74, notes that the Royal Geographical Society offered instruments for loan to travellers from 1834.

support and grants, most famously to ambitious global magnetic surveys.[15] The Royal Institution undertook laboratory work for the state and the Royal Society continued to advise government.[16] After its own period of reform, the Royal Society shifted away from old regime patronage and, from 1849, distributed limited government funds for research. It is possible to understand these developments as part of the gradual formalisation of specialist expertise within the administrative state.[17] Nevertheless, key individuals, as always, continued to move across and influence a range of institutional and informal networks. While the Board of Longitude ceased to exist, its functions – as offered to government, the public and scientific workers – were displaced.

Nevertheless, in the 1870s the Royal Commission on Scientific Instruction and Advancement of Science still found the Board's example relevant as it deliberated on the role of government support for scientific education and research. Giving evidence to the Commission in spring 1872, Airy concluded that, while the Board's late leadership had made it unfit, its services to 'navigation were very great'. It was apt and predictable that he should offer this summary obituary. The Astronomer Royal lauded the Board's publication of the *Nautical Almanac* and a host of navigational and astronomical tables; remarkable improvements in chronometers' accuracy; the establishment of the Cape Observatory; a range of survey and navigational instruments; its monitoring of expeditions into the Arctic; and its sponsorship of global pendulum surveys. But Airy could not resist a swipe at Thomas Young's refusal to 'consent to any change'. He pointedly noted the uselessness and falsehood, as he saw it, of many schemes and the exhaustion of the Board by the time of its disappearance, carefully cataloguing exactly how its significant functions had passed to other organisations after, of course, carefully cataloguing the Board's archives.[18] It had certainly offered a range of support unavailable at that conjuncture from markets or clients, whether private or state. Thereafter, government continued to aid the infrastructure and training required to make use of several technologies that the Board had helped usher into practice, and to apply them to the systematic charting of the world's navigable waters. Scientific specialists found increasing options for paid work and advancement, through government or military work, scientific societies and, increasingly, academic institutions. Most scientific practitioners might lobby for increased state support but it is worth recalling that Airy, head of the nation's principal publicly funded

[15] Cawood, 'Magnetic Crusade'; Morrell and Thackray, *Gentlemen of Science*, pp. 523–531; Bulstrode, 'Eye of the Needle', pp. 14–15, 227–231.

[16] On Faraday's advisory work see, e.g. James, 'Michael Faraday's Work' and 'Michael Faraday and Lighthouses'; Gascoigne, 'Royal Society'. See also Jackson, *Scientific Advice*.

[17] Hall, *All Scientists Now*; Alter, *Reluctant Patron*, pp. 61–62.

[18] Airy, 'Remarks'; Waring, 'Thomas Young', pp. 218–221.

scientific establishment, notoriously preferred 'voluntary associations of private persons to organisations of any kind dependent on the State'.[19]

Yet, as this book demonstrates, such systems of expansive support accompanied stern forms of condescension and exclusion within hierarchical social orders of inherited privilege or specialist qualification. As signal example, already introduced in Chapter 4, with notable exceptions such as Mary Somerville, interpreter of Laplacian celestial mechanics and in 1835 made honorary member of the Royal Astronomical Society, and Janet Taylor, instrument-maker, teacher and publisher of lunar tables during the 1830s, institutional recognition of women's expert participation in the astronomical sciences remained very rare.[20] In several ways, involvement in the principal scientific societies by artisan practitioners, especially the eminent scientific instrument makers who played such a decisive role in the activities of the Board, became more fraught and unevenly rewarded. The network defined by the Royal Observatory, the Royal Society and the Board of Longitude in the long eighteenth century had somehow to negotiate a newly complex world of manufacturing systems and industrial capital, with corresponding challenges for enterprises that depended on personal networks of apprenticeship, patronage and acquaintance.[21] The stern declaration issued by the Board of Longitude at the height of its authority that it would only consider schemes certified by 'persons of Science whose Characters are known' was eloquent testimony to the powers of social judgement within the reward systems of science, innovation and empire.[22]

[19] Airy's Presidential Address to the 1851 Ipswich meeting of the British Association for the Advancement of Science, cited in Morrell, 'Individualism', p. 204; Alter, *Reluctant Patron*, pp.80–81.

[20] Kidwell, 'Women Astronomers', pp. 534–535; Desborough and Clifton, 'Science and the City'; Brock, 'Mary Somerville'; Croucher and Croucher, *Mistress of Science*.

[21] Bennett, 'Instrument Makers', pp. 14–15, 24–25.

[22] BoL, confirmed minutes, 13 June 1772, CUL RGO 14/5, p. 228.

Appendix 1
An Act for Providing a Publick Reward for Such Person or Persons as Shall Discover the Longitude at Sea[1]

WH E R E A S it is well known by all that are acquainted with the Art of Navigation, That nothing is so much wanted and desired at Sea, as the Discovery of the Longitude, for the Safety and Quickness of Voyages, the Preservation of ships, and the Lives of Men: And whereas in the Judgment of able Mathematicians and Navigators, several Methods have already been discovered, true in Theory, though very difficult in Practice, some of which (there is Reason to expect) may be capable of Improvement, some already discovered may be proposed to the Publick, and others may be invented hereafter: And whereas such a Discovery would be of particular Advantage to the Trade of *Great Britain*, and very much for the Honour of this Kingdom; but besides the great Difficulty of the Thing itself, partly for the Want of some Publick Reward to be settled as an Encouragement for so useful and beneficial a Work, and partly for want of Money for Trials and Experiments necessary thereunto, no such Inventions or Proposals, hitherto made, have been brought to Perfection; Be it therefore enacted by the Queen's most Excellent Majesty, by and with the Advice and Consent of the Lords Spiritual and Temporal, and Commons, in Parliament assembled, and by the Authority of the same, That the Lord High Admiral of *Great Britain*, or the first Commissioner of the Admiralty, the Speaker of the Honourable House of Commons, the first Commissioner of the Navy, the first Commissioner of Trade, the Admirals of the Red, White, and Blue Squadrons, the Master of the Trinity-House, the President of the Royal Society, the Royal Astronomer of *Greenwich*, the *Savilian, Lucasian* and *Plumian* Professors of the Mathematicks in *Oxford* and *Cambridge*, all for the Time being, the Right Honourable *Thomas* Earl of *Pembroke* and *Montgomery, Philip* Lord Bishop of *Hereford, George* Lord Bishop of *Bristol, Thomas* Lord *Trevor*, the Honourable Sir *Thomas Hanmer* Baronet, Speaker of the Honourable House of Commons, the Honourable *Francis Robarts* Esq; *James Stanhope* Esq; *William Clayton* Esq; and *William Lowndes* Esq; be constituted, and they

[1] 13 Anne c 14 (sometimes also as 12 Anne c 15); listed in the *Chronological Table of Statutes* as 'Discovery of Longitude at Sea Act, 1713'; copy at CUL RGO 14/1, pp. 10r–12r. The Act received royal assent on 9 July 1714 Old Style (20 July New Style).

are hereby constituted Commissioners for the Discovery of the Longitude at Sea, and for examining, trying, and judging of all Proposals, Experiments, and Improvements relating to the same; and that the said Commissioners, or any five or more of them, have full Power to hear and receive any Proposal or Proposals that shall be made to them for discovering the said Longitude; and in case the said Commissioners, or any five or more of them, shall be so far satisfied of the Probability of any such Discovery, as to think it proper to make Experiment thereof, they shall certify the same, under their Hands and Seals, to the Commissioners of the Navy for the Time being, together with the Persons Names, who are the Authors of such Proposals; and upon producing such Certificate, the said Commissioners are hereby authorized and required to make out a Bill or Bills for any such Sum or Sums of Money, not exceeding two thousand Pounds, as the said Commissioners for the Discovery of the said Longitude, or any five or more of them, shall think necessary for making the Experiments, payable by the Treasurer of the Navy; which Sum or Sums the Treasurer of the Navy is hereby required to pay immediately to such Person or Persons as shall be appointed by the Commissioners for the Discovery of the said Longitude, to make those Experiments, out of any Money that shall be in his Hands, unapplied for the Use of the Navy.

II. And be it further enacted by the Authority aforesaid, That after Experiments made of any Proposal or Proposals for the Discovery of the said Longitude, the Commissioners appointed by this Act, or the major Part of them, shall declare and determine how far the same is found practicable, and to what Degree of Exactness.

III. And for a due and sufficient Encouragement to any such Person or Persons as shall discover a proper Method for finding the said Longitude, Be it enacted by the Authority aforesaid, That the first Author or Authors, Discoverer or Discoverers of any such Method, his or their Executors, Administrators, or Assigns, shall be intitled to, and have such Reward as herein after is mentioned; that is to say, to a Reward, or Sum of ten thousand Pounds, if it determines the said Longitude to one Degree of a great Circle, or sixty Geographical Miles; to fifteen thousand Pounds, if it determines the same to two Thirds of that Distance; and to twenty thousand Pounds, if it determines the same to one Half of the same Distance; and that one Moiety or Half-Part of such Reward or Sum shall be due and paid when the said Commissioners, or the major Part of them, do agree that any such Method extends to the Security of Ships within eighty Geographical Miles of the Shores, which are Places of the greatest Danger, and the other Moiety or Half-Part, when a Ship by the Appointment of the said Commissioners, or the major Part of them, shall thereby actually said over the Ocean, from *Great Britain* to any such Port in the *West Indies*, as those Commissioners, or the major Part of them, shall choose or nominate for the Experiment, without losing their Longitude beyond the Limits before mentioned.

IV. And be it further enacted by the Authority aforesaid, That as soon as such Method for the Discovery of the said Longitude shall have been tried and found practicable and useful at Sea, within any of the Degrees aforesaid, That the said Commissioners, or the major Part of them, shall certify the same accordingly, under their Hands and Seals, to the Commissioners of the Navy for the Time being, together with the Person or Persons Names, who are the Authors of such Proposal; and upon such Certificate the said Commissioners are hereby authorized and required to make out a Bill or Bills for the respective Sum or Sums of Money, to which the Author or Authors of such Proposal, their Executors, Administrators, or Assigns, shall be intitled by virtue of this Act; which Sum or Sums the Treasurer of the Navy is hereby required to pay to the said Author or Authors, their Executors, Administrators, or Assigns, out of any Money that shall be in his Hands unapplied to the Use of the Navy, according to the true Intent and Meaning of this Act.

V. And it is hereby further enacted by the Authority aforesaid, That if any such Proposal shall not, on Trial, be found of so great Use, as aforementioned, yet if the same, on Trial, in the Judgment of the said Commissioners, or the major Part of them, be found of considerable Use to the Publick, that then in such Case, the said Author or Authors, their Executors, Administrators or Assigns, shall have and receive such less Reward therefore, as the said Commissioners, or the major Part of them, shall think reasonable, to be paid by the Treasurer of the Navy, on such Certificate, as aforesaid.

Appendix 2
The Board of Longitude Finances

The Longitude Act of 1714 (Appendix 1) established a group of 'Commissioners for the Discovery of the Longitude at Sea' empowered to assess, put to trial, judge and potentially reward any schemes proposed to them. To enable this, the Act established guidelines not only for judging schemes but also for authorising the payment of rewards, whether to test promising ideas or as a result of successful trials. In either case, the rewards were to be funded from 'any Money … unapplied for the Use of the Navy'. For a reward to be paid, the Commissioners of Longitude had to apply to the Commissioners of the Navy. Once approved, the funds were released by the Treasurer of the Navy, a civilian officer who was one of the Navy's principal commissioners.[1]

There were limits on the funds available: a total allocation of £2000 for bringing schemes to trial and up to £20,000 for rewards for successful schemes. As a result, many subsequent Longitude Acts included clauses that renewed funds that had been spent on bringing schemes to trial or paying lesser rewards. The Act of 1753, for example, noted that £1750 had been paid in rewards to John Harrison and William Whiston, leaving only £250 for further 'Experiments'. The new Act therefore allocated an additional £2000 for future payments to bring schemes to trial.[2] Later Acts similarly renewed this allocation, initially £2000 each time, then £5000 from 1770 and £10,000 from 1806 (see Timeline). What remained in this pot sometimes limited what the Commissioners could pay out, as in 1761, when they resolved to pay £500 to

[1] Discovery of Longitude at Sea Act 1713 (12 Anne c 15), copy at CUL RGO 14/1, pp. 10r–12r. Examples of applications to the Navy Board include: BoL to Navy Board, 13 October 1761, NMM ADM/A/2528 (£500 to John Harrison); BoL to Navy Board, 17 August 1762, NMM ADM/A/2539 (£1000 to John Harrison, £500 to Christopher Irwin); BoL to the Navy Board, 1763, NMM ADM/A/2551 (£100 to Christopher Irwin, £50 to Charles Green, payments to Commissioners and other matters); BoL to Navy Board, 13 June 1765, NMM ADM/A/2572 (£3000 to Maria Mayer, £300 to Leonhard Euler, £100 to Catherine Price); BoL to Navy Board, 11 June 1795, NMM ADM/A/2869 (£200 to Ralph Walker); BoL to Navy Board, 2 March 1771 (£200 to John Arnold), TNA ADM 49/65, pp. 103–104; BoL to Navy Board, 12 December 1805 (£2500 to Thomas Earnshaw; £1678 to John Roger Arnold), TNA ADM 49/65, pp. 144–145.

[2] Discovery of Longitude at Sea Act 1753 (26 Geo 2 c 25), copy at CUL RGO 14/1, pp. 15r–18r.

John Harrison to prepare for the first sea trial of H4. With only £200 available, the remainder could not be paid until the following year.[3]

It is also important to note that lesser rewards were, in principle, to be set against any future greater reward – sometimes explicitly referred to as a 'great reward' or 'great premium' – as a result of a successful trial under the Longitude Act.[4] Between 1771 and 1781, for instance, watchmaker John Arnold was awarded six such advances, totalling £1700. Of this sum, £378 was reckoned as the cost of six timekeepers purchased in the early 1770s for voyages the Board supported, leaving £1322 still considered as an advance.[5] Thus, when a posthumous reward of £3000 was agreed in 1805, that figure was subtracted from the total, leaving £1678 to be paid to Arnold's son, John Roger.[6]

The appointment of a Secretary in 1763 marked an important change. What was soon often called the Board of Longitude was now a bureaucratic entity with regular expenses. It came under greater parliamentary scrutiny thereafter, with financial reporting to the House of Commons often required.[7] It was presumably partly in response to this increased scrutiny that in 1780 the Board set up a standing committee – comprising the president of the Royal Society (Joseph Banks) and any other Commissioners who wished to attend – to examine and audit its own accounts.[8]

The Longitude Act of 1818 changed the procedures even more markedly. While funds still came from the Navy and required an application to the Navy Board, much more of the Board of Longitude's expenditure was put onto an annual rather than an ad hoc basis. The Act allowed the Commissioners to spend up to £1000 each year on trials and publishing, the same 'in ascertaining the Latitude and Longitude of Places'. In addition, the salaries of three newly nominated Commissioners and of the Secretary and the new roles of Superintendent of the Nautical Almanac and Superintendent of Timekeepers were to be paid from the 'Ordinary Estimate of the Navy'. Thereafter, the Commissioners were required to submit an estimate of expenses at the beginning of each year.[9]

Derek Howse's published account of the Board's finances provides an overview and lists the individual rewards given by the Board, as well its

[3] BoL, confirmed minutes, 13 October 1761 and 3 June 1762, CUL RGO 14/5, pp. 33, 35–36.
[4] For example, Banks, *Protest Against a Vote*, p. 1.
[5] Arnold, *An Account Kept During Thirteen Months*, p. 4, notes the situation in 1780, by which time the Board had advanced £1400. Arnold lists six timekeepers purchased for 60 guineas each (total cost £378), leaving £1022 in advances. Another advance of £300 was granted in March 1781.
[6] Howse, 'Britain's Board of Longitude', p. 406.
[7] Admiralty, papers relating to longitude, 1763–1819, TNA ADM 49/65, includes copies of reports from the Treasurer of the Navy to Parliament. Many of these reports use the name 'Board of Longitude'. See also Barrett, 'Explaining Themselves', p. 148. Additional papers concerning expenditure are in BoL, accounts, 1765–1828, CUL RGO 14/2, with more detailed paperwork in CUL RGO 14/15 to RGO 14/21.
[8] BoL, confirmed minutes, 15 July 1780, pp. 2:8–2:9; BoL, reports from the committees on the examination of accounts, 1782–1823, CUL RGO 14/2, pp. 297v–344r.
[9] Discovery of Longitude at Sea, etc. Act 1818 (58 Geo 3 c 20), copy at CUL RGO 14/1, pp. 79r–82v.

publications and the relevant Acts of Parliament from 1714 to 1828. Howse grouped the Board's expenses under four headings:

- rewards;[10]
- publications, including computing and printing costs;
- expeditions and experiments, comprising payments for those preparing for and participating in sea trials and other expeditions, including associated salaries such as those of expeditionary astronomers, as well as payments for purchasing, maintaining and storing instruments;
- other expenses, including the salaries of the Secretary and other officers, allowances for attending meetings and other miscellaneous payments.

These headings have been retained for the analyses below, which also draw on the data Howse gathered in unpublished research.[11]

Broadly speaking, the story of the Board's finances mirrors its history. Between 1737 and 1828, the total expenditure was just over £184,000: approximately £53,000 for rewards; £71,000 for publications; £24,000 for expeditions and experiments; £36,000 for other expenses. This equates to an average expenditure of £2000 per year between 1737 and 1828, although this figure masks the sporadic nature of payments in the first twenty-six years (when payments averaged just under £260 per year). For the period from 1763, when the Board appointed a salaried Secretary, it spent an average of £2687 annually, although the yearly average in the 1820s (over £4990) was more than double that of the previous six decades (£2324).

To put these figures in some comparative context, John Harrison's wealth at death in 1776 was £6500 (plus remaining clocks, watches and scientific instruments and the contents of a studio workshop), while Jesse Ramsden's estate was valued at under £5000 when he died in 1800.[12] In 1799, the basic pay for a Royal Navy lieutenant on active service was between about £91 and £100 annually, for a captain between £146 (for a sixth-rate ship) and £365 (for a first-rate ship).[13] In other words, the financial outgoings of the Board were modest, particularly when compared with its funding body, the Royal Navy – annual naval expenditure was £5.43 million at the beginning of the wars with France in 1792, peaked at £26.93 million in 1797, was still

[10] The £2500 reward to Thomas Mudge in 1793 is included in the analysis, although it does not appear in the BoL accounts; Howse, 'Britain's Board of Longitude', p. 411.

[11] Howse, 'Britain's Board of Longitude'; Howse, 'The Board of Longitude Accounts'. Howse drew data from the confirmed minutes and BoL, accounts, 1765–1828, CUL RGO 14/2; Admiralty, papers relating to longitude, 1763–1819, TNA ADM 49/65. Minor amendments have been made to refine Howse's data.

[12] King, 'Harrison, John'; Chapman, 'Ramsden, Jesse'.

[13] *Steel's Original and Correct List of the Royal Navy* for 1799, pp. 21, 29, lists captains' pay between 8*s* (sixth rate) and £1 (first rate) per day; lieutenants' pay 5*s* to 5*s* 6*d* (when under an admiral) per day; Rodger, *Command of the Ocean*, pp. 622–627, documents pay from 1807.

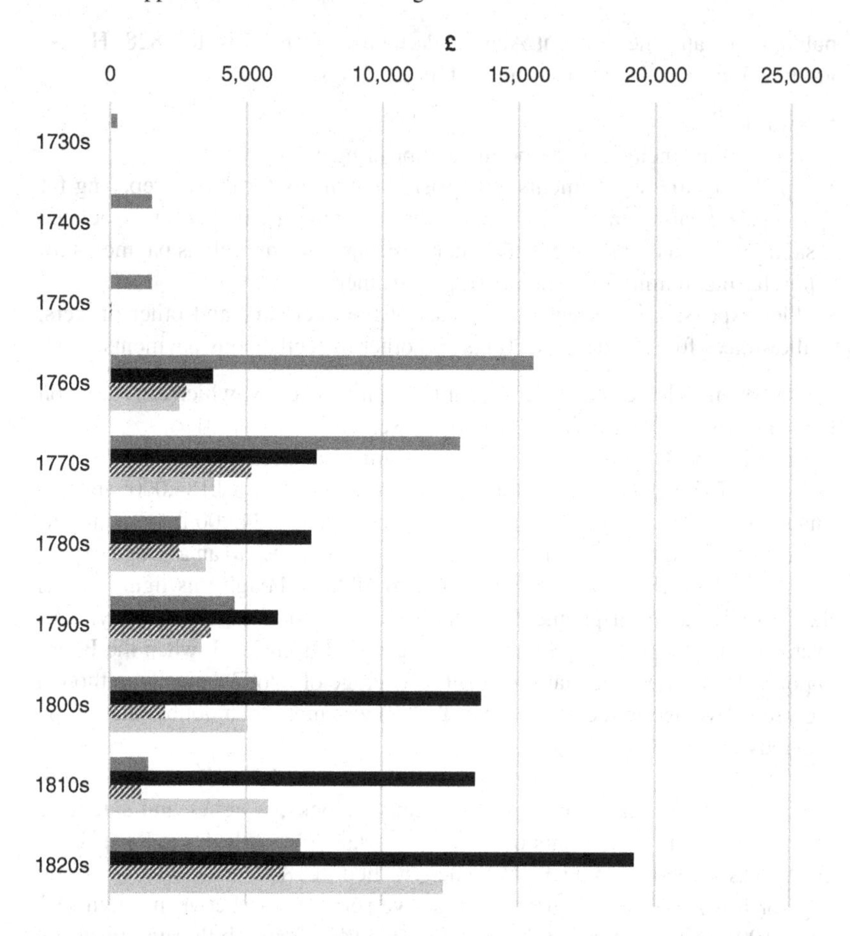

Figure A2.1 Board of Longitude expenditure 1737–1828

£16.37 million in 1815 and dropped to £5.67 million in 1828.[14] The Board's income was even more modest: sales of publications between 1769 and 1828 brought in just under £26,000 (almost £648 per year on average) and never came close to covering the costs incurred.[15] Profit was never the reason for publishing.

Figure A2.1 gives an overview using Howse's headings of how the Board's expenditure changed between 1737, when it first authorised a payment (to John

[14] Salavrakosp, 'A Reassessment', p. 8; Modelski and Thompson, *Seapower in Global Politics*, p. 339.
[15] In addition, a repeating circle was sold to Naples Observatory in 1825 for £210.

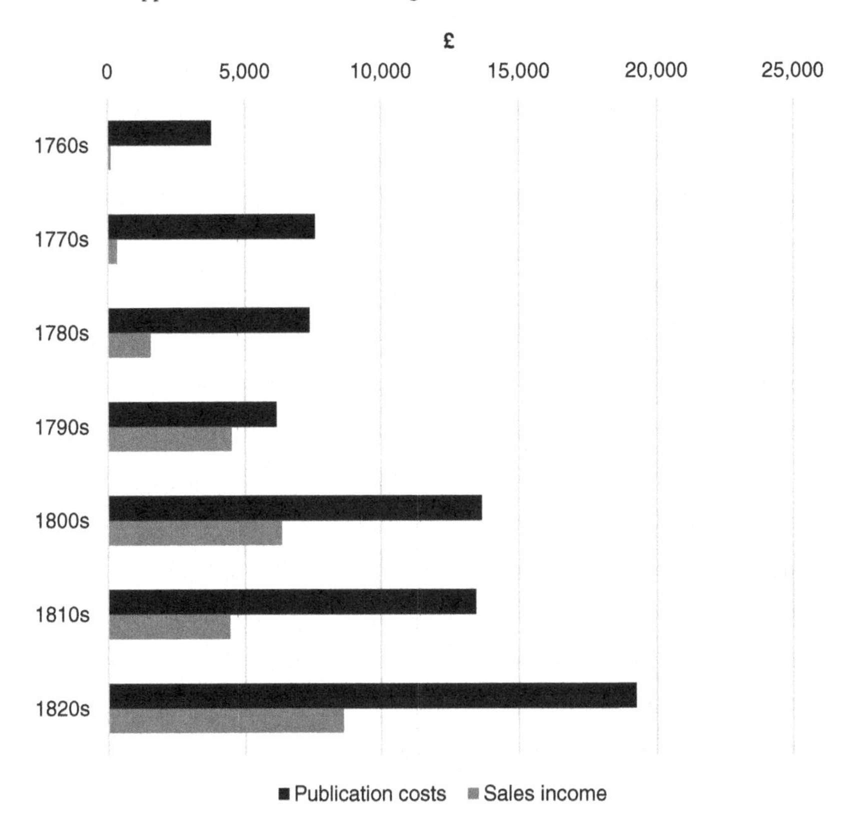

Figure A2.2 Board of Longitude publications – costs and sales income

Harrison), and its closure in 1828. Until 1763, the only costs were rewards given on an ad hoc basis, mostly to Harrison (see also Appendix 3), and some expenses on clocks and instruments. From 1763, however, as a recognisable Board of Longitude emerged as a standing body with regular meetings, it had regular expenses, including its Secretary. Following a new Act in 1765, the Board became a publisher of nautical tables and other works related to its purposes, leading to a substantial increase in expenditure.[16] Publication became the largest area of outlay within twenty years and remained so until the Board's closure. Other expenses, including the salaries of two new officers from 1818, also increased as publication and other activities grew. Figure A2.2 compares publication costs and income from sales.

[16] Discovery of Longitude at Sea Act 1765 (5 Geo 3 c 20), copy at CUL RGO 14/1, pp. 29r–33v.

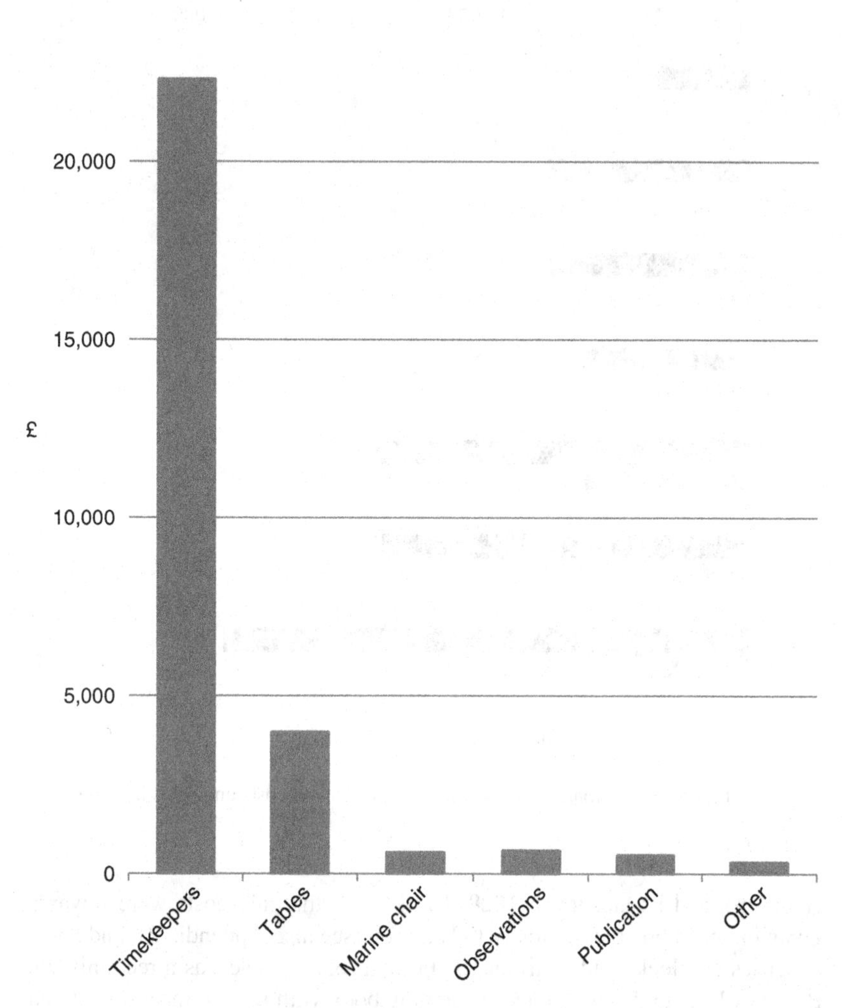

Figure A2.3 Board of Longitude rewards 1737–1773

Rewards, the Board's only expense in the first decades, hit their peak in the 1770s, mostly due to the final award of £8750 to Harrison, after which they never surpassed expenditure on publication. Figures A2.3 and A2.4 compare the methods and activities to which rewards were allocated in the periods 1737–1773 (i.e. until the resolution of dealings with Harrison) and 1774–1828, following the Longitude Act 1774, which significantly changed the rewards available, as did the 1818 Act. As these tables show, while rewards for timekeepers were the largest group in both periods, particularly in the earlier,

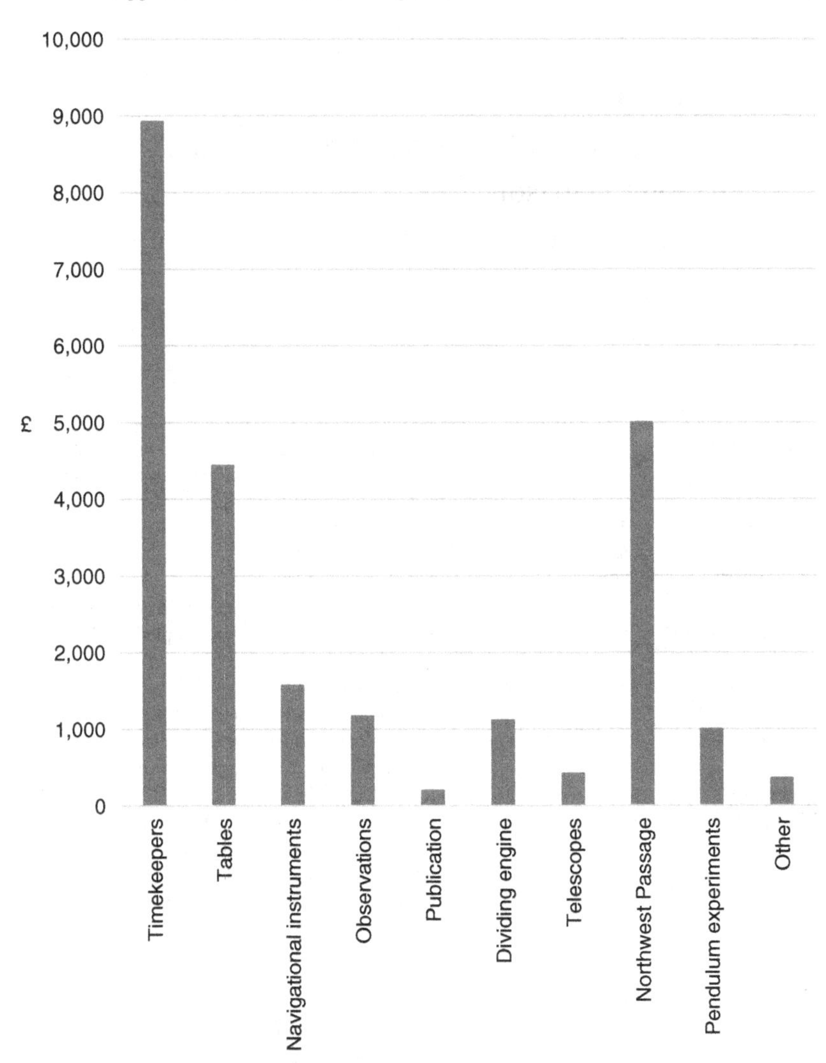

Figure A2.4 Board of Longitude rewards 1774–1828

the range of activities rewarded broadened significantly from 1774. In both periods, however, rewards to a small number of recipients dominated: Harrison in the earlier period; three watchmakers (Arnold, Earnshaw and Mudge) and the recipients of the reward relating to the Northwest Passage in the later one. Lastly, it is worth noting that the Board's support for astronomical methods centred on the compilation and publication of the *Nautical Almanac* and related works rather than on the granting of rewards.

Appendix 3
Payments from the Board of Longitude to John Harrison

Meeting	Reward (£)	Other (£)	Notes
30 June 1737	250		Construction of H2; £500 granted but the remaining £250, dependent on a successful trial, was never paid
16 January 1742	500		Construction of H3
4 June 1746	500		Construction of H3
17 July 1753	500		Construction of H3
19 June 1755	500		Adjustment of H3 and completion of two watches
28 November 1757	500		Final adjustment of H3 and completion of watches
18 July 1760	500		Completion of H4
12 March 1761		250	Costs relating to first sea trial of H4
13 October 1761		200	A further £500 agreed for costs relating to the sea trial, but only £200 was then available
3 June 1762		300	Remainder of the £500 agreed on 13 October 1761
17 August 1762	1500		Reward following first sea trial of H4
9 August 1763		300	Expenses for second sea trial of H4
18 September 1764	1000		Reward following second sea trial
9 February 1765		15	Expenses incurred for computations following second sea trial
28 October 1765	7500		First half of agreed reward of £10,000, minus £2500 paid in 1762 and 1764
19 June 1773	8750		Final payment of remaining £10,000 of the full reward, minus £1250 previously paid in 1737, 1742 and 1746
Total	**£22,000**	**£1065**	

Glossary

Cross-references in **bold**

Académie Royale des Sciences/Académie des Sciences The Académie Royale des Sciences, founded 1666, was the principal learned scientific society of the French state. During most of the eighteenth century it was based in the Louvre and subject to statutes of 1699, often charged with overseeing patent rights. From 1702 it became responsible for the ***Connoissance des Temps***, as well as **ephemerides** released every decade, the ***Ephemerides des Mouvemens Célèstes***, under Nicolas-Louis de Lacaille's direction from 1743. The Académie was abolished during the French Revolution in 1793 and made a section of the new Institut de France in 1795. From 1816 it was restored as the Académie Royale des Sciences

Act of 1714 'An Act for Providing a Publick Reward for such Person or Persons as shall Discover the Longitude at Sea'. The legislation appointed a number of **Commissioners of Longitude** and established rewards of £10,000 for a method that found longitude at sea to one degree of a great circle or 60 miles; £15,000 for a method accurate to 40 minutes or 40 miles; and £20,000 for an accuracy of 30 minutes or 30 miles. See Appendix 1 for the full wording

Act of 1740 'An Act for surveying the Chief Ports and Head Lands on the Coasts of Great Britain and Ireland, and the Islands and Plantations thereto belonging, in order to the more exact Determination of the Longitude and Latitude thereof' permitted the **Commissioners of Longitude** to pay up to £2000 for surveys of British, Irish and colonial coasts that established their latitude and longitude

Act of 1753 Legislation of January 1753 designed to make the **Act of 1714** 'more effectual with regard to the making experiments or proposals made for discovering the Longitude'. The Act increased the number of **Commissioners of Longitude** and permitted a further £2000 to fund schemes

Act of 1763 'An Act for the encouragement of John Harrison, to publish and make known his Invention of a Machine or Watch, for the Discovery of the Longitude at Sea', set out directions for Harrison to make a 'full and clear discovery of the principles' of his **sea watch (H4)** to a group of witnesses including Fellows of the **Royal Society** and London clockmakers. The **Treasurer of the Navy** was then to pay Harrison £5000. Were Harrison later to gain a longitude reward, sums already paid would be deducted from it

Act of 1765 Two Acts were passed in early 1765. The first permitted further sums of up to £2000 for trial costs of schemes the **Commissioners of Longitude** judged viable. The second directed that if within the next six months John Harrison discovered and explained his watch's principles to a majority of the **Commissioners**

and assigned the property in three of these watches to the **Commissioners** for public use, he would be paid £10,000, as half of the original reward announced in 1714. The remainder would be paid when 'other Time Keepers of the same kind shall be made' and shown on trial to determine longitude within 30 miles. The Act authorised payment of up to £300 to Leonhard Euler and up to £3000 to the widow of Tobias Mayer, as reward for work to establish lunar tables and for assignation of Mayer's tables to the public. A further £5000 was offered for future table improvements; and the **Board of Longitude** was authorised to produce the ***Nautical Almanac***

Act of 1774 'An Act for the Repeal of all former Acts concerning the Longitude at Sea' reaffirmed the appointment of the **Commissioners of Longitude** and their licence to release the ***Nautical Almanac*** and other publications; and allowed the **Treasurer of the Navy** to clear the **Board of Longitude**'s debts. It confirmed the Commissioners' right to try projects and grant smaller rewards for navigationally useful schemes. The Act replaced all previous conditions on major rewards and halved the value of the maximum rewards available

Act of 1818 'An Act for more effectually discovering the Longitude at Sea, and encouraging Attempts to find a Northern Passage between the Atlantic and Pacific Oceans, and to approach the Pole' overhauled financial management and membership of the **Board of Longitude** to include three fellows of the **Royal Society** and 'three other persons well versed in the Sciences of Mathematics, Astronomy, or Navigation' resident in or near London. Initially, William Hyde Wollaston, Thomas Young and Henry Kater were appointed to these salaried positions, until January 1820. The Royal Society's Foreign Secretary was named **Secretary of the Board** and **Superintendent of the Nautical Almanac**. The Act also established rewards for navigating the **Northwest Passage**, from £20,000 for reaching the Pacific, £5000 for 110 degrees west or 89 degrees north and £1000 for 83 degrees north

Act of 1828 'An Act for repealing the Laws now in force relating to the Discovery of Longitude at Sea' abolished the **Board of Longitude**, the positions of the Commissioners and all rewards. A clause guaranteed the continuation of the ***Nautical Almanac*** under the direction of the **Lord High Admiral**

Admiral, Lord High The Lord High Admiral was the title given the head of the Navy. The **Act of 1714** named the Lord High Admiral as a **Commissioner of Longitude** ***ex officio***. At that date and for the next century the powers of the Lord High Admiral were in fact exercised by a group of Lords Commissioners

Admirals of Red, White and Blue The British fleet was in principle divided into three large squadrons, each distinguished by the coloured flags associated with their commanders. By the period of the **Act of 1714**, the Red squadron was the senior group, followed by White then Blue. This order governed promotion within the higher levels of the Navy

Admiralty In 1709 the Admiralty was defined as a department of the British government responsible for naval affairs, administered under a group of Lords Commissioners [see **Board of Admiralty**] responsible for the Navy's military direction. During the eighteenth century, other responsibilities, such as financial control, shipbuilding and design, management of naval stores, supplies and administrative organisation, were formally conducted by a group of officials who formed the **Navy Board**. The term ***Admiralty*** was also used for the buildings completed from 1726 housing the Board Room and offices for the Lords of the Admiralty, situated between the northern end of Whitehall and St James's Park, London. From the later 1730s the **Commissioners of Longitude** often held meetings there

Admiralty Board [see **Board of Admiralty**]

Admiralty, First Lord of the [see **First Lord of the Admiralty**]

Admiralty, High Court of [see **High Court of Admiralty**]

almanac a calendar, often produced annually, containing significant dates and statistical information such as astronomical data and tide tables [see also **ephemeris**, ***Nautical Almanac***]

apparent time the time based on the position of the Sun without any adjustment for the variation in its progress during the year

astrolabe [see **mariner's astrolabe**]

Astronomer Royal The position of Astronomer Royal or King's Astronomer, established in 1675, involved the direction of the **Royal Observatory, Greenwich**, founded that year. The Astronomer Royal, who typically held the position for life, was named ***ex officio*** in the **Act of 1714** as a **Commissioner of Longitude**. The Astronomers Royal during the period covered by the Longitude Acts were John Flamsteed (until 1719), Edmond Halley (1720–1742), James Bradley (1742–1762), Nathaniel Bliss (1762–1764), Nevil Maskelyne (1765–1811) and John Pond (from 1811)

astronomical quadrant a **quadrant** used mainly for determining the altitude of a celestial body above the horizon; see also **mural quadrant**

Astronomical Society [see **Royal Astronomical Society**]

atmospheric refraction the deviation of light from a straight line as it passes through the atmosphere, causing an error in observed celestial angles

Attorney General a law officer, normally a Member of Parliament who represented the Crown in legal affairs, in particular in any litigation. Sir Philip Yorke, Attorney General under Robert Walpole 1724–1734, advised on the later interpretation of the **Act of 1714**

backstaff an instrument used for measuring the Sun's altitude, so called because the user had their back to the Sun when observing [see also **quadrant**]

balance an oscillating wheel with a weighted rim that, with an appropriate balance spring, governs the **rate** of a **chronometer**

barometer an instrument for measuring atmospheric pressure

bearing horizontal direction, measured in degrees or compass points

Board of Admiralty The Admiralty Board was composed of Commissioners who collectively performed the office of **Lord High Admiral**, under the presidency of the **First Lord of the Admiralty**. During the eighteenth century most members were civilians, political appointments normally replaced at a change of government. Throughout this period the **Secretary to the Admiralty** attended meetings of the Board but was not officially a member

Board of Longitude The **Act of 1714** established a group of **Commissioners of Longitude**; it did not refer to a Board. The use of the phrase 'Board of Longitude' to refer to any established or permanent institution concerned with the conduct of the Longitude Acts does not seem to predate 1756 and came into common use only from the 1760s, when business became more regular, meetings were held at least annually, and the Commissioners requested a salary for their **Secretary**. The Board was abolished by the **Act of 1828**

Board of Ordnance a government body charged with managing the state's military arms, fortifications and arsenals, principally concerned with supply of weapons, ammunition and equipment for the Army and Navy. The Board was headed by a Master-General, in the later eighteenth century a cabinet appointment. Other members included the Surveyor-General, responsible for inspection of gunnery and later for mapping. The Board built and funded the **Royal Observatory, Greenwich**, until this responsibility was passed to the **Admiralty** in 1818

Board of Visitors, Royal Observatory The **Royal Observatory** was initially built and funded through the **Board of Ordnance** from 1675. In December 1710 a Board of Visitors was set up under the president of the **Royal Society**, then Isaac Newton. The Visitors were supposed to oversee the **Astronomer Royal**'s conduct and act as a means through which the Royal Observatory could lobby the state. The Board was an important agent in the transfer of responsibility for the Royal Observatory from the Ordnance Board to the **Admiralty** in 1818

box chronometer a **chronometer** intended to be used within a protective wooden box and usually mounted in **gimbals**

Bureau des Longitudes The Bureau was established in Paris by a decree of the National Convention in June 1795 (Messidor year III), charged with the improvement of longitude determination at sea and the publication of the ***Connoissance des Temps***, previously the responsibility of the **Académie Royale des Sciences**. The Bureau also took charge of the **Observatoire de Paris**, the Observatory of the former École Militaire, and all state-owned astronomical establishments in France

Christ's Hospital [see **Royal Mathematical School at Christ's Hospital**]

chronometer, box [see **box chronometer**]

chronometer, marine a mechanical timekeeper adjusted to keep accurate time in all variations of temperature; usually used to refer to accurate portable timekeepers that employ the **detent escapement**

clearing the process of applying mathematical corrections to an observed celestial angle in order to account for the effects of **parallax** and **atmospheric refraction**

Commissioners for the Discovery of Mr Harrison's Watch The **Act of 1763** nominated eleven people as witnesses of John Harrison's disclosure of the details of his **sea watch** (H4) and to ensure they were published for others to copy. The group comprised the astronomer James Douglas, Earl of Morton, then president of the **Royal Society**, the Society's vice-presidents Charles Cavendish and Hugh Willoughby, and three other Fellows of the Royal Society, natural philosopher John Michell, writer and mathematician George Lewis Scott and instrument-maker James Short. Also nominated were four clockmakers: Alexander Cumming, William Frodsham, Andrew Dickie and James Green. If a majority of the Commissioners agreed, Harrison would be paid £5000 by the Treasurer of the Navy. In May 1765 the **Board of Longitude** nominated a new group to witness Harrison's planned disclosure. These were 'three Gentlemen skilled in Mechanics': Charles Cavendish, John Michell and Cambridge mathematician William Ludlam, with instrument-maker John Bird as reserve; and three clockmakers, Thomas Mudge, William Mathews and Larcum Kendall, with Justin Vulliamy as reserve. Bird replaced Cavendish, and was present at John Harrison's explanation of the **sea watch** in August 1765 along with Kendall, Ludlam, Mathews, Michell and Mudge

Commissioners of Longitude The **Act of 1714** named a number of Commissioners of Longitude drawn mainly from politics, naval affairs and astronomy and mathematics. Some were described by office: **Lord High Admiral**, **Speaker of the House of Commons**, First Commissioner of the Navy [see **First Lord of the Admiralty**], First Commissioner of Trade, **Admirals of Red, White and Blue Squadrons**, Master of **Trinity House**, president of the **Royal Society**, **Astronomer Royal**, the **Savilian professor** at Oxford (including both Geometry and Astronomy chairs), and the **Lucasian** and **Plumian** professors at Cambridge. The others were explicitly named, and formed a bipartisan list: Thomas Herbert, Earl of Pembroke, formerly Lord High Admiral (1708–1709); Philip Bisse, Bishop of Hereford and Fellow of the Royal Society; the pro-Jacobite George Smalridge, Bishop of Bristol; the Tory lawyer and peer Thomas Trevor; Sir Thomas Hanmer, then Speaker; Francis Robartes, Fellow of the Royal Society and MP; Whig MP and former military commander James Stanhope, who became leader of the government in 1717–1721; Liverpool MP and merchant William Clayton; and William Lowndes MP, **Secretary of the Treasury**. There was no recorded meeting of the Commissioners until 1737 and for the next two decades small groups of Commissioners met from time to time at the **Admiralty**. The **Act of 1753** increased the number of Commissioners, in part to replace named individuals who had died since 1714, to include the Governor of the **Royal Hospital for Seamen**, the Judge of the **High Court of Admiralty**, the Secretary of the Treasury, the **Secretary of the Admiralty** and the **Comptroller of the Navy**. By the 1760s Commissioners' activities became more formal. In 1762 a salaried position was established for a **Secretary**, and travel costs offered to the university professors and the Astronomer Royal. From the later eighteenth century, the Commissioners' activities were increasingly dominated by those of Astronomer Royal Nevil Maskelyne and the president of the Royal Society, Joseph Banks. The **Act of 1828** abolished the Commissioners along with the **Board of Longitude**

compass, magnetic (or mariner's) an instrument indicating direction, using the north-seeking properties of a magnetised needle allowed to turn on a pivot

Comptroller of the Navy The Comptroller chaired the **Navy Board** and was in principle responsible for managing the naval budget including payment of wages. During the later eighteenth century, the Comptroller's office assumed more control over surveillance and management of naval stores and repair costs. The Comptroller was named an ***ex officio*** **Commissioner of Longitude** in the **Act of 1714**

computer In the eighteenth century a computer referred to someone who performed calculations; it was not used for a machine until the later nineteenth century. Nevil Maskelyne's *British Mariner's Guide* (1763) described rules for calculating longitude from lunar positions that only required 'care in the computer'. In January 1765 the **Board of Longitude** agreed to reward persons named as 'computers', such as John Bevis and George Witchell, who had calculated the results of the trial of Harrison's **sea watch** (H4) to Barbados, while six months later the Board confirmed the scheme of Maskelyne, now **Astronomer Royal**, for a ***Nautical Almanac***, naming the 'computers' who would calculate it. From then on, the term was used by the Astronomer Royal and the Board to describe these calculators

Connoissance des Temps In the 1670s astronomers at the **Académie Royale des Sciences** proposed the publication of astronomical tables based on heliocentric astronomy. In 1679 academician Joachim Dalencé and astronomer Jean Picard launched the ***Connoissance des Temps***, containing an annual calendar, tide tables, and ephemerides of stellar and navigational positions. Successive volumes often included essays and analyses of specific astronomical or meteorological topics. The role was a responsibility of an adjunct astronomer at the Académie. From 1760, under the editorship of Jérôme Lalande, the tables offered resources for computing longitude from lunar positions, calculated by a group of **computers**. Between 1762 and 1767 the title was altered to ***Connoissance des Mouvemens Célestes***, then returned to ***Connoissance des Temps*** from 1768. The publication was precedent for and in close relation with the ***Nautical Almanac***. From 1797 its production became a responsibility of the **Bureau des Longitudes**

cross-staff a hand-held instrument, generally of wood, consisting of a graduated staff and one or more perpendicular vanes moving over it, used to measure angles such as altitudes

Davis quadrant an instrument, usually wooden, used to measure the Sun's altitude above the horizon by casting a shadow onto a vane through which the horizon is also viewed; often called a **backstaff**

dead reckoning the navigational method of finding current position from direction and distance covered (calculated from time travelled and regular measurements of speed) since the last known position

deck watch a precision watch used on deck for navigational purposes to avoid moving and potentially disturbing the **chronometer**, which was normally kept below decks

detached escapement an **escapement** in which the motion of the **balance** is free from any contact with the escapement, except during unlocking and impulse

detent escapement a type of **detached escapement** in which the detent (a piece designed to arrest and release another component) locks and is caused to unlock the escape wheel, which enables it to deliver impulse to the **balance**

deviation the angular displacement of a compass needle from magnetic north caused by local magnetic disturbances, such as a ship's iron

Devonshire Commission The Royal Commission on Scientific Instruction and the Advancement of Science was set up by William Gladstone's Liberal administration in 1870 under the chairmanship of William Cavendish Duke of Devonshire, Chancellor of Cambridge University. Its secretary was scientific activist and astronomer Norman Lockyer. It took evidence for the next five years on matters of interest in public science, education and research funding. In his evidence to the Commission, **Astronomer Royal** George Airy recalled his experiences as a **Commissioner of Longitude** in the 1820s

dip (of horizon) the angular depression of the visible horizon below the true or natural horizon; the angle at the eye of an observer between a horizontal line and a tangent drawn from the eye to the surface of the ocean

dip circle an instrument for measuring **magnetic dip**

discovery the action of disclosing or revealing something hidden or previously unseen or unknown

dividing engine a mechanical device used to mark the graduations on the scales of measuring instruments such as rules, sextants and octants

double altitude (method) position finding from two **ex-meridian** altitudes of the Sun or a star (or two stars simultaneously)

East India Company During the period of the Longitude Acts, the East India Company, which held titular if unenforceable monopoly over British trade in the Indian and Pacific oceans, expanded its power over large territories in southern

Asia, initially in Bengal and the other presidencies centred on the port cities of Madras and Bombay. Its finances relied on extracting revenue from Indian commerce and agriculture, backed by fiscal administration and military force, and on profits from trade with China. The Company's vessels, so-called East Indiamen, were commonly the largest merchant ships of the period. **Chronometers** were carried on East Indiamen from the 1770s. Some were owned by the Company, others by ships' officers. From 1768 the Company required officers and mates on East Indiamen to hold a certificate of competence to determine longitude by lunar methods. On occasion candidates were ordered to train at the **Royal Mathematical School at Christ's Hospital**. The Company employed Hydrographers to produce charts and determine positions in the Indian Ocean, and established observatories at St Helena and Madras whose function included longitude determination and **rating** chronometers

eclipse (method) a method for determining longitude by comparing the **local times** at which an eclipse (usually lunar) is observed at two different places

ephemeris/ephemerides a set of tables of the positions of celestial bodies

equal altitudes (method) a method for determining **local time** by noting the times (by a timekeeper) at which the Sun is at the same altitude in the morning and afternoon, local noon determined as the mid-point of these times

equator great circle on Earth in the plane of the celestial equator, 90 degrees from the poles

escapement the mechanism in a timekeeper's **movement** that communicates power from the train (the wheels and pinions that transfer power from the mainspring) to the oscillator, the vibrating element that is the movement's main source of timekeeping [see also **balance, detached escapement, detent escapement, spring-detent escapement**]

ex-meridian description of celestial bodies not on the observer's **meridian**

First Lord of the Admiralty a political appointment, presided over the **Board of Admiralty** and the government's principal adviser on naval affairs. Under the **Act of 1714**, the First Lord was an ***ex officio*** **Commissioner of Longitude**. The position was sometimes held by eminent political figures including John Montagu, Earl of Sandwich (First Lord 1748–1751, 1763, 1771–1782) and Robert Dundas, Lord Melville (served 1812–1827), or senior mariners, including George Anson (served 1757–1762) and Edward Hawke (served 1766–1771)

gimbals rings with pairs of inner and outer supporting pivots at right angles, arranged to keep a **compass**, **chronometer** or other instrument level

Glass Committee The 'Joint Committee of the **Board of Longitude** and the **Royal Society** for the Improvement of Glass for Optical Purposes' was proposed at a meeting of the Board of Longitude in April 1824. Its aim was to fund and manage experimental trials to aid glass production in response to the challenge of optical glass produced by Bavarian artisan Josef von Fraunhofer. It has been suggested it was a means of channelling funds to the ailing Board. Experiments began in 1825 and had fallen under Michael Faraday's control by 1827. He continued this work after the Board's dissolution in summer 1828

Godfrey quadrant an instrument for measuring angles up to around 90 degrees by using mirrors to bring the image of one target into coincidence with a second, directly viewed target, such as the horizon; first devised by Thomas Godfrey in the early 1730s [see also **Hadley quadrant**]

Greenwich Hospital [see **Royal Hospital for Seamen**]

Greenwich Meridian the **meridian** passing through the **Royal Observatory, Greenwich**, used as the baseline for data in the ***Nautical Almanac*** and British charts [see also **Prime Meridian**]

Greenwich Premium Trials competitive **chronometer** trials, run at the **Royal Observatory, Greenwich**, between 1822 and 1835, with winning chronometers purchased for the Royal Navy

Greenwich Observations The **Astronomer Royal** was charged with conducting, correcting and publishing observations made at the **Royal Observatory** of lunar, planetary and stellar positions and times. These provided important data for the development of lunar theory and the efficacy of the **lunar-distance method**. The observations often generated controversy over rightful ownership, variable quality and delay in publication. After many conflicts, Flamsteed's were published posthumously in three volumes in 1725, while Halley's appeared posthumously in 1749. Maskelyne's ***Astronomical Observations made at the Royal Observatory at Greenwich*** were released annually, then compiled in volumes from 1776 until the appearance of a fourth volume in 1811. Copies were distributed among those listed by the **Royal Society**. The observations of Maskelyne's predecessors James Bradley and Nathaniel Bliss eventually appeared, after legal conflict, in 1798 and 1805. Pond continued Maskelyne's series after 1811, published annually until 1824, and then every three months

H1, H2, H3, H4, H5 the appellations, coined in the 1950s, for the five marine timekeepers made by John Harrison, the later ones with help from his son, William. H1–H4 are now in the collection of the National Maritime Museum; H5 is in the collection of the Worshipful Company of Clockmakers

Hadley quadrant an observing instrument of which the frame is an eighth of a circle, based on the same principle as the **Godfrey quadrant**; first devised by John Hadley in the early 1730s

High Court of the Admiralty The High Court included a Prize Court to oversee the seizure and fate of captured ships and their contents and an Instance Court to rule on disputes over collisions, salvage and payments. The Court exercised jurisdiction over serious criminal charges against mariners and shipping workers. Elder Brethren of **Trinity House** sat as Court assessors

Hydrographic Office The Office of Hydrographer to the Admiralty was established in 1795 to publish information from **Admiralty** charts and surveys as well as that gathered from naval voyages. Over the next four decades it gradually became committed to the organisation of survey voyages and the production of marine charts, the first of which appeared in 1800. Charts went on public sale from 1808, as did its *Sailing Directions* by 1829. The Office was headed by the **Hydrographer of the Navy**

Hydrographer of the Navy The first Hydrographer of the Navy was the civilian Alexander Dalrymple, appointed in 1795 after similar service from 1779 to the **East India Company**. His responsibility was to manage chart acquisition and distribution, a charge soon extended to managing dedicated surveys and the publication of charts. In 1808 Dalrymple was succeeded by naval officer Thomas Hurd, who was also **Secretary of the Board of Longitude** from 1810 and **Superintendent of Timekeepers** from 1819. In 1823 Hurd was replaced by Arctic navigator William Edward Parry

Jupiter's satellites (method) a method for determining **longitude** from the motions of the four most visible satellites of Jupiter, requiring tables of the predicted times of their eclipses (at a given reference **meridian**, such as that of Greenwich), which could be compared with the **local time** at which they were observed

K1, K2, K3 the modern names given to the three marine timekeepers made for the Board of Longitude by Larcum Kendall, K1 being an exact copy of **H4**, K2 and K3 being simplified versions. All three are in the collection of the National Maritime Museum

latitude angular distance north–south from the **equator** of a location on the Earth's surface

local time the time at a particular place, reckoned from the instant of the passage of the mean Sun over the **meridian** at that place (which defines local noon)

log and line an instrument for measuring the speed of a ship, consisting of a small board attached to a knotted line

longitude angular distance east-west on Earth from a chosen **meridian**, measured parallel to the **equator**

Longitude Board [see **Board of Longitude**]

Longitude Commissioners [see **Commissioners of Longitude**]

Lord High Admiral [see **Admiral, Lord High**]

Lowndean professorship The Lowndean chair of Astronomy and Geometry was established at Cambridge University in 1749 under an endowment from Cheshire astronomer Thomas Lowndes. Its responsibilities included astronomical observations and lectures. The Lowndean professor was confirmed as ***ex officio*** member of the **Board of Longitude** in the **Act of 1818**

Lucasian professorship The Lucasian chair in Mathematics was established at Cambridge University in 1663 by MP Henry Lucas. The Lucasian professor was named ***ex officio*** **Commissioner of Longitude** in the **Act of 1714**, when the chair was held by Nicholas Saunderson

lunar-distance method (or lunars) a method for finding **longitude** based on a measurement of the angular distance between the Moon and the Sun or another star, and the altitudes of both bodies. The observed position is reduced to a position as viewed from the centre of the Earth to give a true lunar distance, used to find a reference time (e.g. at Greenwich) from appropriate tables, such as those in the *Nautical Almanac*. Comparison of this time to the **local time** allows the navigator to determine the longitude from the reference **meridian**

magnetic dip the angle of Earth's magnetic field lines to the horizontal, which varies according to location

magnetic variation the angle between true north and magnetic north (the angle indicated by a **compass** needle at a particular location), which varies at different locations and changes over time

magnetic variation (method) a method of plotting a ship's position from a measurement of **magnetic variation** at the current location, then plotting that position using the observed **latitude** and a chart showing lines of equal magnetic variation; some attempts were also made to do the same with **magnetic dip**

marine chair an apparatus designed to steady an observer to allow them to make accurate shipboard observations, often of Jupiter's satellites [see also **Jupiter's satellites (method)**]

marine chronometer [see **chronometer, marine**]

mariner's astrolabe an observing instrument used at sea from the sixteenth century for measuring the altitude in degrees of the Sun and stars

mariner's quadrant an instrument, usually made of metal, comprising a quarter of a circle, with a graduated arc of 90 degrees and sights along one of the straight edges, used to measure angles relative to the vertical as defined by a plumb bob

mean time or mean solar time time by the calculated motion of the mean Sun, averaging out the annual variations in the Sun's apparent motion; the time shown by an ordinary clock

meridian defined in geography and geodesy as half a great circle on the Earth's surface, terminating at the North and South Poles; often used when referring to the observer's own meridian

movement (of a timekeeper) the term properly used to describe a timekeeper's mechanism as a whole

mural quadrant a large **astronomical quadrant** fixed to a wall

Nautical Almanac In 1765 **Astronomer Royal** Nevil Maskelyne won approval from the **Board of Longitude** to compile and publish a 'Nautical Ephemeris' to aid the determination of longitude at sea using the **lunar-distance method** by offering tabulations of angles between the Moon and nearby stars or the Sun. The **Act of 1765** approved this plan. The first ***Nautical Almanac and Astronomical Ephemeris*** was available in print in January 1767. With improved management and faster publication schedules, by the 1780s the ***Nautical Almanac*** was available five years in advance. See Box 7.1 for its contents and structure. Users of the ***Nautical Almanac*** also relied on instructions and data in the ***Tables Requisite*** together with tables of **atmospheric refraction** and **parallax**. The accuracy and production of the ***Nautical Almanac*** was widely held to have suffered under Maskelyne's successor John Pond, and the **Act of 1818** transferred responsibility to Thomas Young as **Superintendent of the Nautical Almanac**. A separate Nautical Almanac Office was established three years after the dissolution of the Board of Longitude

naval estimates Funding for the Royal Navy was officially fixed by a parliamentary Committee of Supply, based on estimates that set out expenditure required. These estimates had three components. The largest, the sea service estimate, based on a notional calculation of the number of men in service, supplied funds for wages and maintenance alongside finance for the **Board of Ordnance**. The ordinary estimate met demands such as the administrative costs of the **Admiralty** and the **Navy Board** and stipends for officers on half-pay. The extraordinary estimate funded shipbuilding and dockyards. The sums agreed were drawn by the **Treasurer of the Navy** from credit on a Treasury account, though the estimates never fully met naval costs and were supplemented by the issue of Navy bills

Navy Board The Navy Board was responsible for supplying the fleet with provisions, equipment and ordnance, together with managing naval dockyards and shipbuilding. The **Board of Ordnance** was supposed to supply the **Navy Board** with artillery and military supplies. Until 1786 the Board was housed at the Navy Office near Tower Hill, London, when it moved to Somerset House. During the eighteenth century the Board comprised a group of commissioners, normally retired naval officers, chaired by the **Comptroller of the Navy**, and included the Navy Surveyor and the **Treasurer of the Navy**, as well as so-called resident commissioners to oversee each Royal Dockyard. In 1796 the Navy Board was overhauled and a secretary appointed under the Comptroller's chairmanship. It was eventually abolished in 1832

Northwest Passage There was long an interest among European maritime powers to establish a route linking the northern Atlantic to the Pacific through the Arctic, in principle shortening travel between western Europe and China. Considerable myth and fantasy surrounded these ambitions. It was apparently demonstrated from 1772 that no strait linked Hudson's Bay to the Pacific. James Cook's final voyage of 1776–1779 had as one of its main aims exploring the northern Pacific and charting the American coast to test the possibility of such a passage, followed by George Vancouver's intensive surveys 1792–1794. These programmes were supplied with equipment and instructions by the **Board of Longitude**. The **Act of 1818** set out rewards for sailing from the north Atlantic along the northern coast of the Americas, depending on the distance travelled west and north, with the highest sum awarded to mariners who could reach the Pacific. In 1820 the Board awarded £5000 to William Parry for reaching the longitude of 113 degrees west

Observatoire de Paris The Paris Observatory was founded by the French monarchy in 1667, initially to house experimental philosophical work and equipment for the **Académie Royale des Sciences**, as well as astronomical instruments, personnel and observations for surveys of the Moon, planets and stars. Under the direction of Giovanni Domenico Cassini and his dynastic descendants from 1671, the Observatoire defined the **prime meridian** for France and its surveys and charts. After the Revolution, Jérôme Lalande directed the Observatoire from 1795 and engaged in frequent exchanges with Maskelyne around the production and use of the ***Greenwich Observations*** and ***Nautical Almanac***

octant [see **Hadley quadrant**]

Ordnance Board [see **Board of Ordnance**]

Ordnance Survey In 1790 the **Board of Ordnance** inaugurated a programme to triangulate the positions of numerous landmarks in the British Isles as points for constructing local topographic maps. The scheme owed much to work by the military surveyor William Roy, who had taken part in such projects in Scotland in the wake of Jacobite revolt and in a junction effected from 1784 between the **Royal Observatory, Greenwich** and the **Observatoire de Paris**. The impetus for the 1790–1791 project was military, and started with a triangulation survey of the southern coast facing apparent threats of French invasion. In 1801 the Survey's map of Kent was published and by 1810 most of southern England had been mapped. Under directors William Mudge (from 1798) and Thomas Colby (from 1820), the Ordnance Survey based at the Board of Ordnance headquarters in the Tower of London launched a complete survey of England and Wales

parallax the difference in an observed position or direction due to a difference in the viewpoint of the observer or observers

pendulum a rod, cord or wire with a weight or bob at or near one end, suspended from a fixed point so as to swing or oscillate freely under the influence of gravity; often used to refer to a weighted rod used to regulate a clock; also an instrument based on the same principle for geodetic and gravimetric measurements

Pendulum Committee The length of a **pendulum** beating seconds was long contemplated as offering a viable standard of length. It was understood that exact determination of that length in different places, thus measures of the changing force of gravity across the Earth's surface, would allow estimates of the Earth's shape. In 1816 it was proposed in Parliament, as part of a project to overhaul national weights and measures, that changes in this pendulum length be measured at different **Ordnance Survey** stations. In 1817 the **Royal Society** established a Pendulum Committee to manage experiments on pendulum lengths that would encourage improved standards and determine the figure of the Earth. Committee members included **Astronomer Royal** John Pond, head of the Ordnance Survey William Mudge and **Secretary of the Board of Longitude** Thomas Young. Two **Resident Commissioners of the Board of Longitude**, William Hyde Wollaston and Henry Kater, were also members. Kater launched a programme of pendulum trials in London, while the Committee provided advice on pendulum experiments to a series of scientific voyages, such as those of John Ross in the Arctic in 1818 and of Basil Hall along the Chilean and Peruvian coasts in 1821–1822, for which the **Board of Longitude** provided equipment

perpetual motion the motion of a machine that runs permanently once set going, without any further input of energy; sometimes also used to refer to a mechanism of this kind

philosopher's stone a long-sought-after substance believed to be a source of transformation, most commonly the power to change metals into gold or silver or to cure wounds or diseases and prolong life indefinitely

Plumian professorship The Plumian chair of astronomy and experimental philosophy was established at the University of Cambridge in 1704 under an endowment of former Greenwich vicar Thomas Plume to fund an astronomical observatory, experimental and astronomical equipment and a professorship to maintain and work with these facilities. The Plumian professor was ***ex officio*** a **Commissioner of Longitude** under the **Act of 1714**

pocket chronometer a pocket watch with a **chronometer escapement**; a chronometer in the form of a large watch

Prime Meridian the **meridian** in a geographic coordinate system at which longitude is defined by convention to be 0 degrees. Astronomers and cartographers used many different prime meridians up to the nineteenth century but the meridian running through the **Royal Observatory, Greenwich**, began to be defined as the Prime Meridian of the World following a recommendation at the International Meridian Conference in Washington, DC, in 1884

projector In eighteenth-century English the term projector was used to describe someone who put forward any scheme or venture. This was commonly used in a pejorative sense, with the associations of swindle and ingenious deceit subject to satirical attack and moral censure. **Secretary of the Board of Longitude** Thomas Young used the term in exactly this sense, writing in his lectures of 'superficial projectors of machines'

quadrant an instrument with a graduated arc of 90 degrees, used for measuring the angle between distant bodies or between the horizon and a distant body [see also **astronomical quadrant**; **Davis quadrant**; **Godfrey quadrant**; **Hadley quadrant**; **mariner's quadrant**; **mural quadrant**]

quadrature of the circle [see **squaring the circle**]

rate the difference of a **chronometer**'s timekeeping compared with a time standard (such as **mean solar time**), usually expressed in terms of variance over a twenty-four-hour period (known as daily rate). To set a rate for a chronometer, it was necessary to make comparisons with an astronomical or other time standard over an extended period to work out how much the timekeeper would normally be expected to gain or lose each day. The declared rate could be applied during the course of a voyage or other series of observations to the time indicated by the chronometer in order to calculate the corresponding mean time

reflecting circle a full-circle instrument for measuring the angles between distant bodies, using reflection to bring the images of two targets into coincidence

regulator a clock (usually **pendulum**-controlled) that keeps time accurately, typically used to set other timepieces correctly or to time astronomical observations

Resident Commissioners The **Act of 1818** defined new, salaried positions on the **Board of Longitude** for three Resident Commissioners, described as 'persons well versed in the sciences of mathematics, astronomy, or navigation'. The first were Fellows of the **Royal Society** closely allied with Joseph Banks: chemist and Royal Society secretary William Hyde Wollaston; Henry Kater, former Royal Engineer in India and leading member of the **Pendulum Committee**; and Thomas Young, who was also the **Secretary of the Board of Longitude**. In 1821 the new head of the **Ordnance Survey**, Thomas Colby, was made a Resident Commissioner, while Wollaston was replaced by astronomer and natural philosopher John Herschel

Resident Committee of Scientific Advice When the **Board of Longitude** was abolished in 1828, many of its functions were transferred by the **Admiralty** to a new administratively internal Resident Committee of Scientific Advice, with a slightly larger budget. Admiralty administrator John Barrow invited Henry Kater, John Herschel and Thomas Young, all previously **Resident Commissioners** of the Board of Longitude, to form this new Committee in the same way as the Resident Commissioners had functioned. Kater and Herschel refused and were replaced by Young's recommendations, Michael Faraday and military surveyor Edward Sabine. Barrow explained their role as advising the **Board of Admiralty** on inventions, calculations, discoveries 'and other scientific subjects'. Young died in May 1829, Sabine was posted to Ireland and the committee was abolished in 1831. It was replaced by the Admiralty Scientific Branch under the **Hydrographer of the Navy**, including the **Hydrographic Office**, Nautical Almanac Office, **Royal Observatory Greenwich. Royal Observatory Cape of Good Hope**, and Chronometer Office

Royal Artillery Under the management of the **Board of Ordnance**, the Royal Artillery was formally established in 1720 as a group of field artillery companies, which by 1722 became a regiment based at the Royal Arsenal, Woolwich, and from 1770 at its own barracks on Woolwich Common. Entrance and training were competitive and based at the **Royal Military Academy**

Royal Astronomical Society The Astronomical Society of London was established in early 1820 by a group of businessmen, lawyers and other astronomical practitioners to promote and define the practice of scientific astronomy, especially the pursuit of precision measurement, refinement of instrumentation and determination of planetary and stellar positions. The Society was greeted with hostility by president of the **Royal Society** Joseph Banks, and promoted by mathematicians such as Charles Babbage and John Herschel. The Society met at rented rooms in Covent Garden and then in Lincoln's Inn Fields. It gained a royal charter in 1831

Royal Engineers The **Board of Ordnance** set up an engineer officer corps at Woolwich in 1716, which in 1787 became the Royal Engineers with a corps of military artificers under its command. Their functions included surveys and chart making, drainage and fortification, siege warfare and communications construction.

Trained at the **Royal Military Academy**, they played a decisive role in the launch of the **Ordnance Survey** in the 1790s

Royal Hospital for Seamen The Greenwich Hospital was founded in 1692 as a naval mariners' retirement home. It included a chapel and infirmary, and was overseen by a governor, named an ***ex officio*** **Commissioner of Longitude** in the **Act of 1753**. From 1712, it supported the education of sons of poor Greenwich Pensioners, initially at Thomas Weston's Academy in Greenwich until the Hospital opened its own school in 1758. It became one of the largest navigation schools in Britain during this period

Royal Institution In March 1799, hosted by Joseph Banks, the experimenter and entrepreneur Count Rumford and colleagues proposed a new institution to pursue useful experimental and natural philosophical inquiry, especially on questions of mechanical invention, economy and agriculture, and to organise public lectures. Funds came from the Society for Bettering the Conditions of the Poor. In summer 1799 the Institution was created in Albemarle Street, Mayfair, gaining a royal charter within a year. Between 1801–1812 its laboratory was directed by chemist Humphry Davy, while in 1812 the Institution became a public organisation with a fee-paying membership and group of managers. Its public lectures, especially those launched in the 1820s, were major events in the London social season. In 1821 the pre-eminent natural philosopher and lecturer Michael Faraday, having previously worked as assistant, became superintendent and from 1825 laboratory director

Royal Mathematical School at Christ's Hospital The School was set up in summer 1673 within the charity school Christ's Hospital at Newgate, London. Its promoters appealed to the precedent of French state patronage of navigation schools. The aim was to train boys in the mathematics needed for navigation. These pupils, with support from the **Navy Board**, were apprenticed to ships' masters. James Hodgson, master of the school 1709–1755, sometimes advised on schemes presented to the **Commissioners of Longitude**. Astronomer and **computer** William Wales was employed as mathematics teacher at the School from 1775 and **Secretary of the Board of Longitude** from 1795

Royal Military Academy, Woolwich The **Board of Ordnance** established an Academy at Woolwich in 1741 to train cadet engineers and artillery officers, principally in mathematics, ballistics and fortification. From 1764 the Academy was divided into lower and upper schools, placed under the direction of a lieutenant governor, a position held by William Mudge from 1809. Teaching was led by a group of professors and assistant masters, several eminent practical mathematicians. The Academy also trained military engineers for the **East India Company**. In 1806 dedicated buildings were provided for the Academy at Woolwich; by 1810 East India Company cadets were transferred to a new college at Addiscombe

Royal Naval Academy, Portsmouth A training institution for young naval recruits was ordered in 1729 and its building at Portsmouth Dockyard completed in 1733. Graduates were initially to be granted accelerated promotion, but this scheme met strong resistance and eventually failed. The Academy's masters included **computer** George Witchell (served 1766–1785) and astronomer William Bayly (served 1785–1807). It was renamed the Royal Naval College in 1806 and abolished in 1837, from the following year being used for officers' scientific training

Royal Observatory, Board of Visitors [see **Board of Visitors, Royal Observatory**]

Royal Observatory, Greenwich The astronomical observatory on Greenwich Hill was established under the **Board of Ordnance** in 1675 for the purpose of correcting astronomical tables and measures of star positions to solve the problem of longitude at sea. It was put under the charge of John Flamsteed, the first **Astronomer Royal**, whose charge was to survey stellar transits and motions of the Moon and planets, and compile and publish tables of their positions and times. The Astronomer Royal was ***ex officio*** a **Commissioner of Longitude**. Nevil Maskelyne, who held the position from 1765 to his death in 1811, played a major role on the **Board of Longitude** and in 1766 used the Royal Observatory for trials of **H4**. The **Act of 1774** stipulated that copies of marine timekeepers offered for reward be tested for a year at the Royal Observatory, while from 1822 annual **Greenwich premium trials** of sea clocks were held there under the responsibility of the **Superintendent of Timekeepers**

Royal Observatory, Cape of Good Hope The Cape Colony was seized by the British from the Dutch in 1806. In late 1819 Davis Gilbert, newly appointed a **Commissioner of Longitude** under the provisions of the **Act of 1818**, in co-ordination with president of the **Royal Society** Joseph Banks, lobbied to establish a new astronomical observatory there under the management of the **Board of Longitude**. The observatory was approved at the Board in mid-1820. With the backing of the Commissioner John Herschel, his fellow Cambridge mathematician Fearon Fallows was appointed Astronomer in late 1820 and reached the Cape in August 1821. The Board set out detailed plans for the observatory's work, including **rating chronometers** and compiling a southern star catalogue, and commissioned astronomical instruments. A series of unsatisfactory assistants were hired during the 1820s, the Observatory buildings were not completed until 1827 and Fallows died in 1831, by which time responsibility had passed to the **Admiralty**'s Scientific Branch

Royal Society The Royal Society for Improving Natural Knowledge was established in London in the early 1660s. It comprised Fellows, elected for life, some of them eminent natural philosophers, physicians, mathematicians and instrument-makers, several elected through their connections and influence in church and state. The Society met regularly, heard reports of experimental, mathematical, natural historical and other inquiries, and occasionally witnessed demonstrations or displays. Its affairs were managed by a council composed of the president and vice-president, the two secretaries, the treasurer and a number of other Fellows. Its chair was the president, elected by the Fellows for an indefinite period. From 1710 the Society was housed at Crane Court off Fleet Street, before moving in 1780 to rooms in Somerset House. From 1731 the Society awarded the Copley medal for success in discovery or experiment: in 1749 John Harrison won the medal for his timekeepers. The president was ***ex officio*** a leading member of the **Board of Visitors of the Royal Observatory** and a **Commissioner of Longitude**. Joseph Banks, president 1778–1820, took a principal role on the **Board of Longitude** and sought to bring its work close to that of the Society. Under the **Act of 1818**, three Royal Society fellows were to be appointed as Commissioners, to be nominated by the president and Council. The Society established several committees to organise inquiry and initiative in specific fields, such as the **Pendulum Committee** (1817) and the **Glass Committee** (1824)

Savilian professorship In 1619 the college warden, learned scholar and mathematical authority Henry Savile established two professorships at Oxford University in geometry and in astronomy. The Savilian professor of geometry was to lecture on the classical authorities as well as offering English-language teaching in introductory mathematics; the Savilian professor of astronomy was to combine teaching on the classical tradition with lectures on Copernicus, and conduct regular astronomical observations. The **Act of 1714** named the Savilian professor as an ***ex officio* Commissioner of Longitude**; it was eventually clarified that this included both Savilian chairs

Scientific Advice, Resident Committee of [see **Resident Committee of Scientific Advice**]

sea clock a mechanical clock used for measuring time on a ship

sea watch term used by John Harrison for his fourth and fifth marine timekeepers, **H4** and **H5**

Secretary of the Board of Longitude The **Act of 1714** specified neither administrative nor secretarial role to support the **Commissioners of Longitude**, and their meetings from the 1730s were normally held irregularly at the invitation of the **First Lord of the Admiralty**. In 1762, the Commissioners lobbied for funds for a secretary, a request agreed in summer 1763. The Secretary's tasks included co-ordinating the **Board of Longitude**'s meetings, maintaining and cataloguing its papers and records, and responding to and directing correspondence. The early secretaries were recruited from among senior **Admiralty** clerks, such as John Ibbetson (in post 1763–1782) and Harry Parker (1782–1795). Astronomer and mathematician William Wales, while master at the **Royal Mathematical School**, was Secretary from 1795 to his death in 1798, and was replaced by his former assistant George Gilpin, then also secretary to the **Royal Society**. On Gilpin's death in 1810 he was replaced by the **Hydrographer** Thomas Hurd. The **Act of 1818** overhauled the administrative structure of the Board and Thomas Young became Secretary and manager of the affairs of the Board

Secretary to the Admiralty This political appointment, normally held by a government MP, had financial oversight of naval business, including costs of dockyards and stores. The Secretary of the Admiralty was made an ***ex officio* Commissioner of Longitude** under the **Act of 1753**. In 1763 the title was changed to First Secretary to the Admiralty alongside a Second Secretary, both confirmed as ***ex officio*** Commissioners of Longitude by the Longitude Act of 1790. The Secretary normally attended the **Board of Admiralty** but was not formally a member. As effective head of budgetary surveillance in the Admiralty, the Secretary occupied a significant rôle in the affairs of the Commissioners of Longitude, particularly Josiah Burchett, who served until 1742; and in the later conduct of the **Board of Longitude** the Tory John Wilson Croker, First Secretary from 1809, and John Barrow, Second Secretary from 1804

Secretary of the Treasury During the eighteenth century there were two offices of Secretary of the Treasury, senior and junior, with the understanding from the later eighteenth century that the junior would replace the senior on retirement. The office was a political appointment, with responsibility for managing Treasury business, notably the affairs of the meetings of the Treasury Board and its vast correspondence, dealing principally with contracts and sales, revenue and taxation. It was not until 1782 that the posts received a salary; before then, they were expected to live off fees received. The Secretaries supervised the increasingly large group of Treasury clerks and as political agents frequently managed parliamentary votes and elections of MPs. John Robinson, Secretary 1770–1782 and recipient of lobbies on behalf of John Harrison, was notable for his management of the House of Commons

sextant, marine an observing instrument of which the frame is one sixth of a circle, for measuring angles up to 120 degrees or more, using reflection to bring the images of two targets into coincidence

sidereal time time from the stars, a sidereal day being defined as the time between two successive **meridian** transits of a single star, representing a true 360-degree revolution of the Earth

signalling (method) a method for determining a ship's position by observing the distance and/or **bearing** of the explosions of rockets fired from a known location, such as a ship moored at sea or a place on land

solar time time from the Sun's movement, a solar day being defined as the time between two successive **meridian** transits of the Sun

Solicitor General A government law officer who acted as deputy to the **Attorney General**. In the eighteenth century the position was almost always held by a senior barrister

Speaker of the House of Commons The Speaker acted as presiding officer in meetings of the House of Commons. During the eighteenth century the position was normally held by an MP loyal to the government and often held other ministerial positions. Under the **Act of 1714**, the Speaker was ***ex officio*** a **Commissioner of Longitude**

spring-detent a principal component of a **chronometer escapement**, with a flexible spring blade at its foot, and which is caused to lock and unlock the escape wheel

squaring the circle the challenge of constructing a square with the same area as a given circle, a problem incapable of geometrical solution; used as a metaphor for trying to do the impossible

Superintendent of the Nautical Almanac The reputation of the ***Nautical Almanac*** under **Astronomer Royal** John Pond suffered, and in 1818 its management was transferred to the new Secretary of the **Board of Longitude**, Thomas Young, appointed the first Superintendent of the Nautical Almanac, with general responsibility for oversight of computation and publication. When the Board was abolished in 1828, responsibility for the *Nautical Almanac* passed to the **Admiralty**, Young remaining as Superintendent. Young died the following year, and Pond temporarily resumed control despite fierce criticisms. In 1831 William Stratford, secretary of the **Royal Astronomical Society** was made Superintendent and a new Nautical Almanac Office set up as part of the **Admiralty**'s Scientific Branch

Superintendent of Timekeepers At the start of the nineteenth century the **Navy Board** had a responsibility for the distribution of **marine chronometers** to naval officers. This task was transferred to the **Hydrographic Office** in 1809 under the official direction of Thomas Hurd, who became **Secretary of the Board of Longitude** the following year. In 1821 the responsibility for chronometer distribution and testing was transferred again from the **Hydrographer** to the **Astronomer Royal**, John Pond, who was appointed Superintendent of Timekeepers. The following year he launched premium trials at the **Royal Observatory, Greenwich**, to ascertain the performance of marine chronometers on behalf of the **Admiralty**

Tables Requisite In 1766 the **Board of Longitude** published the first *Tables requisite to be used with the Astronomical and Nautical Ephemeris*. These were used for longitude calculations in tandem with the ***Nautical Almanac***. They offered schematic examples of calculations, as well as logarithm tables, conversion of degrees, and tables to correct for **parallax** and **atmospheric refraction**

time [see **apparent time**, **solar time**, **Greenwich Mean Time**, **local time**, **mean time**, **sidereal time**]

timekeeper (method) a method for finding longitude at sea by using a mechanical timekeeper able to keep the reference time (**local time** at a known location, such as Greenwich) with sufficient accuracy that it could be compared with local time on board ship to establish the difference in longitude between the known location and the ship's current position

Transit of Venus the astronomical event during which the planet Venus passes directly between the Sun and the Earth, becoming visible as a small disc moving across the Sun's face. Observations of the event from different places on Earth can be used to determine the Earth-Sun distance and hence the scale of the solar system

Treasurer of the Navy a political appointment in charge of the Royal Navy's finances and member of the **Navy Board**. By the eighteenth century the position was effectively separated from the official quarters at the Navy Office. Under the **Act of 1763** it was the responsibility of the Treasurer of the Navy to make a payment to John Harrison following the **discovery** of his **sea watch**

Trinity House The Corporation of Trinity House was established to oversee pilotage and provide charitable support for seamen. During the eighteenth century, under a Court of Elder Brethren, responsibilities extended to regulating navigational skills, examining pupils of the **Royal Mathematical School** as well as pilots, those wishing to be masters on His Majesty's ships and naval schoolmasters, and the maintenance of legal order in maritime affairs, including acting as assessors at the **High Court of Admiralty**. The **Act of 1714** named the Master of Trinity House as a **Commissioner of Longitude** *ex officio*. In early correspondence as Commissioner of Longitude, Isaac Newton wrote that Trinity House's advice on the practical utility of navigational schemes was essential. Its lasting legacy was the construction and maintenance of lighthouses and other coastal beacons

variation [see **magnetic variation**]

Woolwich [see **Royal Military Academy, Woolwich**]

Bibliography

* Complete holding available via Cambridge Digital Library (cudl.lib.cam.ac.uk)
† Items relating to Board of Longitude available via Cambridge Digital Library

Archival Sources

Australia

State Library of Victoria
*H17809 'Harrison Journal'

France

Observatoire de Paris
MS 1090 Letters from Franz Xaver von Zach to Jérôme Lalande

Germany

Universitätsarchiv Göttingen
Cod. MS. Mich. Papers of Johann David Michaelis

Republic of Ireland

King's Inns, Dublin
N3/7/4/2 Collection of Indentures and Precedents, 1680–1826

United Kingdom

British Library
Add MS 37182 Correspondence of Charles Babbage
Add MS 39822 Papers of Samuel Pepys and Larcum Kendall

IOR/E	India Office Records: East India Company General Correspondence
IOR L/MIL	India Office Records: Records of the Military Department
RP	Copies of papers exported from the United Kingdom
Sloane MS 4055	Hans Sloane: Original correspondence

Cambridge University Library

*Mm.6.48	Letters, Memoranda and Journal Containing the History of Mr William Gooch
*MS Add.3972	Portsmouth Collection: Papers on Finding the Longitude at Sea
RGO 1	Royal Greenwich Observatory Archives: Papers of John Flamsteed
†RGO 4	Royal Greenwich Observatory Archives: Papers of Nevil Maskelyne
†RGO 5	Royal Greenwich Observatory Archives: Papers of John Pond
RGO 6	Royal Greenwich Observatory Archives: Papers of George Airy
*RGO 14	Royal Greenwich Observatory Archives: Board of Longitude Archive
RGO 15	Royal Greenwich Observatory Archives: Papers of the Cape Observatory
RGO 16	Royal Greenwich Observatory Archives: Papers of the Nautical Almanac Office
UA Obsy	Archives of the University Observatory

Doncaster Archives

CWM	Cooke of Wheatley Muniments

King's College London Archives

K/MUS 1	Astronomical and meteorological observations compiled mainly at the Royal Observatory, Kew

London Metropolitan Archives

MJ/SR	Middlesex Sessions of the Peace: Court in Session

The National Archives, Kew

ADM 1	Admiralty, and Ministry of Defence, Navy Department: Correspondence and Papers
ADM 7	Admiralty: Miscellanea
ADM 49	Navy Board, and Admiralty, Accountant General's Department: Miscellaneous Accounting Records
C 11	Court of Chancery: Six Clerks Office: Pleadings 1714 to 1758
PRIS 1	Fleet Prison: Commitment Books
PROB 11	Prerogative Court of Canterbury and Related Probate Jurisdictions: Will Registers
SP 36	Secretaries of State: State Papers Domestic, George II
SP 78	Secretaries of State: State Papers Foreign, France
TS 21	Treasury Solicitor: Deeds, Evidences and Miscellaneous Papers

National Maritime Museum, Greenwich

†ADM/A	Navy Board: In-letters from the Admiralty, 1689–1815
ADM/B	Board of Admiralty: In-letters from the Navy Board, 1738–1809
†ADM/L	Navy Board: Lieutenants' Logs
†AGC	Letters
*BGN	Papers of William Wildman Barrington-Shute
†FIS	Papers of George Fisher
†MKY	Papers of Andrew Mackay
*MSK	Papers of Nevil Maskelyne
†MSS	Uncatalogued Manuscripts
†NVT	Navigation: Theory
†SAN	Sandwich Papers

Royal Society, London

CB	Papers of Charles Blagden
CLP	Classified Papers of the Royal Society
CMO	Council Minutes Original
EL	Early Letters: Original Manuscripts of Letters to the Royal Society
HS	Herschel Papers
LBO	Letter Book of the Royal Society: 'Original' Copies
MM	Miscellaneous Manuscripts by, about or Belonging to the Fellows of the Royal Society
MS	Manuscripts General
RBO	Register Book of the Royal Society: 'Original' Copies

Spalding Gentleman's Society, Spalding

William Stukeley, 'Memoirs of the Royal Society'

University of Southampton Library

MS61	Wellington Papers

Worshipful Company of Clockmakers, London

6026/1	John Harrison, 'Description for Making Clocks for Use on Land and at Sea'

United States of America

Houghton Library, Harvard University, Cambridge, MA

MS Eng 719	Christopher Smart, 'Jubilate agno' manuscript

McFarlin Library of the University of Tulsa

2007-006	Legal Documents Relating to the Talbot Family (Earls of Shrewsbury)

US National Archives and Records Administration

RG 45	Miscellaneous Letters Received by the Secretary of the Navy

Primary Printed Sources

George Airy, 'Report on the Progress of Astronomy during the Present Century', in *Report of the First and Second Meetings of the British Association for the Advancement of Science*, 2nd ed. (London, 1835), pp. 125–189

George Airy (ed.), *A Catalogue of Circumpolar Stars, Deduced from the Observations of Stephen Groombridge* (London, 1838)

George Airy, 'Report of the Astronomer Royal, to the Board of Visitors, Read at the Annual Visitation of the Royal Observatory, Greenwich, June 1, 1839', in *Astronomical Observations made at the Royal Observatory, Greenwich, in the Year 1838* (London, 1840), pp. G1–6

George Airy, 'Results of the Observations made by the Rev. Fearon Fallows at the Royal Observatory, Cape of Good Hope in the Years 1829, 1830, 1831', *Monthly Notices of the Royal Astronomical Society*, 19 (1851), 1–102

George Airy, 'Report of the Astronomer Royal to the Board of Visitors of the Royal Observatory Greenwich, Read at the Annual Visitation of the Royal Observatory, 1858, June 5', in *Astronomical and Magnetical and Meteorological Observations Made at the Royal Observatory, Greenwich, in the Year 1857* (London, 1859), pp. G1–G19

George Airy, 'Remarks on the History and Position of the (Now Abolished) Board of Longitude', 18 May 1872, in *Royal Commission on Scientific Instruction and the Advancement of Science, Minutes of Evidence, Appendices, and Analyses of Evidence, Vol. II* (London, 1874), Appendix VIII

Wilfrid Airy, *Autobiography of Sir George Biddell Airy* (Cambridge, 1896)

Dorotheo Alimari, *Longitudinis aut terra aut mari investiganda methodus* (London, 1715)

A.M. in Hydrostat, *Thoughts of a Project for Draining the Irish Channel* (Dublin, 1722)

Guillaume Amontons, *Remarques et expériences phisiques sur la construction d'une nouvelle clepsidre, sur les baromètres, termomètres, & higromètres* (Paris, 1695)

Anonymous, *The Life and Glorious Actions of Sir Cloudesly Shovel, Kt.* (London, 1707)

Anonymous, *An Essay Towards a New Method to Shew the Longitude at Sea; Especially Near the Dangerous Shores* (London, 1714)

Anonymous, *Will-with-a-Wisp; or, the Grand Ignis Fatuus of London Being a Lay-Man's Letter to a Country-Gentleman, Concerning the Articles Lately Exhibited Against Mr. Whiston* (London, 1714)

Anonymous, *The History and Proceedings of the House of Commons*, 14 vols (London, 1742)

Anonymous, *A Calculation Shewing the Result of an Experiment Made by Mr. Harrison's Time-Keeper* (n.p., 1762)

Anonymous, *Memorial Concerning Mr. Harrison's Invention for Measuring the Time at Sea* (n.p., 1762)

Anonymous, *Proposal for Examining Mr. Harrison's Time-Keeper at Sea* (n.p., 1762)

Anonymous, *An Account of the Proceedings in Order to the Discovery of the Longitude in a Letter to the Right Honourable ******, Member of Parliament*, 1st ed. (London, 1763)

Anonymous, *An Account of the Proceedings in Order to the Discovery of the Longitude in a Letter to the Right Honourable ******, Member of Parliament*, 2nd ed. (London, 1763)

Anonymous, *A Narrative of the Proceedings Relative to the Discovery of the Longitude at Sea; By Mr. John Harrison's Time-Keeper; Subsequent to those Published in the Year 1763* (London, 1765)

Anonymous, *Some Particulars Relative to the Discovery of the Longitude; Mentioning Several Foreign Premiums, and Exactly Narrating the Particulars of the British Acts of Parliament, Respecting that Affair* (London, 1765)

Anonymous, *Tables Requisite to Be Used with the Astronomical and Nautical Ephemeris for Finding the Latitude and Longitude at Sea. Published by Order of the Commissioners of Longitude* (London, 1766)

Anonymous, *The Case of Mr. John Harrison*, 1st ed. (n.p., c. 1766)

Anonymous, *The Case of Mr. John Harrison*, 2nd ed. (n.p., 1770)

Anonymous, *The Case of Mr. John Harrison*, 3rd ed. (n.p., 1773)

Anonymous, 'The Principles of Mr. Harrison's Time-Keeper', *The Gentleman's Magazine and Historical Chronicle*, 37 (1767), 156

Anonymous, *Tables Requisite to Be Used with the Nautical Ephemeris for Finding the Latitude and Longitude at Sea. Published by Order of the Commissioners of Longitude*, 2nd ed. (London, 1781)

Anonymous, *Objections to Mr. Mudge's Pretensions to a Reward from Parliament. By a Committee of the Board of Longitude* (London, 1793)

Anonymous, *Tables Requisite to Be Used with the Nautical Ephemeris, for Finding the Latitude and Longitude at Sea. Published by Order of the Commissioners of Longitude*, 3rd ed. (London, 1802)

Anonymous, *Selections from the Additions that Have been Occasionally Annexed to the Nautical Almanac: From Its Commencement to the Year 1812* (London, 1813)

John Arbuthnot, *To the Right Honourable the Mayor and Aldermen of the City of London: The Humble Petition of the Colliers, Cooks, Cook-Maids, Black-Smiths* (London, 1716)

John Arnold, *An Account Kept during Thirteen Months in the Royal Observatory at Greenwich, of the Going of a Pocket Chronometer Made on a New Construction* (London, 1780)

John Arnold, *Certificates and Circumstances Relative to the Going of Mr. Arnold's Chronometers* (London, 1791)

John Roger Arnold, *Explanation of Time-Keepers, Constructed by Mr. Arnold. Delivered to the Board of Longitude by Mr. Arnold, March 7th, 1805,* (London, 1805)

Charles Babbage, *Reflections on the Decline of Science in England* (London, 1830)

Charles Babbage, *The Exposition of 1851* (London, 1851)

Charles Babbage, *Passages from the Life of a Philosopher* (London, 1864)

Francis Baily, *Astronomical Tables and Remarks for the Year 1822* (London, 1822)

Francis Baily, *Remarks on the Present Defective State of the Nautical Almanac* (London, 1822)

Francis Baily, *Further Remarks on the Present Defective State of the Nautical Almanac* (London, 1829)

Joseph Banks, *Observations on Mr. Mudge's Application to Parliament for a Reward for His Timekeepers, which, Agreeably to the Act of Parliament of 14th of George III. had been Unanimously Denied him by the Board of Longitude* (London, 1793)

Joseph Banks, *Protest Against a Vote of the Board of Longitude Granting to Mr. Earnshaw a Reward for the Merit of His Time-keepers* (London, 1804)

Joseph Banks, 'Some Account of the Cape of Good Hope', in *The Endeavour Journal of Joseph Banks, 1768–1771*, ed. by J. C. Beaglehole, 2 vols (Melbourne, 1962), II, pp. 250–260

John Barrow, *Account of Travels into the Interior of Southern Africa,* 2 vols (London, 1801 and 1804)

[John Barrow], 'The Cape of Good Hope', *Quarterly Review*, 22 (1819), 203–246

Jeremy Bentham, *The Rationale of Reward* (London, 1825)

Johann Bernoulli [as John Bernoulli], *A Sexcentenary Table; Exhibiting, at Sight, the Result of Any Proportion, Where the Terms Do Not Exceed 600 Seconds or 10 Minutes; with Precepts and Examples* (London, 1779)

Jacques Besson, *Le Cosmolabe ou instrument universel, concernant toutes observations qui se peuvent faire par les sciences mathematiques* (Paris, 1567)

Case Billingsley, 'Since the publication of my late proposal, I find it is the opinion of many learned in astronomy and navigation, that the longitude at sea, will never be discovered to any degree of certainty …' ([London], 1714)

Case Billingsley, *The Longitude at Sea, not to be Found by Firing Guns, nor by the Most Curious Spring-Clocks or Watches* (London, 1714)

John Bird, *The Method of Dividing Astronomical Instruments* (London, 1767)

John Bird, *The Method of Constructing Mural Quadrants* (London, 1768)

Nathaniel Bowditch, *The New American Practical Navigator* (Newburyport, 1802)

William Broughton, *A Voyage of Discovery to the North Pacific Ocean* (London, 1804)

Robert Browne, *Methods, Propositions and Problems, for Finding the Latitude; With the Degree and Minute of the Equator upon the Meridian. And the Longitude at Sea. By Caelestial Observations only. And also by Watches, Clocks, &c. and to Correct them and Know their Alterations* (London, 1714)

Robert Browne, *A System of Theology Revealed from God by the Angels in the British Language* (London, 1728)

Robert Browne, *To the Honourable the Commons of Great Britain in Parliament Assembled, the Case of Robert Browne Humbly Sheweth* ([London], 1732)

Digby Bull, *An Exhortation to Trust in God, and Not to Despair of His Help and Fall from Him, in this Dark Time of Popery that Is Coming Upon the Church* (London, 1695)

Digby Bull, *A Letter of Advice to all the Worthy and Ingenuous Merchants of the City of London, and Elsewhere in England, Scotland, and Ireland … Shewing an Exact, Easie and Speedy Way to Know the Longitude of all Places in the World, Where the European Merchants have their Agents to make Observations; and Also How the Longitude of Places may be Better Known upon Ship-Board* (London, 1706)

Digby Bull, *A Farther Warning of Popery* (London, 1710)

Bureau des Longitudes, *Tables astronomiques publiées par le Bureau des longitudes* (Paris, 1806)

Antonio Cagnoli, *Memoir on a New and Certain Method of Ascertaining the Figure of the Earth by Means of Occultations of the Fixed Stars*, trans. by Francis Baily (London, 1819)

Elizabeth Carter, *A Series of Letters between Mrs Elizabeth Carter and Miss Catherine Talbot, from the Year 1741 to 1770*, 4 vols (London, 1809)

Alexis Clairaut, 'Copy of a Letter from M. Clairault to Dr Bevis, dated Paris, 11 April, 1765; from the English Original, in His Own Hand', *The Gentleman's Magazine, and Historical Chronicle*, 35 (1765), 208

James Clarke, *An Essay, Wherein a Method Is Humbly Propos'd for Measuring Equal Time with the Utmost Exactness; Without the Necessity of Being Confin'd to Clocks, Watches, or any Other Horological Movements; in Order to Discover the Longitude at Sea* (London, 1714)

James Clarke, *The Mercurial Chronometer Improv'd: or, a Supplement to a Book Entituled, An Essay, Wherein a Method is Humbly Propos'd for Measuring Equal Time with the Utmost Exactness* (London, 1715)
William Cobbett, *Cobbett's Parliamentary History of England*, 36 vols (London, 1806–1820)
Zerah Colburn, *A Memoir of Zerah Colburn Written by Himself* (Springfield, 1833)
Commissioners of Longitude, *Minutes of the Proceedings of the Commissioners Appointed by Act of Parliament for the Discovery of the Longitude at Sea, at Their Meetings on the 25th, 28th, and 30th of May, and 13th of June, 1765* ([London], 1765)
James Cook, James King and William Bayly, *The Original Astronomical Observations made in the Course of a Voyage to the Northern Pacific Ocean, for the Discovery of a North East or North West Passage wherein the North West Coast of America and North East Coast of Asia were Explored in His Majesty's Ships the Resolution and Discovery, in the Years MDCCLXXVI, MDCCLXXVII, MDCCLXXVIII, MDCCLXXIX, and MDCCLXXX* (London, 1782)
John Coster, *A Practicable Method for Finding the Longitude at Sea, by a Marmeter, or Instrument, for Measuring the Exact Run of the Ship* (London, 1714)
Alexander Dalrymple (attr.), *Some Notes Useful to Those Who Have Chronometers at Sea* ([London], [1779/1780])
Alexander Dalrymple, *Longitude: A Full Answer to the Advertisement Concerning Mr. Earnshaw's Timekeeper in the Morning Chronicle, 4th Feb. and Times, 13th Feb. 1806* (London, 1806)
Jean Baptiste Joseph Delambre, 'Notice sur la vie et les travaux de M. Maskelyne', in *Mémoires de la classe des sciences mathématiques et physiques de l'Institut Impérial de France. Année 1811* (Paris, 1812), pp. lix–lxxviii
William Derham, *The Artificial Clock-Maker* (London, 1714)
John Theophilus Desaguliers, 'Remarks on Some Attempts Made Towards a Perpetual Motion, by the Reverend Dr. Desaguliers', *Philosophical Transactions*, 31 (1720), 234–239
John Theophilus Desaguliers, *A Course of Experimental Philosophy*, 2 vols (London, 1744–1745)
Henry de Saumarez, *An Account of the Proceedings of Henry de Saumarez of the Island of Guernesey, Gent: Concerning His Discovery of an Invention, by which the Course of a Ship at Sea May be Better Ascertained Than by Logg-line* (London, 1717)
Henry de Saumarez, 'An Account of a New Machine, called the Marine Surveyor', *Philosophical Transactions*, 33 (1725), 411–432
Thomas Earnshaw, *Explanation of Time-Keepers, Constructed by T. Earnshaw. Three of Them Having Been Tried Under the Present Act of Parliament. Delivered to the Board of Longitude by the Astronomer Royal, on the Part of Mr. Earnshaw, March 7th, 1805*, (London, 1805)
Thomas Earnshaw, *Longitude, an Appeal to the Public Stating Mr. Thomas Earnshaw's Claim to the Original Invention of the Improvements in his Timekeepers, their Superior Going in Numerous Voyages, and also as Tried by the Astronomer Royal by Orders of the Commissioners of Longitude and his Consequent Right to National Reward* (London, 1808)
Thomas Earnshaw and John Roger Arnold, *Explanations of Time-Keepers, Constructed by Mr. Thomas Earnshaw and the late Mr. John Arnold. Published by Order of the Commissioners of Longitude* (London, 1806)

William Emerson, *A System of Astronomy* (London, 1769)

William Emerson, *Miscellanies* (London, 1776)

Leonhard Euler, *Lettres à une Princesse d'Allemagne*, 3 vols (Berne, 1775)

Fearon Fallows, 'Communication of a Curious Appearance Lately Observed upon the Moon', *Philosophical Transactions*, 112 (1822), 237–238

Fearon Fallows, 'A Catalogue of Nearly All the Principal Fixed Stars Between the Zenith of Cape Town, Cape of Good Hope, and the South Pole', *Philosophical Transactions*, 114 (1824), 457–470

Jean-Henri-Samuel Formey, *The Life of John Philip Baratier: The Prodigy of this Age for Genius and Learning* (London, 1745)

George Gordon, *A Compleat Discovery of a Method of Observing the Longitude at Sea* (London, 1724)

Richard Graham, 'The Description and Use of an Instrument for Taking the Latitude of a Place at Any Time of the Day', *Philosophical Transactions*, 38 (1733), 450–457

Robert Greene, *The Principles of Natural Philosophy* (Cambridge, 1712)

Robert Greene, *The Principles of the Philosophy of the Expansive and Contractive Forces* (Cambridge, 1727)

John Hadley, 'The Description of a New Instrument for Taking Angles', *Philosophical Transactions*, 37 (1731–1732), 147–157

John Hadley, 'An Account of Observations Made on Board the Chatham-Yacht, August 30th and 31st, and September 1st, 1732', *Philosophical Transactions*, 37 (1731–1732), 341–356

Francis Haldanby, *An Attempt to Discover the Longitude at Sea, Pursuant to What Is Proposed in a Late Act of Parliament* (London, 1714)

William Hall, *A New and True Method to Find the Longitude, Much More Exacter than that of Latitude by Quadrant* (London, 1714)

Edmond Halley, *Astronomical Tables with Precepts Both in English and Latin, for Computing the Places of the Sun, Moon, Planets, and Comets* (London, 1752)

Samuel Hardy, *The Theory of the Moon made Perfect; So Far at Least as to Determine the Longitude, Both at Sea and Land, within the Limits Required by Act of Parliament* (London, 1751)

Samuel Hardy, *A Translation of Scherffer's Treatise on the Emendation of Dioptrical Telescopes ... To Which Are Added, Explanatory Notes; and a Description of a Telescope to be Used at Sea* (London, 1768)

Edward Harrison, *Idea longitudinis* (London, 1696)

John Harrison, *The Principles of Mr. Harrison's Time-Keeper with Plates of the Same* (London, 1767)

John Harrison, *Principes de la montre de Mr. Harrison avec les planches relatives a la même montre*, trans. by Esprit Pezenas (Avignon, 1767)

John Harrison, *Remarks on a Pamphlet Lately Published by the Rev. Mr. Maskelyne, Under the Authority of the Board of Longitude*, 1st ed. (London, 1767)

John Harrison, *Remarks on a Pamphlet Lately Published by the Rev. Mr. Maskelyne, Under the Authority of the Board of Longitude*, 2nd ed. (London, 1767)

John Harrison, *A Description Concerning Such Mechanism as Will Afford a Nice, or True Mensuration of Time; Together with Some Account of the Attempts for the Discovery of the Longitude by the Moon: As Also an Account of the Discovery of the Scale of Musick*, 1st ed. (London, 1775)

John Harrison, *A Description Concerning Such Mechanism as Will Afford a Nice, or True Mensuration of Time; Together with Some Account of the Attempts for the Discovery of the Longitude by the Moon: As Also an Account of the Discovery of the Scale of Musick*, 2nd ed. (London, 1775)

Isaac Hawkins, *An Essay for the Discovery of the Longitude at Sea, by Several New Methods Fully and Particularly Laid before the Publick* (London, 1714)

Robert Heath, *A Supplement to the Royal Astronomer and Navigator. Supplying (by Reduction and Equation) Accurate Astronomical Tables* (London, 1768)

Robert Heath, *The Seaman's Guide to the Longitude: or a Key to the Nautical Almanac and Astronomical Ephemeris* (London, 1770)

John Herschel, 'The Following Paper was Read at a Meeting of the Board of Longitude, April 5th 1827', in 'Copies of any Memorials or Reports presented to the Government since 1st January 1828, on the Subject of the Nautical Almanack, or the Board of Longitude', *House of Commons Papers*, 21 (1829), paper no. 91, 3–6

John Herschel, *Essays from the Edinburgh and Quarterly Reviews* (London, 1857)

William Hobbs, *A New Discovery for Finding the Longitude* (London, 1714)

Józef Maria Hoene-Wroński, *Address of M Hoene Wronski to the British Board of Longitude, Upon the Actual State of the Mathematics, their Reform, and Upon the New Celestial Mechanics, Giving the Definitive Solution of The Problem of Longitude. Translated from the Original in French, by W. Gardiner* (London, 1820)

Johan Horrins, *Memoirs of a Trait in the Character of George III ... Authenticated by Official Papers and Private Letters in Possession of the Author* (London, 1835)

John Horsley, 'An Account of his Observations at Sea for Finding out the Longitude by the Moon', *Philosophical Transactions*, 54 (1764), 329–332

House of Commons, 'Estimates of Ordinary of the Navy; Building and Repair of Ships; and Wages for 22,000 Men; for the Year 1821', *House of Commons Papers*, 15 (1821), paper no. 41

House of Commons, 'Navy Estimate no. II. For the Year 1832–3; with Estimate of the Expense of the Victualling and Medical Departments from 1 April 1832 to 31 March 1833', *House of Commons Papers*, 27 (1832), paper no. 116

House of Commons, 'Navy Estimates for the Year 1857–58', *House of Commons Papers*, 26 (1857), paper no. 0.13

House of Commons Committee, *Report from the Committee on Mr. Earnshaw's Petition* (London, 1809)

House of Commons Select Committee, *Report from the Select Committee of the House of Commons, to Whom It was Referred to Consider of the Report which was Made from the Committee to Whom the Petition of Thomas Mudge, Watch-maker, was Referred; and Who Were Directed to Examine Into the Matter Thereof, and Also to Make Enquiry Into the Principles on which Mr. Mudge's Time-keepers Have Been Constructed* (London, 1793)

House of Lords Select Committee, *Report and Minutes of Evidence Taken Before the Select Committee of the House of Lords Appointed to Consider of the Bill, Intituled, 'An Act Further to Amend the Law Touching Letters Patent for Inventions;' and also of the Bill, Intituled, 'An Act for the Further Amendment of the Law Touching Letters Patent for Inventions;' and to Report Thereon to the House. Session 1851. Ordered, by the House of Commons, to be Printed, 4 July, 1851* (London, 1851)

Henry Hunter (trans.), *Letters of Euler to a German Princess*, 2 vols (London, 1795)

Charles Hutton, *Tables of the Products and Powers of Numbers* (London, 1781)

Christopher Irwin, *A Summary of the Principles and Scope of a Method, Humbly Proposed, for Finding the Longitude at Sea* (London, 1760)

Benjamin Habakkuk Jackson, *Some New Thoughts Founded upon new Principles, Concerning a Threefold Motion of the Earth ... And Facilitating the Discovery of the Longitude* (London, 1714)

Elizabeth Johnson (attr.), *The Explication of the Vision to Ezekiel* (London, 1781)

Elizabeth Johnson (attr.), *The Astronomy and Geography of the Created World, and of Course the Longitude* (Oxford, 1785)

Jorge Juan, 'Informe sobre el reloj de Harrison', in *Colección Documentos inéditos para la Historia de España*, ed. by Miguel Salvá and Pedro Sainz de Baranda, 113 vols (Madrid, 1842–1895), XXI, pp. 224–238

George Keith, *An Easie Method not to be Found Hitherto in any Author, or History, Whereby the Longitude of any Places at Sea or Land, West or East from any First Meridian, E.g. London, May be Found at any Distance, Great or Small, by Certain Fixed Stars* (London, 1713)

Thomas Kirby, *The Seaman's Almanac for the Year 1802* (London, 1801)

Jérôme Lalande, *Tables Astronomiques de M. Halley* (Paris, 1759)

Jérôme Lalande, *Astronomie*, 2nd ed. (Paris, 1771)

Jérôme Lalande, *Astronomie*, 3rd ed. (Paris, 1792)

Jérôme Lalande, *Journal d'un voyage en Angleterre*, ed. by Hélène Monod-Cassidy (Oxford, 1980)

William Lax, *Tables to be Used with the Nautical Almanac for Finding the Latitude and Longitude at Sea; with Easy and Accurate Methods for Performing the Computations Required for these Purposes* (London, 1821)

Georg Christoph Lichtenberg, *Georg Christoph Lichtenberg: Briefwechsel*, ed. by Ulrich Joost and Albrecht Schöne, 6 vols (Munich, 1983–2004)

Richard Locke, *The Circle Squared. To Which Is Added, a Problem to Discover the Longitude Both at Land and Sea* (London, 1730)

Richard Locke, *A New Problem to Discover the Longitude at Sea, by the Same Observation, and with the Same Certainty, as the Latitude* (London, 1751)

Christian Carl Lous (ed.), *Historien af Mr. Harrisons Forsøg til Længdens Opfindelse formedelst et Uhr eller en Tidmaaler* [The History of Mr. Harrison's Attempt at the Discovery of the Longitude by Means of a Watch or Timekeeper] (Copenhagen, 1768)

John Lund, *The Mirrour: A Poem. In imitation of C. Churchill. To Which Are Added, Three Tales, (in the Manner of Prior)* (London, 1771)

Nevil Maskelyne, *The British Mariner's Guide* (London, 1763)

Nevil Maskelyne, *An Account of the Going of Mr. John Harrison's Watch, at the Royal Observatory* (London, 1767)

Nevil Maskelyne, *An Answer to a Pamphlet Entitled 'A Narrative of Facts', Lately Published by Mr. Thomas Mudge Junior* (London, 1792)

Nevil Maskelyne, *Arguments for Giving a Reward to Mr Earnshaw, for His Improvements on Time-keepers. By the Astronomer-Royal. To the Commissioners of Longitude* (London, 1804)

Tobias Mayer, *Theoria lunae juxta systema Newtonianum* (London, 1767)

Tobias Mayer, *Tabulæ motuum solis et lunæ novæ et correctæ* (London, 1770)

Tobias Mayer and Charles Mason, *Mayer's Lunar Tables, Improved by Mr. Charles Mason* (London, 1787)

Thomas Mudge, *Thoughts on the Means of Improving Watches, and More Particularly those for the Use of the Sea* (London, 1765)

Thomas Mudge junior, *A Narrative of Facts Relating to Some Timekeepers Constructed by Mr. T. Mudge for the Discovery of the Longitude at Sea, Together with Observations upon the Conduct of the Astronomer Royal Respecting them* (London, 1792)

Thomas Mudge junior, *A Reply to the Answer of the Rev. Dr. Maskelyne, Astronomer Royal, to A Narrative of Facts, Relating to Some Time-keepers, Constructed by Mr. Thomas Mudge, for the Discovery of the Longitude at Sea, &c.* (London, 1792)

Thomas Mudge junior, *A Description, with Plates, of the Time-Keeper Invented by the Late Mr. Thomas Mudge* (London, 1799)

[William Nicholson], 'Spider's Webs to Form the Cross Wires in the Eye Piece of Astronomical and Other Instruments', *Journal of Natural Philosophy, Chemistry and the Arts*, 1 (1802), 319–320

Stephen Plank, *An Introduction to the Only Method for Discovering Longitude. Humbly Presented to Both Houses of Parliament, and to those Worthy Persons that Are to Judge of It* ([London], 1714)

Stephen Plank, *An Introduction to a True Method for the Discovery of Longitude at Sea. Humbly Offer'd to the Consideration of Both Houses of Parliament* (London, 1720)

Elias Pledger, *A Brief DESCRIPTION of A NEW INVENTED, SMALL, Yet ACCURATE Astronomical Quadrant: Particularly Calculated for Curious Observations at SEA, But Equally Useful at LAND: AND THAT Either by DAY Or NIGHT* (London, 1731)

Conyers Purshall, *An Essay at the Mechanism of the Macrocosm: Or, the Dependance of Effects upon their Causes. In a new Hypothesis, Accommodated to Our Modern and Experimental Philosophy. In which are Solved Several Phoenomena, hitherto Unaccounted for; as the Cause of Gravitation, Motion, Reflexion, Refraction, &c. With a Method Proposed to Find out the Exact Rate that a Ship Runs, and Consequently the Longitude at Sea*, 1st ed. (London, 1705)

Conyers Purshall, *An Essay on the Mechanical Fabrick of the Universe: Or the Dependence of Effects Upon Their Causes. In a New Hypothesis ... In Which Are Solved Several Phaenomena ... Hitherto Unaccounted for; as the Cause of Gravitation, Motion, Reflection, Refraction, &c. With a Method Proposed, to Find out the Exact Rate that a Ship Runs, and Consequently the Longitude at Sea* (London, 1707)

Conyers Purshall, *To the Lords Commissioners for the Longitude* (London, c.1714)

Jesse Ramsden, *Description of an Engine for Dividing Mathematical Instruments* (London, 1777)

Jesse Ramsden, *Description of an Engine for Dividing Strait Lines on Mathematical Instruments* (London, 1779)

Sebastiano Ricci, *The New Method Proposed by Segnior Dorotheo Alimari to Discover the Longitude* (London, 1714)

P. I. Rocquette, *Nouveau sisteme de la sphere celeste suivi d'un moyen pour trouver les longuitudes sur mer. Par mr. P. I. Rocquette, horologeur de S. M. l'imperatrice de toutes les Russies, etc.* (St Petersburg, 1732)

Jacob Rowe, *Navigation Improved* (London, 1725)

William Roy, 'An Account of the Trigonometrical Operation, Whereby the Distance Between the Meridians of the Royal Observatories of Greenwich and Paris has been Determined', *Philosophical Transactions*, 80 (1790), 111–270, 591–614

Anthony Shepherd, *Tables for Correcting the Apparent Distance of the Moon and a Star from the Effects of Refraction and Parallax* (Cambridge, 1772)

Society for the Encouragement of Arts, Manufactures, and Commerce, *A List of the Society for the Encouragement of Arts, Manufactures and Commerce* (London, 1766)

James South, *Practical Observations on the Nautical Almanac and Astronomical Ephemeris* (London, 1822)

Jane Squire, *A Proposal to Determine our Longitude*, 1st ed. ([London], c.1731)

Jane Squire, *A Proposal for Discovering our Longitude / Proposition pour la decouverte de notre Longitude* (London, 1742)

Jane Squire, *A Proposal to Determine our Longitude*, 2nd ed. (London, 1743)

Henry Sully, *Description abregée d'une horloge d'une nouvelle invention* (Paris, 1726)

Jonathan Swift, *Travels into Several Remote Nations of the World. In Four Parts. By Lemuel Gulliver, First a Surgeon, and then a Captain of Several Ships*, 2 vols (London, 1726)

Michael Taylor, *A Sexagesimal Table, Exhibiting, at Sight, the Result of any Proportion, Where the Terms Do Not Exceed Sixty Minutes* (London, 1780)

Michael Taylor, *Tables of Logarithms of all Numbers, from 1 to 101000; and of the Sines and Tangents to Every Second of the Quadrant* (London, 1792)

Jeremy Thacker, *The Longitudes Examin'd. Beginning with a Short Epistle to the Longitudinarians, and Ending with the Description of a Smart, Pretty Machine of my own, which I am (almost) Sure will do for the Longitude, and Procure me the Twenty Thousand Pounds* (London, 1714)

Johann Wilhelm von Archenholz, *A Picture of England: Containing a Description of the Laws, Customs, and Manners of England* (London, 1789)

William Wales and William Bayly, *The Original Astronomical Observations Made in the Course of a Voyage towards the South Pole and Round the World* (London, 1777)

John Ward, *A Practical Method, To Discover the Longitude at Sea, By a New Contrived Automaton* (London, 1714)

William Whiston, *The Longitude and Latitude Found by the Inclinatory or Dipping Needle* ([London], 1719)

William Whiston, *The Calculation of Solar Eclipses Without Parallaxes* (London, 1724)

William Whiston, *The Longitude Discovered by the Eclipses, Occultations and Conjunctions of Jupiter's Planets* (London, 1738)

William Whiston, *Memoirs of the Life and Writings of Mr. William Whiston*, 2 vols (London, 1749)

William Whiston and Humphry Ditton, *A New Method for Discovering the Longitude Both at Sea and Land, Humbly Proposed to the Consideration of the Publick* (London, 1714)

William Whiston and Humphry Ditton, *A Petition about the Longitude* (London, 1714)

William Whiston and Humphry Ditton, *Reasons for a Bill, Proposing a Reward for the Discovery of the Longitude* (London, 1714)

Robert Wright, *An Humble Address to the Right Honourable the Lords and the Rest of the Honourable Commissioners* (London, 1728)

Robert Wright, *New and Correct Tables of the Lunar Motions, According to the Newtonian Theory ... with a Description of a New Instrument for Taking Altitudes at Sea* (Manchester, 1732)

[Thomas Young], 'A Reply to Mr. Baily's Remarks on the Nautical Almanac', *Quarterly Journal of Science, Literature and the Arts*, 13 (1822), 201–208

Thomas Young, 'Report on a Memorandum of a Plan for Reforming thc Nautical Almanack', in 'Copies of any Memorials or Reports presented to the Government since 1st January 1828, on the Subject of the Nautical Almanack, or the Board of Longitude', House of Commons Papers, 21 (1829), paper no. 91, 6–11

Periodicals

The British Palladium
Busy Body
Cape Town Gazette
Country Journal or The Craftsman
The Critical Review, or Annals of Literature
Daily Advertiser
Daily Courant
Daily Journal
Daily Post
The Dublin Journal
The Englishman
Evening Post
Gazetteer and New Daily Advertiser
General Advertiser
General Evening Post
The Gentleman's Magazine and Historical Chronicle
The Guardian
Hansard
Journal Littéraire
Journals of the House of Commons
Journals of the House of Lords
Lloyd's Evening Post and British Chronicle
London Chronicle
London Evening Post
London Gazette
London Journal
The London Magazine, or, Gentleman's Monthly Intelligencer
Monthly Chronicle
Monthly Notices of the Royal Astronomical Society
Monthly Review
Morning Chronicle
Morning Post
The Nautical Almanac and Astronomical Ephemeris
Old Whig or The Consistent Protestant
Original Weekly Journal
Parliamentary Register
Post Boy
Post Man and the Historical Account
St. James's Chronicle or the British Evening Post
Steel's Original and Correct List of the Royal Navy

The Times
Weekly Journal or British Gazetteer
Weekly Journal or Saturday's Post
Weekly Journal with Fresh Advices Foreign and Domestick
Weekly Packet
Whitehall Evening Post or London Intelligencer

Secondary Sources

Akademie der Wissenschaften zu Göttingen (ed.), *Georg Christoph Lichtenberg. Vorlesungen zur Naturlehre, Band 2: Gottlieb Gamauf: 'Erinnerungen aus Lichtenbergs Vorlesungen'. Die Nachschrift eines Hörers* (Göttingen, 2008)

Emily Akkermans, 'Chronometers and Chronometry on British Voyages of Exploration, 1819–1836' (unpublished PhD thesis, University of Edinburgh, 2020)

David Alff, *The Wreckage of Intentions: Projects in British Culture 1660–1730* (Philadelphia, 2017)

Peter Alter, *The Reluctant Patron: Science and the State in Britain, 1850–1920* (Oxford, 1987)

William J. H. Andrewes (ed.), *The Quest for Longitude* (Cambridge, MA, 1996)

William J. H. Andrewes, 'Even Newton Could Be Wrong: The Story of Harrison's First Three Sea Clocks', in *The Quest for Longitude*, ed. by William J. H. Andrewes (Cambridge, MA, 1996), pp. 189–234

William J. H. Andrewes, 'Finding Local Time at Sea, and the Instruments Employed', in *The Quest for Longitude*, ed. by William J. H. Andrewes (Cambridge, MA, 1996), pp. 394–404

Anonymous, *The Shropshire Gazetteer* (Wem, 1824)

Philip Arnott, 'Chronometers on East India Company Ships', *Antiquarian Horology*, 30 (2007), 481–500

William J. Ashworth, 'The Calculating Eye: Baily, Herschel, Babbage and the Business of Astronomy', *British Journal for the History of Science*, 27 (1994), 409–441

William J. Ashworth, *Customs and Excise: Trade, Production, and Consumption in England, 1640–1845* (Oxford, 2003)

Samuel Elliott Atkins and William Henry Overall, *Some Account of the Worshipful Company of Clockmakers of the City of London*, (London, 1881)

Jeanette Atkinson, 'Steampunking Heritage: How Steampunk Artists Reinterpret Museum Collections', in *A Museum Studies Approach to Heritage*, ed. by Sheila Watson, Amy Jane Barnes and Katy Bunning (London, 2018), pp. 205–220

David Aubin, Charlotte Bigg and H. Otto Sibum, 'Introduction: Observatory Techniques in Nineteenth-Century Science and Society', in *The Heavens on Earth: Observatories and Astronomy in Nineteenth-Century Science and Culture*, ed. by David Aubin, Charlotte Bigg and H. Otto Sibum (Durham, NC, 2010), pp. 1–32

Lawrence J. Baack, '"A Practical Skill that Was without Equal": Carsten Niebuhr and the Navigational Astronomy of the Arabian Journey, 1761–7', *Mariner's Mirror*, 99 (2013), 138–152

Alexi Baker, 'Squire, Jane (*bap.* 1686–1743), Scientific Writer', in *Oxford Dictionary of National Biography* (Oxford, 2004) online edition www.oxforddnb.com/view/article/45826 [accessed 8 April 2021]

Alexi Baker, 'The Business of Life: The Socioeconomics of the "Scientific" Instrument Trade in Early Modern London', in *Generations in Towns: Succession and Success in Pre-Industrial Urban Societies,* ed. by F.-E. Eliassen and K. Szende (Newcastle upon Tyne, 2009), pp. 169–191

Alexi Baker, '"This Ingenious Business": The Socio-Economics of the Scientific Instrument Trade in London, 1700–1750' (unpublished DPhil thesis, University of Oxford, 2010)

Alexi Baker, '"Humble Servants", "Loving Friends", and Nevil Maskelyne's Invention of the Board of Longitude', in *Maskelyne: Astronomer Royal*, ed. by Rebekah Higgitt (London, 2014), pp. 203–228

Alexi Baker, 'Interpreting Silences and Overturning Assumptions: Women in the Instrument Trade of 18th-Century London', XLII Symposium of the Scientific Instrument Commission in Palermo, Italy, September 18–22, 2023

David Barrado Navascués, *Cosmography in the Age of Discovery and the Scientific Revolution* (Cham, 2023)

Katy Barrett, '"Explaining" Themselves: The Barrington Papers, the Board of Longitude, and the Fate of John Harrison', *Notes and Records of the Royal Society*, 65 (2011), 145–162

Katy Barrett, *Looking for Longitude: A Cultural History* (Liverpool, 2022)

Katy Barrett and Richard Dunn, 'A Mechanic Art', *Apollo*, CLXXX no. 625 (November 2014), 82–86

Michael K. Barritt, 'Matthew Flinders's Survey Practices and Records', *Journal of the Hakluyt Society* (2014) www.hakluyt.com/downloadable_files/Journal/Barritt_Flinders.pdf [accessed 5 January 2021]

Daniel Baugh, 'Parliament, Naval Spending and the Public: Contrasting Financial Legacies of Two Exhausting Wars, 1689–1713', *Histoire & Mesure*, 30 no. 2 (2015), 23–49

John Cawte Beaglehole (ed.), *The Journals of Captain James Cook on his Voyages of Discovery,* 4 vols (Cambridge, 1955–1974)

John Cawte Beaglehole, *The Life of Captain James Cook*, 2 vols (London, 1974)

Silvio A. Bedini, *Thinkers and Tinkers: Early American Men of Science* (New York, 1975)

John Bendall, *Kendall's Longitude* (London, 2019)

James Arthur Bennett, *The Mathematical Science of Christopher Wren* (Cambridge, 1983)

James Arthur Bennett, 'Instrument Makers and the "Decline of Science in England": The Effect of Institutional Change on the Elite Makers of the Early Nineteenth Century', in *Nineteenth-Century Scientific Instruments and Their Makers*, ed. by Peter de Clercq (Amsterdam, 1985), pp. 13–27

James Arthur Bennett, *The Divided Circle: The History of Instruments for Astronomy, Navigation and Surveying* (Oxford, 1987)

James Arthur Bennett, 'Science Lost and Longitude Found: The Tercentenary of John Harrison', *Journal of the History of Astronomy*, 24 (1993), 281–287

Jim Bennett, 'The Travels and Trials of Mr Harrison's Timekeeper', in *Instruments, Travel and Science: Itineraries of Precision from the Seventeenth to the Twentieth Century*, ed. by Marie-Noëlle Bourguet, Christian Licoppe and Heinz Otto Sibum (London, 2002), pp. 75–95

Jim Bennett, 'Catadioptrics and Commerce in Eighteenth-Century London', *History of Science*, 44 (2006), 246–278

Jim Bennett, '"The Rev. Mr. Nevil Maskelyne, F.R.S. and Myself": The Story of Robert Waddington', in *Maskelyne: Astronomer Royal*, ed. by Rebekah Higgitt (London, 2014), pp. 59–88

Jim Bennett, 'James Short and John Harrison: Personal Genius and Public Knowledge', *Science Museum Group Journal*, 2 (2014) http://dx.doi.org/10.15180/140209 [accessed 13 April 2021]

Jim Bennett, *Navigation: A Very Short Introduction* (Oxford, 2017)

Jim Bennett, 'Adventures with Instruments: Science and Seafaring in the Precarious Career of Christopher Middleton', *Notes and Records of the Royal Society*, 73 (2019), 303–327

Jim Bennett, 'Mathematicians on Board: Introducing Lunar Distances to Life at Sea', *British Journal for the History of Science*, 52 (2019), 65–83

Jim Bennett, 'The First Nautical Almanac and Astronomical Ephemeris', in *The History of Celestial Navigation: Rise of the Royal Observatory and Nautical Almanacs*, ed. by P. Kenneth Seidelmann and Catherine Y. Hohenkerk (Cham, 2020), pp. 145–156

Paola Bertucci, *Artisanal Enlightenment: Science and the Mechanical Arts in Old Regime France* (New Haven, 2017)

Jonathan Betts, 'Introduction & Technical Appraisal', in *Principles and Explanations of Timekeepers*, ed. by Harrison, Arnold and Earnshaw (Upton, 1984)

Jonathan Betts, 'Arnold and Earnshaw: The Practicable Solution', in *The Quest for Longitude*, ed. by William J. H. Andrewes (Cambridge, MA, 1996), pp. 311–328

Jonathan Betts, 'Earnshaw, Thomas (1749–1829), Maker of Watches and Chronometers', in *Oxford Dictionary of National Biography* (Oxford, 2004) online edition https://doi.org/10.1093/ref:odnb/8406 [accessed 8 May 2020]

Jonathan Betts, 'Emery, Josiah (bap. 1725, d. 1794), Watchmaker', in *Oxford Dictionary of National Biography* (Oxford, 2004) online edition https://doi.org/10.1093/ref:odnb/49517 [accessed 28 December 2020]

Jonathan Betts, 'Kendall, Larcum (1719–1790), Watchmaker', in *Oxford Dictionary of National Biography* (Oxford, 2004) online edition https://doi.org/10.1093/ref:odnb/677 [accessed 5 January 2021]

Jonathan Betts, *Time Restored: The Harrison Timekeepers and R. T. Gould, the Man Who Knew (Almost) Everything* (Oxford, 2006)

Jonathan Betts, *Harrison* (London, 2007)

Jonathan Betts, *Marine Chronometers at Greenwich* (Oxford, 2017)

Jonathan Betts, 'The Quest for Precision: Astronomy and Navigation', in *A General History of Horology*, ed. by Anthony Turner, James Nye and Jonathan Betts (Oxford, 2022), pp. 311–339

Jonathan Betts and Andrew King, 'Jeremy Thacker: Longitude Imposter?' *Times Literary Supplement*, 18 March 2009, p. 6

Cyprian Blagden, 'Thomas Carnan and the Almanack Monopoly', *Studies in Bibliography*, 14 (1961), 23–43

Guy Boistel, 'Esprit Pezenas (1692–1776), jésuite, astronome et traducteur: un acteur méconnu de la diffusion de la science anglaise en France au XVIIIe siècle', in *Échanges franco-britanniques entre savants depuis le XVIIe siècle,* ed. by Robert Fox and Bernard Joly, *Cahiers de Logique et de philosophie*, 7 (2010), 135–157

Guy Boistel, 'From Lacaille to Lalande: French Work on Lunar Distances, Nautical Ephemerides and Lunar Tables, 1742–85', in *Navigational Enterprises in Europe and its Empires, 1730–1850*, ed. by Richard Dunn and Rebekah Higgitt (Basingstoke, 2015), pp. 47–64

Guy Boistel, *'Pour la gloire de M. de la Lande': Une histoire matérielle, scientifique, institutionnelle et humaine de la* Connaissance des temps, *de 1679 à 1920* (Paris, 2022)

Jamie Bolker, 'Lost at Sea: *Robinson Crusoe* and the Art of Navigation', *Eighteenth-Century Studies*, 53 (2020), 589–606

William Bowe, 'Short Account of the Life of Mr William Emerson', in *Tracts*, ed. by William Emerson, new edition (London, 1793), pp. i–xxiii

David Brady, *The Contribution of British Writers between 1560 and 1830 to the Interpretation of Revelation 13.16-18* (Tübingen, 1983)

John Brewer, *The Sinews of Power: War Money and the English State, 1688–1783* (London, 1989)

Claire Brock, 'The Public Worth of Mary Somerville', *British Journal for the History of Science*, 39 (2006), 255–272

David J. Bryden, 'Magnetic Inclinatory Needles: Approved by the Royal Society?', *Notes and Records of the Royal Society*, 47 (1993), 17–31

David J. Bryden, 'Georgian Instruments Patents – Some Ghosts and Spectres', *Bulletin of the Scientific Instrument Society*, 112 (2011), 4–23

David J. Bryden, 'William Ross and a Misguided Means of Finding Longitude', *The Antiquarian Astronomer*, 11 (2017), 55–68

David J. Bryden, 'William Shires (1780–1858): Astronomer, Mathematician and Instrument Designer', *Bulletin of the Scientific Instrument Society*, 137 (2018), 2–9

David J. Bryden, *Innovation in the Design of Scientific Instruments in the Georgian Era: The Role of the Society of Arts* (London, 2019)

Jenny Bulstrode, 'Cetacean Citations and the Covenant of Iron', *Notes and Records of the Royal Society*, 73 (2019), 167–185

Jenny Bulstrode, 'The Eye of the Needle: Magnetic Survey and the Compass of Capital in the Age of Revolution and Reform' (unpublished PhD thesis, University of Cambridge, 2020)

M. Diane Burton and Tom Nicholas, 'Prizes, Patents and the Search for Longitude', *Explorations in Economic History*, 64 (2017), 21–36

Ronald S. Calinger, *Leonhard Euler: Mathematical Genius in the Enlightenment* (Princeton, 2016)

Geoffrey Cantor, *Quakers, Jews, and Science: Religious Responses to Modernity and the Sciences in Britain, 1650–1900* (Oxford, 2005)

Bernard Capp, *Astrology and the Popular Press: English Almanacs 1500–1800* (London and Boston, 1979)

Audrey T. Carpenter, *John Theophilus Desaguliers: A Natural Philosopher, Engineer and Freemason in Newtonian England* (London, 2011)

Victoria Carroll, *Science and Eccentricity: Collecting, Writing and Performing Science for Early Nineteenth-Century Audiences* (London, 2008)

William Carter and Merri Carter, 'The Age of Sail: A Time When the Fortunes of Nations and Lives of Seamen Literally Turned with the Winds Their Ships Encountered at Sea', *Journal of Navigation*, 63 (2010), 717–731

John Cawood, 'The Magnetic Crusade: Science and Politics in Early Victorian Britain', *Isis* 70 (1979), 493–518

Paul Ceruzzi, *GPS* (Cambridge, MA, 2018)

Neil Chambers (ed.), *The Letters of Sir Joseph Banks: A Selection, 1768–1820* (London, 2000)

Neil Chambers (ed.), *Scientific Correspondence of Joseph Banks*, 6 vols (London, 2007)

Allan Chapman, *Dividing the Circle: The Development of Critical Angular Measurement in Astronomy 1500–1850* (Chichester, 1990)

Allan Chapman, 'The Astronomical Revolution', in *Möbius and His Band: Mathematics and Astronomy in Nineteenth Century Germany*, ed. by John Fauvel, Raymond Flood and Robin Wilson (Oxford, 1993), pp. 32–77

Allan Chapman, 'Dunkin, Edwin (1821–1898)', in *Oxford Dictionary of National Biography* (Oxford, 2004) online edition www.oxforddnb.com/view/article/57900 [accessed 19 May 2017]

Allan Chapman, 'Ramsden, Jesse (1735–1800), Maker of Scientific Instruments', in *Oxford Dictionary of National Biography* (Oxford, 2004) online edition https://doi.org/10.1093/ref:odnb/23105 [accessed 4 February 2021]

Allan Chapman, 'Airy's Greenwich Staff', *The Antiquarian Astronomer*, 6 (2012), 4–18

Edwin Chappell (ed.), *The Tangier Papers of Samuel Pepys* (London, 1935)

Davida Charney, 'Lone Geniuses in Popular Science: The Devaluation of Scientific Consensus', *Written Communication*, 20 (2003), 215–241

David Clarke, *Reflections on the Astronomy of Glasgow* (Edinburgh, 2013)

Agnes M. Clerke, 'A Southern Observatory', *Contemporary Review* 55 (1889), 380–392

Agnes M. Clerke, 'Bevis, John (1695–1771)', rev. Anita McConnell, *Oxford Dictionary of National Biography* (Oxford, 2004) online edition www.oxforddnb.com/view/article/2330 [accessed 14 May 2017]

Gloria Clifton, 'The Adoption of the Octant in the British Isles', in *Koersvast. Vijf eeuwen navigatie op zee*, ed. by Remmelt Daalder, Frits Loomeijer and Diederick Wildemann (Zaltbommel, 2005), pp. 85–94

Gareth Cole, *Arming the Royal Navy, 1793–1815: The Office of Ordnance and the State* (London, 2012)

Thomas Edward Colebrooke, *Life of H T. Colebrooke* (London, 1873)

Quintin Colville and James Davey (eds), *A New Naval History* (Manchester, 2019)

Alan H. Cook, *Edmond Halley: Charting the Heavens and the Seas* (Oxford, 1997)

Alan H. Cook, 'Edmond Halley and the Magnetic Field of the Earth', *Notes and Records of the Royal Society*, 55 (2001), 473–490

Andrew Cook, 'Alexander Dalrymple and John Arnold: Chronometers and the Representation of Longitude on East India Company Charts', *Vistas in Astronomy*, 28 (1985), 189–195

Andrew Cook, 'Establishing the Sea-Routes to India and China', in *The Worlds of the East India Company*, ed. by Huw V. Bowen, Margarette Lincoln and Nigel Rigby (Woodbridge, 2002), pp. 119–136

Charles H. Cotter, *A History of Nautical Astronomy* (London, 1968)

Charles H. Cotter, 'A History of Nautical Astronomical Tables' (unpublished PhD thesis, University of Wales Institute of Science and Technology, 1974)

Charles H. Cotter, 'John Hamilton Moore and Nathaniel Bowditch', *Journal of Navigation*, 30 (1977), 323–326

William Cotton, *Sir Joshua Reynolds, and His Works: Gleanings from His Diary, Unpublished Manuscripts, and from Other Sources* (London, 1856)

Mary Croarken, 'Providing Longitude for All', *Journal for Maritime Research*, 4 no. 1 (2002), 106–126

Mary Croarken, 'Astronomical Labourers: Maskelyne's Assistants at the Royal Observatory, Greenwich', *Notes and Records of the Royal Society*, 57 (2003), 285–298

Mary Croarken, 'Mary Edwards: Computing for a Living in 18th-Century England', *Annals of the History of Computing*, 25 (2003), 9–15

Mary Croarken, 'Tabulating the Heavens: Computing the *Nautical Almanac* in 18th-Century England', *IEEE Annals of the History of Computing*, 23 (2003), 48–61

Mary Croarken, 'Nevil Maskelyne and His Human Computers', in *Maskelyne: Astronomer Royal*, ed. by Rebekah Higgitt (London, 2014), pp. 130–161

Mary Croarken, 'Greenwich, Nevil Maskelyne and the Solution to the Longitude Problem', in *Mathematics at the Meridian: The History of Mathematics at Greenwich*, ed. by Raymond Flood, Tony Mann and Mary Croarken (Boca Raton, 2020), pp. 45–62

John S. Croucher and Rosalind F. Croucher, *Mistress of Science: The Story of the Remarkable Janet Taylor, Pioneer of Sea Navigation* (Stroud, 2016)

Patrick Curry, 'Andrews, Henry (1774–1820), Astronomer and Astrologer', in *Oxford Dictionary of National Biography* (Oxford, 2004) online edition https://doi.org/10.1093/ref:odnb/523 [accessed 9 December 2022]

Krzysztof Czaplewski and Dana Goward, 'Global Navigation Satellite Systems – Perspectives on Development and Threats to System Operation', *International Journal on Marine Navigation and Safety of Sea Transportation*, 10 (2016), 183–192

John Darwin, *Unfinished Empire: The Global Expansion of Britain* (London, 2013)

Lorraine Daston, 'Calculation and the Division of Labor, 1750–1950', *Bulletin of the German Historical Institute*, 62 (2018), 9–30

Andrew David, 'Vancouver's Survey Methods and Surveys', in *From Maps to Metaphors. The Pacific World of George Vancouver*, ed. by Robin Fisher and Hugh Johnston (Vancouver, 1993), pp. 51–69

Karel Davids, 'Finding Longitude at Sea by Magnetic Declination on Dutch East-Indiamen, 1596–1795', *The American Neptune* 50 (1990), 281–290

Karel Davids, 'Dutch and Spanish Global Networks of Knowledge in the Early Modern Period: Structures, Connections, Changes', in *Centres of Accumulation in the Low Countries in the Early Modern Period*, ed. by Lissa Roberts (Münster, 2011), pp. 29–52

Karel Davids, 'The Longitude Committee and the Practice of Navigation in the Netherlands, c. 1750–1850', in *Navigational Enterprises in Europe and Its Empires, 1730–1850*, ed. by Richard Dunn and Rebekah Higgitt (Basingstoke, 2015), pp. 32–46

Simon C. Davidson, 'Box Chronometers for India 1800–1936', *Antiquarian Horology*, 40 (2014), 76–91

Simon C. Davidson, 'The Use of Chronometers to Determine Longitude on East India Company Voyages', *Mariner's Mirror*, 102 (2016), 344–348

Simon C. Davidson, 'Marine Chronometers: The Rapid Adoption of New Technology by East India Captains in the Period 1770–1792 on over 580 Voyages', *Antiquarian Horology*, 40 (2019), 76–91

Alun C. Davies, 'The Life and Death of a Scientific Instrument: The Marine Chronometer, 1770–1920', *Annals of Science*, 35 (1978), 509–525

Alun C. Davies, 'Horology and Navigation: The Chronometers on Vancouver's Expedition, 1791–95', *Antiquarian Horology*, 21 (1993), 244–255

Alun C. Davies, 'Vancouver's Chronometers', in *From Maps to Metaphors. The Pacific World of George Vancouver*, ed. by Robin Fisher and Hugh Johnston (Vancouver, 1993), pp. 70–84

Alun C. Davies, 'Arnold's Chronometer No. 176 and Vancouver's Expedition, 1791–95', *Antiquarian Horology*, 32 (2010), 255–262

Warren R. Dawson (ed.), *The Banks Letters: A Calendar of the Manuscript Correspondence of Sir Joseph Banks, Preserved in the British Museum, the British Museum (Natural History) and Other Collections in Great Britain* (London, 1958)

Suzanne Débarbat, 'L'évolution administrative du Bureau des longitudes: une approche par les textes officiels', in *Pour une histoire du Bureau des longitudes (1795–1932)*, ed. by Martina Schiavon and Laurent Rollet (Nancy, 2017), pp. 23–40

Richard de Grijs, *Time and Time Again: Determination of Longitude at Sea in the 17th Century* (Bristol, 2017)

Richard de Grijs, 'A (Not So) Brief History of Lunar Distances: Lunar Longitude Determination at Sea before the Chronometer', *Journal of Astronomical History and Heritage*, 23 (2020), 495–522

Richard de Grijs, 'European Longitude Prizes. I: Longitude Determination in the Spanish Empire', *Journal of Astronomical History and Heritage*, 23 (2020), 465–494

Richard de Grijs, 'European Longitude Prizes. 2: Astronomy, Religion and Engineering Solutions in the Dutch Republic', *Journal of Astronomical History and Heritage*, 24 (2021), 405–439

Richard de Grijs, 'European Longitude Prizes. 3: The Unsolved Mystery of an Alleged Venetian Longitude Prize', *Journal of Astronomical History and Heritage*, 24 (2021), 728–738

Richard de Grijs, 'European Longitude Prizes. 4: Thomas Axe's Impossible Terms', *Journal of Astronomical History and Heritage*, 24 (2021), 739–750

Johan de Jong, 'Navigating through Technology: Technology and the Dutch East India Company VOC in the Eighteenth Century' (unpublished PhD thesis, University of Twente, 2016)

Greg Dening, *The Death of William Gooch: A History's Anthropology* (Honolulu, 1995)

Jane Desborough and Gloria Clifton, 'Science and the City: The Role of Women in the Science City: London 1650–1800', *Science Museum Group Journal*, 15 (2021) https://dx.doi.org/10.15180/211502/001 [accessed 19 February 2024]

Nicholas Dew, '*Vers la ligne*: Circulating Measurements around the French Atlantic', in *Science and Empire in the Atlantic World*, ed. by James Delbourgo and Nicolas Dew (London, 2008), pp. 53–72

Nicholas Dew, 'Scientific Travel in the Atlantic World: The French Expedition to Gorée and the Antilles, 1681–1683', *British Journal for the History of Science*, 43 (2010), 1–17

David Dewhirst, 'Meridian Astronomy in the Private and University Observatories of the United Kingdom: Rise and Fall', *Vistas in Astronomy*, 28 (1985), 147–158

Harry W. Dickinson, *Educating the Navy: Eighteenth and Nineteenth Century Education for Officers* (Abingdon, 2007)

Henry Dircks, *Perpetuum Mobile: Or, Search for Self-Motive Power, During the 17th, 18th, and 19th Centuries* (London, 1st ed., 1861; 2nd ed., 1870)

Graham Dolan, 'People: John Pond, the Sixth Astronomer Royal', *The Royal Observatory Greenwich* www.royalobservatorygreenwich.org/articles.php?article=1301 [accessed 8 March 2023]

Richard Drayton, *Nature's Government: Science, Imperial Britain and the 'Improvement' of the World* (New Haven, 2000)

John Louis Emile Dreyer, 'The Decade 1830–1840', in *History of the Royal Astronomical Society 1820–1920*, ed. by John Louis Emile Dreyer and Herbert Hall Turner (London, 1923), pp. 50–81
Saul Dubow, *A Commonwealth of Knowledge: Science, Sensibility and White South Africa, 1820–2000* (Oxford, 2006)
Jonathan Duncan, *The History of Guernsey: With Occasional Notices of Jersey, Alderney, and Sark, and Biographical Sketches* (London, 1841)
Edwin Dunkin, *A Far Off Vision: A Cornishman at the Greenwich Observatory: Auto-Biographical Notes by Edwin Dunkin*, ed. by Peter Hingley and Tamsin Daniel (Truro, 1999)
William Dunkin, 'Notes on Some Points Connected with the Early History of the "Nautical Almanac"', *The Observatory*, 262 (1898), 49–53, and 264 (1898), 123–127
Richard Dunn, 'Collecting and Interpreting Navigation at Greenwich', in *Sextants at Greenwich*, ed. by Willem Mörzer Bruyns (Oxford, 2009), pp. 71–82
Richard Dunn, 'The Impudent Mr Jennings and his Insulating Compass', *Bulletin of the Scientific Instrument Society*, 111 (2011), 6–9
Richard Dunn, 'Scoping Longitude: Optical Designs for Navigation at Sea', in *From Earth-Bound to Satellite. Telescopes, Skills and Networks*, ed. by Alison Morrison-Low, Sven Dupré, Stephen Johnston and Giorgio Strano (Leiden, 2011), pp. 141–154
Richard Dunn, 'Heaving a Little Ballast: Seaborne Astronomy in the Late-Eighteenth Century', in *Scientific Instruments in the History of Science: Studies in Transfer, Use and Preservation*, ed. by Marcus Granato and Marta C. Lourenço (Rio de Janeiro, 2014), pp. 79–100
Richard Dunn, 'James Cook and the New Navigation', in *Arctic Ambitions: Captain Cook and the Northwest Passage*, ed. by James Barnett and David Nicandri (Anchorage/Seattle, 2015), pp. 89–107
Richard Dunn, 'A Bird in the Hand, or, Manufacturing Credibility in the Instruments of Enlightenment Science', in *The Material Cultures of Enlightenment Arts and Sciences*, ed. by Adriana Craciun and Simon Schaffer (London, 2016), pp. 73–90
Richard Dunn, *Navigational Instruments* (Oxford, 2016)
Richard Dunn, 'Longitude Found: Innovation and Navigational Practice, 1750–1860', in *The Routledge History of the Modern Maritime World*, ed. by Kenneth Morgan (Abingdon, 2025), pp. 381–401
Richard Dunn and Rebekah Higgitt, *Finding Longitude: How Ships, Clocks and Stars Helped Solve the Longitude Problem* (Glasgow, 2014)
Richard Dunn and Rebekah Higgitt (eds), *Navigational Enterprises in Europe and Its Empires, 1730–1850* (Basingstoke, 2015)
Richard Dunn and Eóin Phillips, 'Of Clocks and Cats', *Antiquarian Horology*, 34 (2013), 88–93
Paul Elliott, 'The Birth of Public Science in the English Provinces: Natural Philosophy in Derby, c. 1690–1760', *Annals of Science*, 57 (2000), 61–100
Alan Ereira, 'The Voyages of H1', *Mariner's Mirror*, 87 (2001), 144–149
Amy Louise Erickson, *Women and Property in Early Modern England*, 2nd ed. (London, 2002)
Amy Louise Erickson, 'Married Women's Occupations in Eighteenth-Century London', *Continuity and Change*, 23 (2008), 267–307
Thomas Ertman, *Birth of the Leviathan: Building States and Regimes in Medieval and Early Modern Europe* (Cambridge, 1997)

David Faris, *Plantagenet Ancestry of Seventeenth-Century Colonists*, 2nd ed. (Boston, 1999)

Maureen Farrell, *William Whiston* (New York, 1981)

Danielle Fauque, 'Testing Longitude Methods in Mid-Eighteenth Century France', in *Navigational Enterprises in Europe and Its Empires, 1730–1850*, ed. by Richard Dunn and Rebekah Higgitt (Basingstoke, 2015), pp. 159–179

Timothy Feist, 'The Stationers' Voice: The English Almanac Trade in the Early Eighteenth Century', *Transactions of the American Philosophical Society*, 95 no. 4 (2005), i–129

Jean-Marie Feurtet, 'Le Bureau des longitudes (1795–1854): De Lalande à Le Verrier' (unpublished thesis, École nationale des chartes, Paris, 2005)

Fernando B. Figueiredo, 'Astronomical and Nautical Ephemerides in late 18th Century Portugal', in *Pour une histoire du Bureau des longitudes (1795–1932)*, ed. by Martina Schiavon and Laurent Rollet (Nancy, 2017), pp. 147–173

Fernando B. Figueiredo and Guy Boistel, 'Monteiro da Rocha and the International Debate in the 1760s on Astronomical Methods to Find the Longitude at Sea: His Proposals and Criticisms to Lacaille's Lunar-Distance Method', *Annals of Science*, 79 (2022), 215–258

Paula Findlen, 'Calculations of Faith: Mathematics, Philosophy, and Sanctity in 18th-Century Italy' (New Work on Maria Gaetana Agnesi)', *Historia Mathematica*, 38 (2011), 248–291

Robin Fisher and Hugh Johnston (eds), *From Maps to Metaphors: The Pacific World of George Vancouver* (Vancouver, 1993)

Eric G. Forbes, 'The Foundation and Early Development of the *Nautical Almanac*', *Journal of the Institute of Navigation*, 18 (1965), 391–401

Eric Forbes, 'Index of the Board of Longitude Papers at the Royal Greenwich Observatory', *Journal of the History of Astronomy*, 1 (1970), 169–179, 2 (1971), 58–70, and 3 (1971), 133–145

Eric Forbes, 'Tobias Mayer (1723–62): A Case of Forgotten Genius', *British Journal for the History of Science*, 5 (1970), 1–20

Eric Forbes, 'Schultz's Proposal for Finding Longitude at Sea', *Journal for the History of Astronomy*, 2 (1971), 35–41

Eric Forbes, *The Birth of Navigational Science: The Solving in the 18th Century of the Problem of Finding Longitude at Sea* (London, 1974)

Eric Forbes, *Greenwich Observatory: The Royal Observatory at Greenwich and Herstmonceux, 1675–1975. Vol. I. Origins and Early History (1675–1835)* (London, 1975)

Eric Forbes, 'Tobias Mayer's Claim for the Longitude Prize: A Study in 18th Century Anglo-German Relations', *Journal of Navigation*, 28 (1975), 77–90

Eric Forbes, *Tobias Mayer (1723–1762), Pioneer of Enlightened Science in Germany* (Göttingen, 1980)

Eric Forbes, Lesley Murdin and Frances Wilmoth (eds), *The Correspondence of John Flamsteed, The First Astronomer Royal*, 3 vols (London, 1995–2001)

Rudolf Freiburg, 'Our Never Failing Guide, the Watch: Der "Längengrad" in der englischen Literatur und Kultur des achzehntehn Jarhunderts', in *Zwischen Literatur und Naturwissenschaft: Debatten – Probleme – Visionen 1680–1820*, ed. by Rudolf Freiburg, Christian Lubkoll and Harald Neymeyer (Berlin, 2017), pp. 33–62

Alan Frost, *The Global Reach of Empire: Britain's Maritime Expansion in the Indian and Pacific Oceans, 1764–1815* (Carlton, 2003)

Tim Fulford, 'The Role of Patronage in Early Nineteenth-Century Science, as Evidenced in Letters from Humphry Davy to Joseph Banks', *Notes and Records of the Royal Society*, 73 (2019), 457–475
Tim Fulford and Sharon Ruston (eds), *Collected Letters of Sir Humphry Davy,* 4 vols (Oxford, 2020)
John Gascoigne, *Science in the Service of Empire: Joseph Banks, the British State and the Uses of Science in the Age of Revolution* (Cambridge, 1998)
John Gascoigne, 'The Royal Society and the Emergence of Science as an Instrument of State Policy', *British Journal for the History of Science*, 32 (1999), 171–184
John Gascoigne, '"Getting a Fix": The *Longitude* Phenomenon', *Isis*, 98 (2007), 760–778
John Gascoigne, *Science and the State: From the Scientific Revolution to World War II* (Cambridge, 2019)
Caroline Gillan, 'Lord Bute and Eighteenth-Century Science and Patronage' (unpublished PhD thesis, University of Galway, 2018)
Owen Gingerich, 'Cranks and Opportunists: "Nutty" Solutions to the Longitude Problem', in *The Quest for Longitude*, ed. by William J. H. Andrewes (Cambridge, MA, 1996), pp. 133–148
Rupert T. Gould, *The Marine Chronometer: Its History and Development* (London, 1923)
Rod Gow, 'Letters of William Emerson and Francis Holliday to the Publisher John Nourse', *British Society for History of Mathematics Bulletin*, 21 (2006), 40–50
Aaron Graham and Patrick Walsh, *The British Fiscal-Military States, 1660–c.1783* (Abingdon, 2016)
David Alan Grier, *When Computers Were Human* (Princeton, 2005)
Hartmut Grosser (ed.), *Johann Friedrich Benzenberg. Die Astronomie, Physische Geographie, Meteorologie und Geologie. Georg Christoph Lichtenbergs Vorlesung 1797/1798* (Göttingen, 2004)
Niccolò Guicciardini, 'Hutton, Charles (1737–1823), Mathematician Lieutenant in the Royal Artillery', in *Oxford Dictionary of National Biography* (Oxford, 2004) online edition https://doi.org/10.1093/ref:odnb/14300 [accessed 9 December 2022]
Marie Boas Hall, *All Scientists Now: The Royal Society in the Nineteenth Century* (Cambridge, 1984)
A. Rupert Hall and Laura Tilling (eds), *The Correspondence of Isaac Newton*, 7 vols (Cambridge, 1977)
A. A. Hanham, 'Stanhope, James, first Earl Stanhope (1673–1721)', in *Oxford Dictionary of National Biography* (Oxford, 2004) online edition https://doi.org/10.1093/ref:odnb/26248 [accessed 11 May 2021]
Richard Harding and Sergio Ferri (eds), *The Contractor State and Its Implications, 1659–1815* (Las Palmas, 2012)
Philip Harling, *The Waning of 'Old Corruption': The Politics of Economical Reform in Britain, 1779–1846* (Oxford, 1996)
Vincent T. Harlow, *The Founding of the Second British Empire, 1763–1793*, 2 vols (London, 1952–1964)
William Henry Hart, 'Gleanings from the Records of the Treasury – No. II', *Notes and Queries*, 2nd series, 9 (1860), 297–298
David Boyd Haycock, *William Stukeley: Science, Religion and Archaeology in Eighteenth-Century England* (Woodbridge, 2002)

John Lewis Heilbron, 'A Mathematicians' Mutiny, with Morals', in *World Changes, Thomas Kuhn and the Nature of Science*, ed. by Paul Horwich (Cambridge, MA, 1993), pp. 81–129

Robert D. Hicks, *Voyage to Jamestown: Practical Navigation in the Age of Discovery* (Annapolis, 2011)

Rebekah Higgitt, 'The Royal Observatory, Greenwich, and Its Publics Past and Present', in *Astronomy and Its Instruments before and after Galileo*, ed. by Luisa Pigatto and Valeria Zanni (Padua, 2010), pp. 439–450

Rebekah Higgitt, 'Prize Fights: Animadversions on the Almanac', *Royal Museums Greenwich blog*, 8 November 2012 www.rmg.co.uk/stories/blog/prize-fights-animadversions-on-almanac [accessed 21 April 2022]

Rebekah Higgitt, 'The Projects of Eighteenth-Century Astronomy', in *Maskelyne: Astronomer Royal*, ed. by Rebekah Higgitt (London, 2014), pp. 89–96

Rebekah Higgitt, 'Revisiting and Revising Maskelyne's Reputation', in *Maskelyne: Astronomer Royal*, ed. by Rebekah Higgitt (London, 2014), pp. 22–49

Rebekah Higgitt (ed.), *Maskelyne: Astronomer Royal* (London, 2014)

Rebekah Higgitt, 'Equipping Expeditionary Astronomers: Nevil Maskelyne and the Development of "Precision Exploration"', in *Geography, Technology and Instruments of Exploration*, ed. by Fraser MacDonald and Charles W. J. Withers (Farnham, 2015), pp. 15–36

Rebekah Higgitt, 'Challenging Tropes: Genius, Heroic Invention, and the Longitude Problem in the Museum', *Isis*, 108 (2017), 371–380

Rebekah Higgitt, '"Greenwich near London": The Royal Observatory and Its London Networks in the Seventeenth and Eighteenth Centuries', *British Journal for the History of Science*, 52 (2019), 297–322

Rebekah Higgitt, 'Mathematical Examiners at Trinity House: Teaching and Examining Mathematics for Navigation in London during the Long Eighteenth Century', in *Beyond the Learned Academy: The Practice of Mathematics, 1600–1850*, ed. by Philip Beeley and Christopher Hollings (Oxford, 2024), pp. 80–111

Rebekah Higgitt and Richard Dunn, 'The Bureau and the Board: Change and Collaboration in the Final Decades of the British Board of Longitude', in *Pour une histoire du Bureau des longitudes (1795–1932)*, ed. by Martina Schiavon and Laurent Rollet (Nancy, 2017), pp. 195–219

Rebekah Higgitt and James Wilsdon, 'The Benefits of Hindsight: How History Can Contribute to Science Policy', in *Future Direction for Scientific Advice in Whitehall*, ed. by Robert Doubleday and James Wilsdon (Cambridge, 2013), pp. 79–85

Roger Highfield, 'Calling All Geniuses for the New Longitude Prize', *The Telegraph*, 19 May 2014

Barry William Higman, 'Locating the Caribbean: The Role of Slavery and the Slave Trade in the Search for Longitude', *Journal of Caribbean History*, 54 (2020), 296–317

Liliane Hilaire-Pérez, 'Technical Invention and Institutional Credit in France and Britain in the 18th Century', *History and Technology*, 16 (2000), 285–306

HM Nautical Almanac Office, 'The Nautical Almanac and Its Superintendents' (2016), http://astro.ukho.gov.uk/nao/history/nao_1832.html [accessed 13 June 2021]

Caitlin Homes, 'Friend and Foe: The Tempestuous Relationship between Nevil Maskelyne and Joseph Banks', in *Maskelyne: Astronomer Royal*, ed. by Rebekah Higgitt (London, 2014), pp. 235–264

Hans Hooijmaiers, 'A Claim for Finding the Longitude at Sea by Zumbach de Koesfelt', *Antiquarian Horology*, 30 (2007), 347–364

Julian Hoppit, *Britain's Political Economies: Parliament and Economic Life, 1660–1800* (Cambridge, 2017)

Derek Howse, 'Captain Cook's Marine Timekeepers. Part I – The Kendall Watches', *Antiquarian Horology*, 6 (1969), 190–205

Derek Howse, 'Captain Cook's Marine Timekeepers. Part II – The Arnold Chronometers', *Antiquarian Horology*, 6 (1969), 276–280

Derek Howse, *Greenwich Observatory: The Royal Observatory at Greenwich and Herstmonceux, 1675–1975. Vol. 3: The Buildings and Instruments* (London, 1975)

Derek Howse, 'The Principal Scientific Instruments Taken on Captain Cook's Voyages of Exploration, 1768–80', *Mariner's Mirror*, 65 (1979), 119–135

Derek Howse, *Nevil Maskelyne: The Seaman's Astronomer* (Cambridge, 1989)

Derek Howse, *Greenwich Time and the Longitude* (London, 1997)

Derek Howse, 'Britain's Board of Longitude: The Finances, 1714–1828', *Mariner's Mirror*, 84 (1998), 400–417

Derek Howse, 'The Board of Longitude Accounts, 1737–1828' (unpublished, 1998), copies at CUL, NMM and RS

Derek Howse, 'Campbell, John (*b*. in or before 1720, *d*. 1790)', in *Oxford Dictionary of National Biography* (Oxford, 2004) online edition www.oxforddnb.com/view/article/4518 [accessed 14 May 2017]

Robert D. Hume, 'The Value of Money in Eighteenth-Century England: Incomes, Prices, Buying Power – and Some Problems in Cultural Economics', *Huntington Library Quarterly*, 77 (2014), 373–416

Roger Hutchins, *British University Observatories 1772–1839* (Aldershot, 2017)

George Huxtable and Ian Jackson, 'Journey to Work: James Cook's Transatlantic Voyages in the Grenville 1764–1767', *Journal of Navigation*, 63 (2010), 207–214

Rob Iliffe, 'Mathematical Characters: Flamsteed and Christ's Hospital Royal Mathematical School', in *Flamsteed's Stars: New Perspectives on the Life and Work of the First Astronomer Royal, 1649–1719*, ed. by Frances Wilmoth (Woodbridge, 1997), pp. 115–144

Joanna Innes, 'Legislation and Public Participation, 1760–1830', in *The British and Their Laws in the Eighteenth Century*, ed. by David Lemmings (Woodbridge, 2005), pp. 102–132

Yuto Ishibashi, '"A Place for Managing Government Chronometers": Early Chronometer Service at the Royal Observatory Greenwich', *Mariner's Mirror*, 99 (2013), 52–66

Myles W. Jackson. *Spectrum of Belief: Joseph von Fraunhofer and the Craft of Precision Optics* (Cambridge, MA/London, 2000)

Roland Jackson, *Scientific Advice to the Nineteenth-Century British State* (Pittsburgh, 2023)

Frank James, 'Michael Faraday's Work on Optical Glass', *Physics Education*, 26 (1991), 296–300

Frank James, 'Michael Faraday and Lighthouses', in *The Golden Age: Essays in British Social and Economic History, 1850–1870*, ed. by Ian Inkster, Colin Griffin, Jeff Hill and Judith Rowbotham (London, 2017), pp. 92–104

Lisa Jardine, 'Scientists, Sea Trials and International Espionage: Who Really Invented the Balance-Spring Watch?', *Antiquarian Horology*, 9 (2006), 663–683

Lisa Jardine, *Going Dutch: How England Plundered Holland's Glory* (London, 2008)
Lisa Jardine, 'Accidental Anglo–Dutch Collaborations: Seventeenth-Century Science in London and The Hague', *Sartoniana*, 23 (2010), 15–40
Lisa Jardine, *Temptation in the Archives: Essays in Golden Age Dutch Culture* (London, 2015)
Adrian Johns, *Piracy: The Intellectual Property Wars from Gutenberg to Gates* (Chicago and London, 2009)
Peter Johnson, 'The Board of Longitude 1714–1828', *Journal of the British Astronomical Association*, 99 (1989), 63–69
Art Roeland Theo Jonkers, 'Finding Longitude at Sea: Early Attempts in Dutch Navigation', *De Zeventiende Eeuw*, 12 no. 1 (1996), 186–197
Art Roeland Theo Jonkers, *Earth's Magnetism in the Age of Sail* (London, 2003)
Art Roeland Theo Jonkers, 'Rewards and Prizes', in *The Oxford Encyclopedia of Maritime History*, ed. by John Hattendorf, 4 vols (Oxford, 2007), III, pp. 433–436
Ludmilla Jordanova, 'On Heroism', *Science Museum Group Journal*, 1 (2014) http://dx.doi.org/10.15180/140107 [accessed 30 April 2021]
Vincent Jullien (ed.), *Le calcul des longitudes: un enjeu pour les mathématiques, l'astronomie, la mesure de temps et la navigation* (Rennes, 2002)
Sarah Tindal Kareem, 'Forging Figures of Invention in Eighteenth-Century Britain', in *The Age of Projects*, ed. by Maximilian Novak (Toronto, 2008), pp. 344–369
Vera Keller and Ted McCormick, 'Towards a History of Projects', *Early Science and Medicine*, 21 (2016), 423–444
[Patrick Kelly], 'Maskelyne, Nevil', in *The Cyclopaedia; or, Universal Dictionary of Arts, Sciences, and Literature*, ed. by Abraham Rees, 39 vols (London, 1802–1820), XXII, pp. 698–700
Paul Kennedy, *The Rise and Fall of British Naval Mastery* (London, 1976)
Carolyn Kennett, 'An 18th-Century Astronomical Hub in West Cornwall', *Antiquarian Astronomer*, 11 (2017), 45–54
Alvin Kernan, *Printing Technology, Letters and Samuel Johnson* (Princeton, 1987)
B. Zorina Khan, *Inventing Ideas: Patents, Prizes, and the Knowledge Economy* (Oxford, 2020)
Peggy Aldrich Kidwell, 'Women Astronomers in Britain, 1780–1930', *Isis* 75 (1984), 534–546
Andrew King, '"John Harrison, Clockmaker at Barrow; Near Barton upon Humber; Lincolnshire": The Wooden Clocks, 1713–1730', in *The Quest for Longitude*, ed. by William J. H. Andrewes (Cambridge, MA, 1996), pp. 167–187
Andrew King, 'Harrison, John (bap. 1693, d. 1776), Horologist', in *Oxford Dictionary of National Biography* (Oxford, 2004) online edition https://doi.org/10.1093/ref:odnb/12438 [accessed 4 February 2021]
Charles Kirby-Miller (ed.), *Memoirs of the Extraordinary Life, Works and Discoveries of Martinus Scriblerus, Written in Collaboration by the Members of the Scriblerus Club: John Arbuthnot, Alexander Pope, Jonathan Swift, John Gay, Thomas Parnell, and Robert Harley, Earl of Oxford* (Oxford, 1988)
Roger Knight and Martin Wilcox, *Sustaining the Fleet: War, the British Navy and the Contractor State* (Woodbridge, 2010)
Jake V. Th. Knoppers, 'The Visits of Peter the Great to the United Provinces in 1697–98 and 1716–17 as Seen in Light of the Dutch Sources' (unpublished MA thesis, McGill University, 1969)

Wolfgang Köberer, 'On the First Use of the Term "Chronometer"', *Mariner's Mirror*, 102 no. 2 (2016), 203–206

Wolfgang Köberer, 'German Contributions to Solving the Longitude Problem in the Seventeenth and Eighteenth Centuries', *Mariner's Mirror*, 108 (2022), 262–285

Albert Kuhn, 'Dr. Johnson, Zachariah Williams, and the Eighteenth-Century Search for the Longitude', *Modern Philology*, 82 (1984), 40–52

David Landes, 'Finding the Point at Sea', in *The Quest for Longitude*, ed. by William J. H. Andrewes (Cambridge, MA, 1996), pp. 20–30

Paul Langford, *Public Life and the Propertied Englishman 1689–1798* (Oxford, 1994)

Arlen J. Large, 'How Far West Am I? The Almanac as an Explorer's Yardstick', *Great Plains Quarterly*, 13 (1993), 117–131

Philip Laundy, 'Norton, Fletcher, First Baron Grantley (1716–1789)', in *Oxford Dictionary of National Biography*, (Oxford, 2006) www.oxforddnb.com/view/article/20342 [accessed 14 January 2017]

Philip Laurie, 'The Board of Visitors of the Royal Observatory – II: 1830–1965', *Quarterly Journal of the Royal Astronomical Society*, 7 (1967), 334–353

David Lemmings, *Law and Government in England in the Long Eighteenth Century* (Basingstoke, 2011)

John Leopold, 'The Longitude Timekeepers of Christiaan Huygens', in *The Quest for Longitude*, ed. by William J. H. Andrewes (Cambridge, MA, 1996), pp. 102–114

Trevor Levere, *Science and the Canadian Arctic: A Century of Exploration 1818–1918* (Cambridge, 1993)

Elri Liebenberg, 'Unveiling South Africa: John Barrow's Map of 1801', *International Journal of Cartography* 7 (2021), 164–170

Kristen Lippincott, *A Guide to the Royal Observatory, Greenwich* (London, 2007)

Philip Loft, 'A Tapestry of Laws: Legal Pluralism in Eighteenth-Century Britain', *Journal of Modern History*, 91 (2019), 276–310

Longitude Prize, 'Origins of the Longitude Prize' https://longitudeprize.org/the-history/ [accessed 30 April 2021]

Gregory Lynall, *Swift and Science: The Satire, Politics, and Theology of Natural Knowledge, 1690–1730* (Basingstoke, 2012)

Gregory Lynall, 'Scriblerian Projections of Longitude: Arbuthnot, Swift, and the Agency of Satire in a Culture of Invention', *Journal of Literature and Science*, 7 no. 2 (2014), 1–18

David Mackay, *In the Wake of Cook: Exploration, Science and Empire, 1780–1801* (London, 1985)

Christine Macleod, *Inventing the Industrial Revolution: The English Patent System 1660–1800* (Cambridge, 1988)

Dániel Margócsy, 'A Long History of Breakdowns: A Historiographical Review', *Social Studies of Science*, 47 (2017), 307–325

Frédéric Marguet, *Histoire générale de la navigation du XVe au XXe siècle* (Paris, 1931)

Ben Marsden and Crosbie Smith, *Engineering Empires: A Cultural History of Technology in Nineteenth-Century Britain* (New York, 2004)

Peter James Marshall (ed.), *The Oxford History of the British Empire, Vol. 2, The Eighteenth Century* (Oxford, 1998)

Jean Mascart, *La vie et les travaux du chevalier Jean-Charles de Borda (1733–1799)* (Paris, 2000)

William Edward May, 'Naval Compasses in 1707', *Journal of Navigation*, 6 (1953), 405–409

William Edward May, 'The Last Voyage of Sir Clowdisley Shovell', *Journal of Navigation*, 13 (1960), 324–332

William Edward May, 'How the Chronometer Went to Sea', *Antiquarian Horology*, 9 (1974), 638–663

William Edward May, 'The Log-Books Used by Ships of the East India Company', *Journal of Navigation*, 27 (1974), 116–118

William Edward May, 'The Gentleman of Jamaica', *Mariner's Mirror*, 73 (1987), 149–165

John McAleer, '"Stargazers at the World's End": Telescopes, Observatories and "Views" of Empire in the Nineteenth-Century British Empire', *British Journal for the History of Science* 46 (2013), 389–413

John McAleer, *Britain's Maritime Empire: Southern Africa, the South Atlantic and the Indian Ocean, 1763–1820* (Cambridge, 2017)

Katherine McAlpine, 'Ships, Clocks & Stars: The Quest for Impact', *JCOM*, 14 no. 3 (2015), C02 https://doi.org/10.22323/2.14030302 [accessed 3 June 2021]

Peter McBride and Richard Larn, *Admiral Shovell's Treasure and Shipwreck in the Isles of Scilly* (Penryn, 1999)

Anita McConnell, *Instrument Makers to the World: A History of Cooke, Troughton and Simms* (York, 1992)

Anita McConnell, 'Bird, John (1709–1776)', in *Oxford Dictionary of National Biography* (Oxford, 2004) online edition www.oxforddnb.com/view/article/2448 [accessed 14 May 2017]

Anita McConnell, 'Witchell, George (1728–1785)', in *Oxford Dictionary of National Biography* (Oxford, 2004) online edition www.oxforddnb.com/view/article/51607 [accessed 14 May 2017]

Anita McConnell, *Jesse Ramsden (1735–1800): London's Leading Scientific Instrument Maker* (Aldershot, 2007)

Anita McConnell, *A Survey of the Networks Bringing a Knowledge of Optical Glass-working to the London Trade, 1500–1800*, ed. by Jenny Bulstrode (Cambridge, 2016)

Rory McEvoy, 'Maskelyne's Time', in *Maskelyne: Astronomer Royal*, ed. by Rebekah Higgitt (London, 2014), pp. 168–193

Arthur Jack Meadows, *Greenwich Observatory: The Royal Observatory at Greenwich and Herstmonceux, 1675–1975. Vol.2: Recent History* (London, 1975)

Vaudrey Mercer, *John Arnold & Son Chronometer Makers 1762–1843* (London, 1972)

David Philip Miller, 'The Royal Society of London 1800–1835: A Study in the Cultural Politics of Scientific Organization' (unpublished PhD thesis, University of Pennsylvania, 1981)

David Philip Miller, 'The "Sobel Effect": The Amazing Tale of How Multitudes of Popular Writers Pinched All the Best Stories in the History of Science and became Rich and Famous while Historians Languished in Accustomed Poverty and Obscurity, and How This Transformed the World. A Reflection on a Publishing Phenomenon', *Metascience*, 2 (2002), 185–199

David Philip Miller, 'Longitude Networks on Land and Sea: The East India Company and Longitude Measurement "in the Wild", 1770–1840', in *Navigational Enterprises in Europe and Its Empires, 1730–1850*, ed. by Richard Dunn and Rebekah Higgitt (Basingstoke, 2015), pp. 223–247

Jane Miller, *Books Will Speak Plain: A Handbook for Identifying and Describing Historical Bindings* (Ann Arbor, MI, 2010)

Tessa Mobbs and Robert Unwin, 'The Longitude Act of 1714 and the Last Parliament of Queen Anne', *Parliamentary History*, 35 (2016), 152–170

George Modelski and William R. Thompson, *Seapower in Global Politics, 1494–1993* (Basingstoke, 1988)

Kenneth Morgan (ed.), *Australia Circumnavigated: The Voyage of Matthew Flinders in HMS Investigator, 1801–1803*, 2 vols (London, 2015)

Kenneth Morgan, 'Finding Longitude: The *Investigator* Expedition, 1801–1803', *International Journal of Maritime History*, 29 (2017), 771–787

Samuel Eliot Morison, *Admiral of the Ocean Sea: A Life of Christopher Columbus* (New York, 1992)

Jack B. Morrell, 'Individualism and the Structure of British Science in 1830', *Historical Studies in the Physical Sciences*, 3 (1971), 183–204

Jack Morrell and Arnold Thackray, *Gentlemen of Science: Early Years of the British Association for the Advancement of Science* (Oxford, 1981)

Willem Mörzer Bruyns, *Sextants at Greenwich* (Oxford, 2009)

Lori L. Murray and David R. Bellhouse, 'How Was Edmond Halley's Map of Magnetic Declination (1701) Constructed?', *Imago Mundi*, 69 (2017), 72–84

Martin Myrone, 'Drawing after the Antique at the British Museum, 1809–1817: "Free" Art Education and the Advent of the Liberal State', *British Art Studies*, 5 (2017) https://doi.org/10.17658/issn.2058-5462/issue-05/mmyrone [accessed 16 March 2023]

National Maritime Museum, *4 Steps to Longitude: An Exhibition to Mark the Bicentenary of the First Successful Trial in 1762 of John Harrison's Fourth Marine Timekeeper* (London, 1962)

Marjorie Nicolson and Nora M. Mohler, 'The Scientific Background of Swift's Voyage to Laputa', *Annals of Science*, 2 (1937), 299–334

Susanna Nockolds, 'Early Timekeepers at Sea: The Story of the General Adoption of the Chronometer', *Antiquarian Horology*, 4 (1963), 110–113, 148–152

Patrick O'Brien, 'The Triumph and Denouement of the British Fiscal State', in *The Fiscal-Military State in Eighteenth-Century Europe: Essays in Honour of P.G.M. Dickson*, ed. by Christopher Storrs (Farnham, 2009), pp. 167–200

Günther Oestmann, *Auf dem Weg zum "Deutschen Chronometer": Die Einführung von Präzisionszeitmessern bei der deutschen Handels- und Kriegsmarine bis zum Ersten Weltkrieg* [= *Deutsche Maritime Studien*, 21] (Bremerhaven, 2012)

John W. Olmsted, 'The Voyage of Jean Richer to Acadia in 1670: A Study in the Relations of Science and Navigation under Colbert', *Proceedings of the American Philosophical Society*, 104 (1960), 612–634

Jacob Orrje, 'Patriotic and Cosmopolitan Patchworks: Following a Swedish Astronomer into London's Communities of Maritime Longitude, 1759–60', in *Navigational Enterprises in Europe and Its Empires, 1730–1850*, ed. by Richard Dunn and Rebekah Higgitt (Basingstoke, 2015), pp. 89–110

George Peacock, *Life of Thomas Young, M.D., F.R.S., &c.* (London, 1855)

Douglas T. Peck, 'Theory Versus Practical Application in the History of Early Ocean Navigation', *Terrae Incognitae*, 34 (2002), 46–59

Douglas T. Peck, 'The Controversial Skill of Columbus as a Navigator: An Enduring Historical Enigma', *Journal of Navigation*, 62 (2009), 417–425

Nigel Penn, 'Mapping the Cape: John Barrow and the First British Occupation of the Colony, 1795–1803', in *Maps and Africa*, ed. by Jeffrey Stone (Aberdeen, 1994), pp. 108–127

David Penney, 'Thomas Mudge and the Longitude: A Reason to Excel', in *The Quest for Longitude*, ed. by William J. H. Andrewes (Cambridge, MA, 1996), 293–310

Adam J. Perkins, 'Edmond Halley, Isaac Newton and the Longitude Act of 1714', in *The History of Celestial Navigation: Rise of the Royal Observatory and Nautical Almanacs*, ed. by P. Kenneth Seidelmann and Catherine Y. Hohenkerk (Cham, 2020), pp. 69–143

Adam J. Perkins and Steven J. Dick, 'The British and American Nautical Almanacs in the 19th Century', in *The History of Celestial Navigation: Rise of the Royal Observatory and Nautical Almanacs*, ed. by P. Kenneth Seidelmann and Catherine Y. Hohenkerk (Cham, 2020), pp. 157–197

Maureen Perkins, *Visions of the Future: Almanacs, Time, and Cultural Change* (Oxford, 1996)

Reginald Henry Phillimore, *Historical Records of the Survey of India*, 4 vols (Dehra Dun, 1946–1958)

Eóin Phillips, 'Remembering Matthew Flinders', *Journal for Maritime Research*, 14 (2012), 111–119

Eóin Phillips, 'Making Time Fit: Astronomers, Artisans and the State, 1770–1820' (unpublished PhD thesis, University of Cambridge, 2014)

Eóin Phillips, 'Instrumenting Order: Longitude, Seamen, and Astronomers, 1770–1805', in *Geography, Technology and Instruments of Exploration*, ed. by Fraser MacDonald and Charles W. J. Withers (Farnham, 2015), pp. 37–55

Nicola Phillips, *Women in Business* (Woodbridge, 2006)

Nicholas Plumley, 'The Royal Mathematical School within Christ's Hospital: The Early Years – Its Aims and Achievements', *Vistas in Astronomy*, 20 (1976), 51–59

Roy Porter and Willam Hobbs, *The Earth Generated and Anatomized: An Early Eighteenth Century Theory of the Earth* (London, 1981)

James Poskett, 'Sounding in Silence: Men, Machines and the Changing Environment of Naval Discipline, 1796–1815', *British Journal for the History of Science*, 48 (2015), 213–232

Humphrey Quill, *John Harrison: The Man Who Found Longitude* (London, 1966)

William Graham Lister Randles, 'Portuguese and Spanish Attempts to Measure Longitude in the 16th Century', *Vistas in Astronomy*, 28 (1985), 235–241

Martin Rees, 'A Longitude Prize for the Twenty-First Century', *Nature*, 509 (2014), 401

Nicky Reeves, 'Maskelyne the Manager', in *Maskelyne: Astronomer Royal*, ed. by Rebekah Higgitt (London, 2014), pp. 97–123

Michael Reidy, *Tides of History: Ocean Science and Her Majesty's Navy* (Chicago and London, 2008)

Tim Reinke-Williams, *Women, Work and Sociability in Early Modern London* (London, 2014)

Derek Robson, *Some Aspects of Education in Cheshire in the Eighteenth Century*, (Manchester, 1966)

Nicholas Andrew Martin Rodger, *The Command of the Ocean: A Naval History of Britain, 1649–1815* (London, 2005)

Nicholas Andrew Martin Rodger, 'From the "Military Revolution" to the "Fiscal-Naval State"', *Journal for Maritime Research*, 13 (2011), 119–128

Pat Rogers, 'Longitude Forged: How an Eighteenth-Century Hoax Has Taken in Dava Sobel and Other Historians', *Times Literary Supplement*, 12 November 2008, p. 6

Pat Rogers, *Documenting Eighteenth Century Satire: Pope, Swift, Gay, and Arbuthnot in Historical Context* (Cambridge, 2011)

Alex Roland, 'Science, Technology, and War', in *The Cambridge History of Science: Volume 5: The Modern Physical and Mathematical Sciences*, ed. by Mary Jo Nye (Cambridge, 2002), pp. 559–578
Anna Marie Roos, *Martin Folkes (1690–1754): Newtonian, Antiquary, Connoisseur* (Oxford, 2021)
Barrington Rosier, 'The Construction Costs of Eighteenth-Century Warships', *Mariner's Mirror*, 96 (2010), 161–172
George Sebastian Rousseau and David Haycock, 'Voices Calling for Reform: The Royal Society in the Mid-Eighteenth Century – Martin Folkes, John Hill, and William Stukeley', *History of Science*, 37 (1999), 377–406
Valerie Rumbold, 'Scriblerus Club (act. 1714)', in *Oxford Dictionary of National Biography* (Oxford, 2004) online edition www.oxforddnb.com/view/theme/71160 [accessed 11 June 2017]
Steven Ruskin, *John Herschel's Cape Voyage: Private Science, Public Imagination and the Ambitions of Empire* (Aldershot, 2004)
Donald H. Sadler, 'The Bicentenary of the Nautical Almanac', *Journal of the Institute of Navigation*, 21 (1968), 6–18
Donald H. Sadler, *Man Is Not Lost: A Record of Two Hundred Years of Astronomical Navigation with The Nautical Almanac 1767–1967* (London, 1968)
Ioannis-Dionysios Salavrakosp, 'A Reassessment of the British and Allied Economic and Military Mobilization in the Revolutionary and Napoleonic Wars (1792–1815)', *Res Militaris*, 7 (2017) http://resmilitaris.net/index.php?ID=1025453 [accessed 4 February 2021]
Shirley D. Saunders, 'Sir Thomas Brisbane's Legacy to Colonial Science: Colonial Astronomy at the Parramatta Observatory, 1822–1848', *Historical Records of Australian Science*, 15 (2004), 177–209
Ann Savours, 'A Very Interesting Point in Geography: The 1773 Phipps Expedition Towards the North Pole', *Arctic*, 37 (1984), 402–428
Simon Schaffer, 'Joseph Banks', *Cambridge Digital Library* https://cudl.lib.cam.ac.uk/view/ES-LON-00019 [accessed 8 June 2017]
Simon Schaffer, 'Papers of the Board of Longitude', *Cambridge Digital Library* https://cudl.lib.cam.ac.uk/collections/rgo14/2 [accessed 3 May 2021]
Simon Schaffer, 'The Show that Never Ends: Perpetual Motion in the Early Eighteenth Century', *British Journal for the History of Science*, 28 (1995), 157–189
Simon Schaffer, 'Our Trusty Friend the Watch', *London Review of Books*, 31 (1996), 11–12
Simon Schaffer, 'The Disappearance of Useful Sciences', *Cambridge Anthropology*, 22 (2000/2001), 2–24
Simon Schaffer, 'The Bombay Case: Astronomers, Instrument Makers and the East India Company', *Journal for the History of Astronomy* 43 (2012), 151–180
Simon Schaffer, 'Swedenborg's Lunars', *Annals of Science*, 71 (2013), 1–25
Simon Schaffer, 'Chronometers, Charts, Charisma: On Histories of Longitude', *Science Museum Group Journal*, 2 (2014) http://dx.doi.org/10.15180/140203 [accessed 30 May 2021]
Martina Schiavon, 'The Bureau des Longitudes: An Institutional Study', in *Navigational Enterprises in Europe and Its Empires, 1730–1850*, ed. by Richard Dunn and Rebekah Higgitt (Basingstoke, 2016), pp. 65–85
Martina Schiavon and Laurent Rollet (eds), *Pour une histoire du Bureau des longitudes (1795–1932)* (Nancy, 2017)

Martina Schiavon and Laurent Rollet (eds), *Le Bureau des longitudes au prisme de ses procès-verbaux (1795–1932)* (Nancy, 2021)

Robert E. Schofield, *Mechanism and Materialism: British Natural Philosophy in an Age of Reason* (Princeton, 2015)

Margaret Schotte, 'Expert Records: Nautical Logbooks from Columbus to Cook', *Information & Culture*, 48 (2013), 281–322

Margaret Schotte, *Sailing School: Navigating Science and Skill, 1550–1800* (Baltimore, 2019)

Thomas Seccombe and David Penney, 'Mudge, Thomas (1715/16–1794), Horologist', in *Oxford Dictionary of National Biography* (Oxford, 2004) online edition https://doi.org/10.1093/ref:odnb/19486 [accessed 10 January 2021]

Anne Secord, 'Corresponding Interests: Artisans and Gentlemen in Nineteenth-Century Natural History', *British Journal of the History of Science*, 27 (1994), 383–408

P. Kenneth Seidelmann and Catherine Y. Hohenkerk (eds), *The History of Celestial Navigation: Rise of the Royal Observatory and Nautical Almanacs* (Cham, 2020)

Steven Shapin, 'Science and the Public', in *Companion to the History of Modern Science*, ed. by Robert Cecil Olby, Geoffrey N. Cantor, John R. R. Christie and Jonathan Hodge (London, 1996), pp. 990–1007

Jonathan R. Siegel, 'Law and Longitude', *Tulane Law Review*, 84 no. 1 (2009), 1–66

Samuel Smiles, *A Publisher and His Friends: Memoir and Correspondence of John Murray*, 2 vols (London, 1891)

Dava Sobel, *Longitude: The True Story of the Lone Genius Who Solved the Greatest Scientific Problem of His Time* (New York, 1995)

Richard Sorrenson, 'George Graham, Visible Technician', *British Journal for the History of Science*, 32 (1999), 203–221

Robin W. Spencer, 'Open Innovation in the Eighteenth Century', *Research-Technology Management*, 55 no. 4 (2012), 39–43

Tom Stamp and Cordelia Stamp, *William Scoresby: Arctic Scientist* (Whitby, 1975)

Alexander Donald Stewart, 'The British Naval Chronometers of 1821', *Antiquarian Horology*, 37 (2016), 247–252

Larry Stewart, *The Rise of Public Science: Rhetoric, Technology, and Natural Philosophy in Newtonian Britain, 1660–1750* (Cambridge, 1992)

Alan Stimson, 'Some Board of Longitude Instruments in the Nineteenth Century', in *Nineteenth Century Scientific Instruments and Their Makers*, ed. by Peter de Clercq (Leiden, 1985), pp. 93–115

Eva Germaine Rimington Taylor, 'Old Henry Bond and the Longitude', *Mariner's Mirror*, 25 (1939), 162–169

Eva Germaine Rimington Taylor, *The Haven-Finding Art: A History of Navigation from Odysseus to Captain Cook* (London, 1956)

Eva Germaine Rimington Taylor, *The Mathematical Practitioners of Hanoverian England, 1714–1840*, (Cambridge, 1966)

John C. Taylor and Arnold W. Wolfendale, 'John Harrison: Clockmaker and Copley Medalist. A Public Memorial at Last', *Notes and Records of the Royal Society*, 61 (2007), 53–62

George McCall Theal, *Records of the Cape Colony*, 36 vols (Cape Town, 1897–1905)

Tamara Plakins Thornton, *Nathaniel Bowditch and the Power of Numbers: How a Nineteenth-Century Man of Business, Science, and the Sea Changed American Life* (Chapel Hill, 2016)

Norman J. W. Thrower, 'Cartography', in *The Quest for Longitude*, ed. by William J. H. Andrewes (Cambridge, MA, 1996), pp. 49–62
Arthur Cecil Todd, *Beyond the Blaze: A Biography of Davies Gilbert* (Truro, 1967)
Jonathan Topham, 'Publishing Popular Science in Early Nineteenth Century Britain', in *Science in the Marketplace: Nineteenth-Century Sites and Experiences*, ed. by Aileen Fyfe and Bernard Lightman (Chicago, 2007), pp. 135–168
Treasury, *Calendar of Treasury Papers ... Preserved in Her Majesty's Public Record Office*, 6 vols (London, 1868–1889)
A. Alan Treherne, 'Massey Family (per. c. 1760–1891), Makers of Clocks, Watches, and Nautical Instruments', in *Oxford Dictionary of National Biography* (Oxford, 2004) online edition https://doi.org/10.1093/ref:odnb/49519 [accessed 22 January 2021]
Anthony Turner, 'L'Angleterre, la France et la navigation: le contexte historique de l'oeuvre chronométrique de Ferdinand Berhoud', in *Ferdinand Berthoud 1727–1807, horloger mécanicien de roi et de la marine*, ed. by Catherine Cardinal (La Chaux-de-Fonds, 1984), pp. 143–163
Anthony Turner, 'Berthoud in England, Harrison in France: The Transmission of Horological Knowledge in 18th Century Europe', *Antiquarian Horology*, 20 (1992), 219–251
Anthony Turner, 'In the Wake of the Act, but Mainly Before', in *The Quest for Longitude*, ed. by William J. H. Andrewes (Cambridge, MA, 1996), pp. 115–127
Anthony Turner, 'Longitude Finding', in *The Oxford Encyclopedia of Maritime History*, ed. by John Hattendorf, 4 vols (Oxford, 2007), II, pp. 405–415
Gerard L'E. Turner, 'The Government and the English Optical Glass Industry, 1650–1850', *Annals of Science*, 57 (2000), 399–414
Herbert Hall Turner, 'The Decade 1820–1830', in *History of the Royal Astronomical Society 1820–1920*, ed. by John Louis Emil Dreyer and Herbert Hall Turner (London, 1923), pp. 1–49
Alfons Van der Kraan, 'The Dutch East India Company, Christiaan Huygens and the Marine Clock, 1682–95', *Prometheus*, 19 (2001), 279–298
Albert Van Helden, 'Longitude and the Satellites of Jupiter', in *The Quest for Longitude*, ed. by William J. H. Andrewes (Cambridge, MA, 1996), pp. 86–100
Wendy Wales, *Captain Cook's Computer: The Life of William Wales* (York, 2015)
Benjamin Wardhaugh, *Poor Robin's Prophecies: A Curious Almanac, and the Everyday Mathematics of Georgian Britain* (Oxford, 2012)
Sophie Waring, 'Thomas Young, the Board of Longitude and the Age of Reform' (unpublished PhD thesis, University of Cambridge, 2014)
Brian Warner, 'The William Herschel 14-foot Telescope', *Monthly Notices of the Astronomical Society of South Africa*, 46 (1987), 158–163
Brian Warner, *Royal Observatory, Cape of Good Hope 1820–1831: The Founding of a Colonial Observatory* (Dordrecht, 1995)
Deborah Warner, 'Terrestrial Magnetism: For the Glory of God and the Benefit of Mankind', *Osiris*, 2nd Series, 9 (1994), 66–84
Adrian Webb, 'The Expansion of British Naval Hydrographic Administration, 1808–1829' (unpublished PhD thesis, University of Exeter, 2010)
Richard Wendorf, *Sir Joshua Reynolds: The Painter in Society* (Cambridge, MA, 1998)
Steven Wepster, *Between Theory and Observations: Tobias Mayer's Explorations of Lunar Motion, 1751–1755* (Berlin, 2010)

Simon Werrett, *Fireworks: Pyrotechnic Arts and Sciences in European History*, (Chicago, 2010)

Simon Werrett, '"Perfectly Correct": Russian Navigators and the Royal Navy', in *Navigational Enterprises in Europe and its Empires, 1730–1850*, ed. by Richard Dunn and Rebekah Higgitt (Basingstoke, 2016), pp. 111–133

Simon Werrett, '"Both by Sea and Land": William Whiston, Longitude, and the Measurement of Space', in *Spaces of Enlightenment Science*, ed. by Gordon McOuat and Larry Stewart (Leiden, 2022), pp. 193–214

Jane Wess, 'Navigation and Mathematics: A Match Made in the Heavens?', in *Navigational Enterprises in Europe and Its Empires, 1730–1850*, ed. by Richard Dunn and Rebekah Higgitt (London, 2015), pp. 201–222

Jane Wess, 'The Role of Instruments in Exploration: A Study of the Royal Geographical Society 1830–1930' (unpublished PhD thesis, University of Edinburgh, 2017)

Richard S. Westfall, *Never at Rest: A Biography of Isaac Newton* (Cambridge, 1987)

Jeffrey R. Wigelsworth, 'Navigation and Newsprint: Advertising Longitude Schemes in the Public Sphere ca. 1715', *Science in Context*, 21 (2008), 351–376

Jeffrey R. Wigelsworth, *Selling Science in the Age of Newton: Advertising and the Commoditization of Knowledge* (Farnham, 2010)

Chris Wilde, 'Hutchinsonianism, Natural Philosophy, and Religious Controversy in 18th-Century Britain', *History of Science*, 18 (1980), 1–24

George H. Wilkins, 'The Expanding Role of H.M. Nautical Almanac Office, 1818–1975', *Vistas in Astronomy*, 20 (1976), pp. 239–243

George H. Wilkins, 'The History of H.M. Nautical Almanac Office', in *Proceedings: Nautical Almanac Office Sesquicentennial Symposium, U.S. Naval Observatory, March 3–4, 1999*, ed. by Alan D. Fiala and Steven J. Dick (Washington, DC, 1999), pp. 55–81

Glyndwr Williams, *The Great South Sea: English Voyages and Encounters 1570–1750* (New Haven and London, 1997)

Frances Willmoth, *Sir Jonas Moore: Practical Mathematics and Restoration Science* (Woodbridge, 1993)

Hannah Wills, 'The Diary of Charles Blagden: Information Management and the Gentleman of Science in Eighteenth-Century Britain' (unpublished PhD thesis, University College London, 2019)

Rif Winfield, *British Warships in the Age of Sail 1714–1792: Design, Construction, Careers and Fates* (Barnsley, 2007)

Rif Winfield, *British Warships in the Age of Sail 1603–1714: Design, Construction, Careers and Fates* (Barnsley, 2009)

C. Winton, 'Steele, Sir Richard (*bap.* 1672, *d.* 1729)', in *Oxford Dictionary of National Biography* (Oxford, 2004) online edition https://doi.org/10.1093/ref:odnb/26347 [accessed 11 May 2021]

Alexander Wood, *Thomas Young: Natural Philosopher 1773–1829* (Cambridge, 1954)

Bennet Woodcroft, *Subject-Matter Index of Specifications of Patents* (London, 1857)

Worshipful Company of Clockmakers, *Register of Apprentices of the Worshipful Company of Clockmakers of the City of London from Its Incorporation in 1631 to Its Tercentenary in 1931* (London, 1931)

Ulrike Zimmermann, 'John Harrison (1693–1776) and the Heroics of Longitude', *helden. heroes. héros*, 2 no. 2 (2014), 119–129

Broadcast Media

Coast (television series, UK, 2005–)
'The Clock that Changed the World' (television documentary, UK, 2010)
Great Britons (television series, UK, 2002)
Horizon (television series, UK, 1964–)
Inside Science (radio series, UK, 2013–)
Longitude (television series, Charles Sturridge, UK/USA, 2000)
Science Britannica (television documentary, UK, 2013)
The One Show (television series, UK, 2006–)

Online Sources

Cambridge Digital Library cudl.lib.cam.ac.uk
OED Online (Oxford: Oxford University Press, 2021) www.oed.com
Oxford Dictionary of National Biography www.oxforddnb.com
The Royal Observatory Greenwich www.royalobservatorygreenwich.org
Spalding Gentlemen's Society Stukeley Memoirs Project sgsprojects.omeka.net

Index

Page numbers/folios in *italics* refers to box/figures/tables.